FIBER OPTICS AND OPTOELECTRONICS

R.P. Khare
Associate Professor of Physics/Instrumentation
Birla Institute of Technology and Science
Pilani, Rajasthan

OXFORD
UNIVERSITY PRESS

YMCA Library Building, Jai Singh Road, New Delhi 110001

Oxford University Press is a department of the University of Oxford.
It furthers the University's objective of excellence in research, scholarship and education by publishing worldwide in

Oxford New York

Auckland Bangkok Buenos Aires Cape Town Chennai
Dar es Salaam Delhi Hong Kong Istanbul Karachi Kolkata
Kuala Lumpur Madrid Melbourne Mexico City Mumbai Nairobi
São Paulo Shanghai Taipei Tokyo Toronto

Oxford is a registered trade mark of Oxford University Press
in the UK and in certain other countries.

Published in India
by Oxford University Press

First published 2004

ISBN 0-19-566930-4

Typeset in Times New Roman
by The Composers, New Delhi 110063
Printed in India by Ram Book Binding House, New Delhi 110020
and published by Manzar Khan, Oxford University Press
YMCA Library Building, Jai Singh Road, New Delhi 110001

To

my spiritual parents,

Vandaniya Mataji Bhagwati Devi Sharma

and

Pandit Sriram Sharma Acharya

Preface

In the present 'information era', where efficient transfer of information is highly relevant to our well-being, fiber-optic systems—owing to their better suitability—are poised to form the very means of such information transfer, and hence find a very important role, direct or indirect, in almost every sphere of life. This fact has been well recognized by the planners and course-development teams of almost all engineering institutions/universities. As a result, most of them are now offering this subject under varied titles such as 'fiber optics and optoelectronics', 'fiber-optic communications', 'optical communications', 'optoelectronic devices and systems', etc., where the course content is more or less the same. Therefore, there is a need for a balanced and up-to-date text that discusses the major building blocks of present-day fiber-optic systems and also presents their uses in both communication and non-communication areas. Apart from theory, there is also a need for stimulating a practical approach so that students can undertake lab-oriented projects in this emerging area. This book aims at achieving these objectives.

Coverage and Structure

Divided into four parts, this book has grown out of the notes developed by me over 10 years of teaching this discipline to undergraduate students of electronics and instrumentation, and electrical and electronics engineering, and postgraduate students of physics at Birla Institute of Technology and Science, Pilani.

Part I starts with the easy-to-understand topic of ray propagation in optical fibers and then proceeds towards the more complex topics of wave propagation in planar and cylindrical waveguides. Special emphasis has been given to the treatment of single-mode fibers—the backbone of present-day fiber-optic communications. Production and characterization methods of optical fibers and the design of cables along with fiber-to-fiber connection techniques and related losses have also been discussed in this part.

Part II offers a detailed treatment of the theory behind optoelectronic sources (light-emitting diodes and injection laser diodes), detectors, modulators, and optical amplifiers.

Part III presents the application areas of fiber optics and optoelectronics in modern optical communications, wavelength-division multiplexing (WDM) and dense WDM, and fiber-optic sensing. Laser-based systems along with their applications in various fields have also been described in this part.

Part IV helps the student/instructor to make his own kit for carrying out PC-based/ manual measurements. It discusses the step-by-step procedures for implementing a variety of projects.

Key Features

The book covers all relevant topics related to modern fiber-optic communications. It includes separate chapters on laser-based systems and their applications in various fields as well as *lab-oriented projects* involving PC-based/manual measurements. In addition, it explains fiber-optic sensors, which will help readers to understand the non-communication applications of fiber optics.

An important aspect of this student-friendly text is that it reinforces theoretical concepts by providing numerous solved examples, along with multiple choice questions and review questions. Therefore, it will be very useful to students pursuing the above-mentioned disciplines.

Acknowledgements

I would like to thank all those who directly or indirectly contributed in making this venture a success. Sincere and specific mention must be made of Professor S. Venkateswaran, Vice Chancellor, BITS, Pilani; Professor L.K. Maheshwari, Director, BITS, Pilani (Pilani Campus); Professor K.E. Raman, Deputy Director (Administration) for his constant encouragement and support; and my colleagues Professor G. Raghurama and Professor S. Gurunarayanan for all the help. I am also thankful to the reviewers whose suggestions helped a lot in improving the presentation and content of the book. Special thanks are due to Mr Budh Ram for word-processing the manuscript and Mr K.N. Sharma for making the diagrams. Thanks are also due to my students, past and present, specially, Rajesh Purohit, K. Sudhir, Shweta Gaur, K. Priya, Sujatha, and Shartha. Finally, I must thank my wife Manjula and my children Gunjan, Ruchir, and Amit for their patience and cooperation during the course of this work.

R.P. KHARE

Contents

PART IV: PROJECTS

1 Introduction

After reading this chapter you will be able to appreciate the following:

- The subject of fiber optics and optoelectronics
- Historical developments in the field
- The configuration of a fiber-optic communication system
- Advantages of fiber-optic systems
- Emergence of fiber optics as a key technology
- Role of fiber optics technology

1.1 FIBER OPTICS AND OPTOELECTRONICS

Fiber optics is a branch of optics that deals with the study of propagation of light (rays or modes) through transparent dielectric waveguides, e.g., optical fibers. Optoelectronics is the science of devices that are based on processes leading to the generation of photons by electrons (e.g., laser diodes) or electrons by photons (e.g., photodiodes). The large-scale use of optoelectronic devices in fiber-optic systems has led to the integration of these two branches of science, and they are now synonymous with each other.

Prior to delving into the subject and its applications, a curious reader would like to know the following:

(i) the emergence of fiber optics as a dominant technology
(ii) the basic configuration of a fiber-optic system
(iii) the merits of such a system
(iv) the role of this technology in sociological evolution

This chapter aims at exploring these and other related issues.

1.2 HISTORICAL DEVELOPMENTS

The term 'fiber optics' was first coined by N.S. Kapany in 1956 when he along with his colleagues at Imperial College of Science and Technology, London, developed an image-transmitting device called the 'flexible fiberscope'. This device soon found application in inspecting inaccessible points inside reactor vessels and jet aircraft engines. The flexible endoscope became quite popular in the medical field. Improved versions of these devices are now increasingly being used in medical diagnosis and surgery.

The next important development in this area was the demonstration of the first pulsed ruby laser in 1960 by T. Maiman at the Hughes Research Laboratory and the realization of the first semiconductor laser in 1962 by researchers working almost independently at various research laboratories. However, it took another eight years before laser diodes for application in communications could be produced.

Almost around the same period, another interesting development took place when Charles Kao and Charles Hockham, working at the Standard Telecommunication Laboratory in England, proposed in 1966 that an optical fiber might be used as a means of communication, provided the signal loss could be made less than 20 decibels per kilometer (dB/km). (The definition of a decibel is given in Appendix A1.1.) At that time optical fibers exhibited losses of the order of 1000 dB/km.

At this point, it is important to know why the need for optical fibers as a transmission medium was felt. In fact, the transfer of information from one point to another, i.e., communication, is achieved by superimposing (or modulating) the information onto an electromagnetic wave, which acts as a carrier for the information signal. The modulated carrier is then transmitted through the information channel (open or guided) to the receiver, where it is demodulated and the original information sent to the destination. Now the carrier frequencies present certain limitations in handling the volume and speed of information transfer. These limitations generated the need for increased carrier frequency. In fiber-optic systems, the carrier frequencies are selected from the optical range (particularly the infrared part) of the electromagnetic spectrum shown in Fig. 1.1.

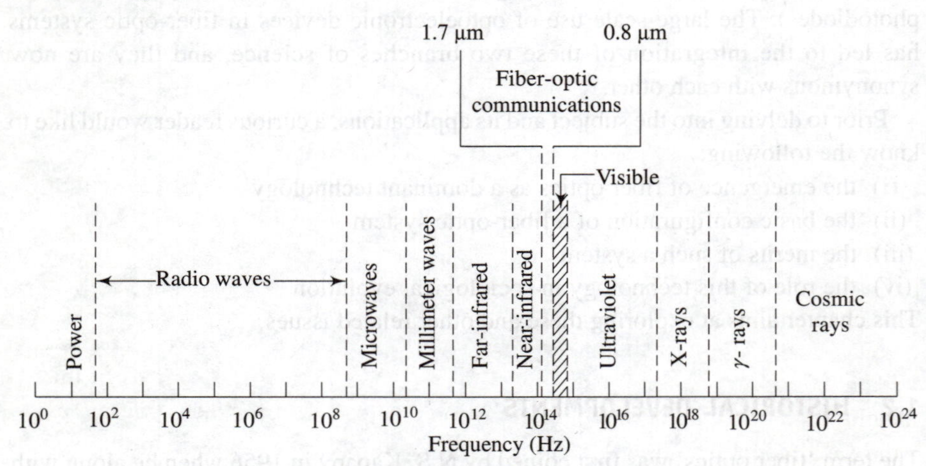

Fig. 1.1 Electromagnetic spectrum

The typical frequencies are of the order of 10^{14} Hz, which is 10,000 times greater than that of microwaves. Optical fibers are the most suitable medium for transmitting these frequencies, and hence they present theoretically unlimited possibilities.

Coming back to Kao and Hockham's proposal, the production of a low-loss optical fiber was required. The breakthrough came in 1970, when Dr Robert Maurer, Dr Donald Keck, and Dr Peter Schultz of Corning Glass Corporation of USA succeeded in producing a pure glass fiber which exhibited an attenuation of less than 20 dB/km. Concurrent developments in optoelectronic devices ushered in the era of fiber-optic communications technology.

1.3 A FIBER-OPTIC COMMUNICATION SYSTEM

Before proceeding further, let us have a look at the generalized configuration of a fiber-optic communication system, shown in Fig. 1.2. A brief description of each block in this figure will give us an idea of the prime components employed in this system.

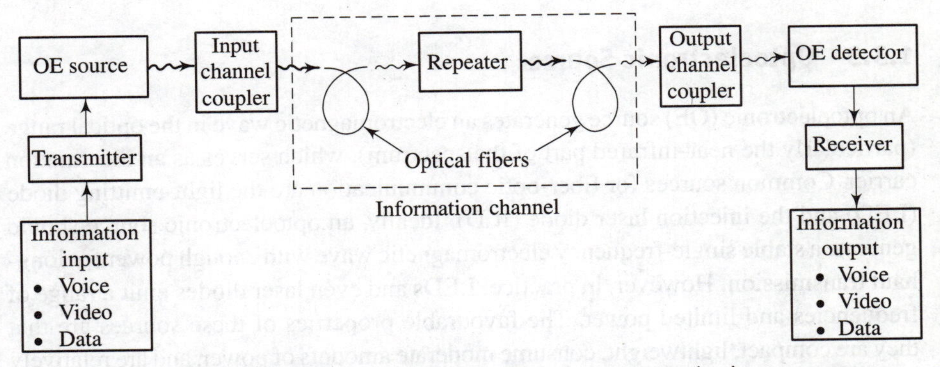

Fig. 1.2 Generalized configuration of a fiber-optic communication system

1.3.1 Information Input

The information input may be in any of the several physical forms, e.g., voice, video, or data. Therefore an input transducer is required for converting the non-electrical input into an electrical input. For example, a microphone converts a sound signal into an electrical current, a video camera converts an image into an electric current or voltage, and so on. In situations where the fiber-optic link forms a part of a larger system, the information input is normally in electrical form. Examples of this type include data transfer between different computers or that between different parts of the same computer. In either case, the information input must be in the electrical form for onward transmission through the fiber-optic link.

1.3.2 Transmitter

The transmitter (or the modulator, as it is often called) comprises an electronic stage which (i) converts the electric signal into the proper form and (ii) impresses this signal onto the electromagnetic wave (carrier) generated by the optoelectronic source.

The modulation of an optical carrier may be achieved by employing either an analog or a digital signal. An analog signal varies continuously and reproduces the form of the original information input, whereas digital modulation involves obtaining information in the discrete form. In the latter, the signal is either on or off, with the *on* state representing a digital 1 and the *off* state representing a digital 0. These are called *binary digits* (or bits) of the digital system. The number of bits per second (bps) transmitted is called the *data rate*. If the information input is in the analog form, it may be obtained in the digital form by employing an analog-to-digital converter.

Analog modulation is much simpler to implement but requires higher signal-to-noise ratio at the receiver end as compared to digital modulation. Further, the linearity needed for analog modulation is not always provided by the optical source, particularly at high modulation frequencies. Therefore, analog fiber-optic systems are limited to shorter distances and lower bandwidths.

1.3.3 Optoelectronic Source

An optoelectronic (OE) source generates an electromagnetic wave in the optical range (particularly the near-infrared part of the spectrum), which serves as an information carrier. Common sources for fiber-optic communication are the light-emitting diode (LED) and the injection laser diode (ILD). Ideally, an optoelectronic source should generate a stable single-frequency electromagnetic wave with enough power for long-haul transmission. However, in practice, LEDs and even laser diodes emit a range of frequencies and limited power. The favourable properties of these sources are that they are compact, lightweight, consume moderate amounts of power, and are relatively easy to modulate. Furthermore, LEDs and laser diodes which emit frequencies that are less attenuated while propagating through optical fibers are available.

1.3.4 Channel Couplers

In the case of open channel transmission, for example, the radio or television broadcasting system, the channel coupler is an antenna. It collects the signal from the transmitter and directs this to the atmospheric channel. At the receiver end again the antenna collects the signal and routes it to the receiver. In the case of guided channel transmission, e.g., a telephone link, the coupler is simply a connector for attaching the transmitter to the cable.

In fiber-optic systems, the function of a coupler is to collect the light signal from the optoelectronic source and send it efficiently to the optical fiber cable. Several designs are possible. However, the coupling losses are large owing to Fresnel reflection and limited light-gathering capacity of such couplers. At the end of the link again a coupler is required to collect the signal and direct it onto the photodetector.

1.3.5 Fiber-optic Information Channel

In communication systems, the term 'information channel' refers to the path between the transmitter and the receiver. In fiber-optic systems, the optical signal traverses along the cable consisting of a single fiber or a bundle of optical fibers. An optical fiber is an extremely thin strand of ultra-pure glass designed to transmit optical signals from the optoelectronic source to the optoelectronic detector. In its simplest form, it consists of two main regions: (i) a solid cylindrical region of diameter 8–100 μm called the core and (ii) a coaxial cylindrical region of diameter normally 125 μm called the cladding. The refractive index of the core is kept greater than that of the cladding. This feature makes light travel through this structure by the phenomenon of total internal reflection. In order to give strength to the optical fiber, it is given a primary or buffer coating of plastic, and then a cable is made of several such fibers. This optical fiber cable serves as an information channel.

For clarity of the transmitted information, it is required that the information channel should have low attenuation for the frequencies being transmitted through it and a large light-gathering capacity. Furthermore, the cable should have low dispersion in both the time and frequency domains, because high dispersion results in the distortion of the propagating signal.

1.3.6 Repeater

As the optical signals propagate along the length of the fiber, they get attenuated due to absorption, scattering, etc., and broadened due to dispersion. After a certain length, the cumulative effect of attenuation and dispersion causes the signals to become weak and indistinguishable. Therefore, before this happens, the strength and shape of the signal must be restored. This can be done by using either a regenerator or an optical amplifier, e.g., an erbium-doped fiber amplifier (EDFA), at an appropriate point along the length of the fiber.

1.3.7 Optoelectronic Detector

The reconversion of an optical signal into an electrical signal takes place at the OE detector. Semiconductor *p-i-n* or avalanche photodiodes are employed for this purpose. The photocurrent developed by these detectors is normally proportional to the incident optical power and hence to the information input. The desirable characteristics of a detector include small size, low power consumption, linearity, flat spectral response, fast response to optical signals, and long operating life.

1.3.8 Receiver

For analog transmission, the output photocurrent of the detector is filtered to remove the dc bias that is normally applied to the signal in the modulator module, and also to block any other undesired frequencies accompanying the signal. After filtering, the photocurrent is amplified if needed. These two functions are performed by the receiver module.

For digital transmission, in addition to the filter and amplifier, the receiver may include decision circuits. If the original information is in analog form, a digital-to-analog converter may also be required.

The design of the receiver is aimed at achieving high sensitivity and low distortion. The signal-to-noise ratio (SNR) and bit-error rate (BER) for digital transmission are important factors for quality communication.

1.3.9 Information Output

Finally, the information must be presented in a form that can be interpreted by a human observer. For example, it may be required to transform the electrical output into a sound wave or a visual image. Suitable output transducers are required for achieving this transformation. In some cases, the electrical output of the receiver is directly usable. This situation arises when a fiber-optic system forms the link between different computers or other machines.

1.4 ADVANTAGES OF FIBER-OPTIC SYSTEMS

Fiber-optic systems have several advantages, some of which were apparent when the idea of optical fibers as a means of communication was originally conceived.

For communication purposes, the transmission bandwidth and hence the information-carrying capacity of a fiber-optic system is much greater than that of coaxial copper cables, wide-band radio, or microwave systems. (The concepts of bandwidth and channel capacity are explained in Appendix A1.2.) Small size and light weight of optical fibers coupled with low transmission loss (typically around 0.2 dB/km) reduces the system cost as well as the need for numerous repeaters in long-haul telecommunication applications.

Optical fibers are insulators, as they are made up of glass or plastic. This property is useful for many applications. Particularly, it makes the optical signal traversing through the fiber free from any radio-frequency interference (RFI) and electromagnetic interference (EMI). RFI is caused by radio or television broadcasting stations, radars, and other signals originating in electronic equipment. EMI may be caused by these sources of radiation as well as from industrial machinery, or from naturally occurring phenomena such as lightning or unintentional sparking. Optical fibers do not pick up or propagate electromagnetic pulses (EMPs). Thus, fiber-optic systems may be employed for reliable monitoring and telemetry in industrial environments, where EMI and EMPs cause problems for metallic cables. In fact, in recent years, a variety of fiber-optic sensor systems have been developed for accurate measurement of parameters such as position, displacement, liquid level, temperature, pressure, refractive index, and so on.

In contrast to copper cables, the signal being transmitted through an optical fiber cannot be obtained from it without physically intruding the fiber. Further, the optical signal propagating along the fiber is well protected from interference and coupling with other communication channels (electrical or optical). Thus, optical fibers offer a high degree of signal security. This feature is particularly suitable for military and banking applications and also for computer networks.

For large-scale exploitation, the system's cost and the availability of raw material are two important considerations. The starting material for the production of glass fibers is silica, which is easily available. Regarding the cost, it has been shown that for long-distance communication, fiber cables are cheaper to transport and easier to install than metallic cables. Despite the fragile nature of glass fiber, these cables are surprisingly strong and flexible.

1.5 EMERGENCE OF FIBER OPTICS AS A KEY TECHNOLOGY

Fiber optics technology has been developed over the past three decades in a series of generations mainly identified by the operating wavelengths they employed. The first-generation fiber-optic systems employed a wavelength of 0.85 μm. This wavelength corresponds to the 'first (low-loss) window' in a silica-based optical fiber. This is shown in Fig. 1.3, where attenuation (dB/km) has been plotted as a function of wavelength for a typical silica-based optical fiber. The region around 0.85 μm was attractive for two reasons, namely, (i) the technology for LEDs that could be coupled to optical fibers had already been perfected and (ii) silicon-based photodiodes could be used for detecting a wavelength of 0.85 μm. This window exhibited a relatively high loss of the order of 3 dB/km. Therefore, as technology progressed, the first window become less attractive.

In fact from the point of view of long-haul communication applications, it is not only the attenuation but also the dispersion of pulses by a fiber that plays a key role in selecting the wavelength and the type of fiber. Therefore second-generation systems used a 'second window' at 1.3 μm with theoretically zero dispersion for silica fibers and also lesser attenuation around 0.5 dB/km.

In 1977, Nippon Telegraph and Telephone (NTT) succeeded in developing a 'third window' at 1.55 μm that offered theoretically minimum attenuation (but non-zero dispersion) for silica fibers. The corresponding loss was about 0.2 dB/km. The same year witnessed the successful commercial deployment of fiber-optic systems by AT&T Corp. (formerly, American Telephone and Telegraph Corporation) and GTE Corp.

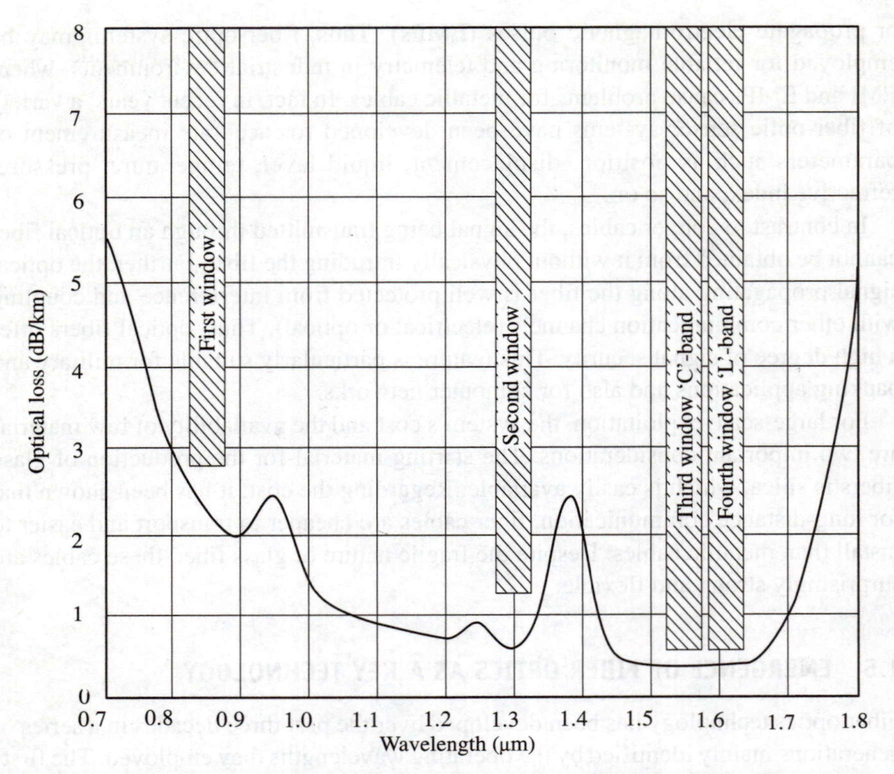

Fig. 1.3 Attenuation versus wavelength for a typical silica-based fiber. Low-loss windows (i.e., wavelength ranges) have been shown by hatched areas.

(formerly, General Telephone and Electronics Corporation). Since then there has been tremendous progress in this technology.

Originally, short-haul fiber-optic links were designed around multimode graded-index fibers; but in the 1980s fiber-optic networks first used single-mode fibers (SMF) operating at 1.31 μm. Later on, use of the dispersion-shifted fiber (DSF) with low dispersion and low attenuation (at 1.55 μm) became the standard practice in long-haul communication links.

In 1990, Bell Laboratories demonstrated the transmission of 2.5 Gbits (10^9 bits) per second over 7500 km without regeneration. This system employed a soliton laser and an EDFA.

In 1998 Bell Laboratories achieved another milestone of transmitting 100 simultaneous optical signals, each at a data rate of 10 Gbits per second, for a distance of nearly 400 km. In this experiment, the concept of dense wavelength-division multiplexing (DWDM), which allows multiple wavelengths to be transmitted as a single optical signal, was employed to increase the total data rate on one fiber to 1 Tbit (10^{12} bits) per second. Since then this technology has been growing fast.

In fact, researches in dispersion management technologies have played a key role in the explosive progress in optical-fiber communications in the past decade. As the demand for more capacity, more speed and more clarity increases, driven mainly by the phenomenal growth of the Internet, the move to optical networking is the focus of new technologies. Therefore the next-generation fiber-optic systems will involve a revamping of existing network architectures and developing all-optical networks. Tunable lasers offer the possibility of connecting any WDM node with a single transmitter, thus enhancing network flexibility and also enabling smooth upgrading of networks. In order to get maximum benefit from a reconfigurable network, one would need not only tunable transmitters but also tunable wavelength add-drop multiplexers (WADM) and tunable receivers. Such configurations are called tunable-transmitter–tunable-receiver (TTTR) networks.

As an optical fiber has immense potential bandwidth (more than 50 THz), there are extraordinary possibilities for future fiber-optic systems and their applications.

1.6 THE ROLE OF FIBER OPTICS TECHNOLOGY

We are living in an 'information society', where the efficient transfer of information is highly relevant to our well-being. Fiber-optic systems are going to form the very means of such information transfer and hence they are destined to have a very important role, directly or indirectly, in the development of almost every sphere of life. Table 1.1 gives just a glance at the areas in which fiber optics technology is going to have a major impact.

Table 1.1 Role of fiber optics technology

Direct	Indirect
Voice Communication	Entertainment
• Inter-office	• High-definition television (HDTV)
• Intercity	• Video on demand
• Intercontinental links	• Video games
Video communication	Power systems
• TV broadcast	• Monitoring of power-generating stations
• Cable television (CATV)	• Monitoring of transformers
• Remote monitoring	Transportation
• Wired city	• Traffic control in high-speed electrified railways
• Videophones	• Traffic control in metropolitan cities
Data transfer	• Monitoring of aircraft
• Inter-office data link	Health care
• Local area networks	• Minimal invasive diagnosis, surgery, and therapy
• Computers	
• Satellite ground stations	

(*contd*)

Table 1.1 (*contd*)

Direct	Indirect
Internet • Email • Access to remote information (e.g., web pages) • Videoconferencing Sensor systems for industrial applications • Point sensors for measurement of position, flow rate, temperature, pressure, etc. • Distributed sensors • Smart structures • Robotics	• Endoscopes • Biomedical sensors • Remote monitoring of patients Military defence • Strategic base communication • Guided missiles • Sensors • Virtual wars Business development • Videoconferencing • Industrial CAD/CAM • Monitoring of manufacturing plants Education • Closed-circuit television (CCTV) • Distance learning • Access to remote libraries

1.7 OVERVIEW OF THE TEXT

An optical fiber is the heart of any fiber-optic system. We, therefore, begin with an easy-to-understand ray model of the propagation of light through optical fibers. The effect of multipath and material dispersion is discussed to assess the limitation of optical fibers. This is presented in Chapter 2.

In fact, light is an electromagnetic wave. Its propagation through optical fibers cannot be properly analysed using a ray model. Therefore, in Chapter 3, we start with Maxwell's equations governing the propagation of electromagnetic waves through any medium and apply them with appropriate boundary conditions to the simplest kind of planar waveguide. This treatment provides the much needed understanding of the concept of modes. Wave propagation in cylindrical dielectric waveguides is treated in Chapter 4. The modal analysis of step-index fibers is given in detail. The relevant parameters of graded-index fibers are also discussed in this chapter.

Modern fiber-optic communication systems use SMFs. Therefore, Chapter 5 is fully devoted to the discussion of such fibers. Thus key parameters of SMF—dispersion, attenuation, and types—have been explained here in detail.

Chapter 6 deals with the manufacturing methods of optical fibers, the optical fiber cable design, and the losses associated with fiber-to-fiber connections. Several methods of characterization of optical fibers have also been given in this chapter.

Any fiber-optic system would require an appropriate source of light at the transmitter end and a corresponding detector at the receiver end. The treatment of optoelectronic sources is given in Chapter 7. Starting with basic semiconductor physics, this chapter explains the theory behind LEDs, their design, and limitations, and then focuses on ILDs, which are commonly used in fiber-optic communication systems. Chapter 8 explains the basic principles of optoelectronic detection. Different types of photodiodes have also been described briefly.

A newer trend in fiber-optic communications is to use optoelectronic modulation rather than electronic modulation. Thus, Chapter 9 includes different types of optoelectronic modulators based on the electro-optic and the acousto-optic effect.

The in-line repeaters, power boosters, or pre-amplifiers of present-day communication systems are all optical amplifiers. Therefore, Chapter 10 is fully devoted to the discussion of different types of optical amplifiers, e.g., semiconductor optical amplifier, EDFA, etc.

Chapters 11 and 12 together present a discussion of the relevant concepts and related components of modern fiber-optic communication systems. Thus, the latest concepts of WDM and DWDM and related components have been described in Chapter 11. System design considerations for digital and analog systems, different types of system architectures, effects of non-linear phenomena, dispersion management, and solitons have been described in Chapter 12.

Optical fibers have not only revolutionized communication systems but also, owing to their small size, immunity from EMI and RFI, compatibility with fully distributed sensing, etc., they have, along with lasers, invaded the field of industrial instrumentation. Thus, Chapter 13 describes various types of point and distributed fiber-optic sensors and their applications in industrial measurements. Chapter 14 focuses mainly on laser-based systems and their applications in various fields.

Finally, in order to obtain hands-on experience and a deeper understanding of the basic principles and also a feel of their applications, some laboratory-oriented projects in this area have been described in Chapter 15. Industrious students can use the material presented in this chapter for making their own kit with the help of their professors for performing various measurements.

APPENDIX A1.1: RELATIVE AND ABSOLUTE UNITS OF POWER

The relative power level between two points along a fiber-optic communication link is measured in decibels (dB). For a particular wavelength λ, if P_0 is the power launched at one end of the link and P is the power received at the other end, the efficiency of transmission of the link is P/P_0. When P and P_0 are both measured in the same units, their ratio in decibels is expressed as follows:

$$\text{dB} = 10\log_{10}(P/P_0)$$

There is always some loss in the communication link. Hence the ratio P/P_0 is always less than 1 and the logarithm of this ratio is negative.

In order to make absolute measurements, P_0 is given a reference value, normally 1 mW. The value of power (say, P) relative to P_0 (= 1 mW) is denoted by dBm. Thus

$$\text{dBm} = 10\log_{10}\frac{P\ (\text{mW})}{P_0\ (1\ \text{mW})} = 10\log_{10}P$$

APPENDIX A1.2: BANDWIDTH AND CHANNEL CAPACITY

The optical bandwidth of a fiber-optic system is the range of frequencies (transmitted by the system) between two points (f_1 and f_2 on a frequency scale) where the output optical power drops to 50% of its maximum value (see Fig. A1.2). This corresponds to a loss of –3 dB. Normally f_1 is taken to be zero.

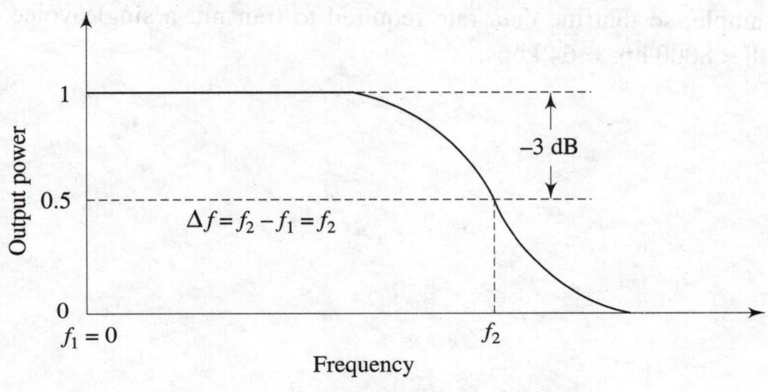

Fig. A1.2

It is important to note that in an all-electrical system, the power is proportional to the square of the root-mean-square value of the current. Thus the 0.5 (or –3 dB) point on a power scale corresponds to 0.707 on a current scale. The electrical bandwidth for such systems (with $f_1 = 0$) is defined as the frequency for which the output current amplitude drops to 0.70 of its maximum value. In a fiber-optic system, the power supplied by the optical source is normally proportional to the current supplied to it (and not to the square of the current as in an all-electrical system). In this case, therefore, the half-power point is equivalent to half-current point. Thus the electrical bandwidth $(\Delta f)_{el}$ of a fiber-optic system is less than its optical bandwidth $(\Delta f)_{opt}$. Actually,

$$(\Delta f)_{el} = (\Delta f)_{opt}(1/2)^{1/2}$$

or

$$(\Delta f)_{el} = 0.707(\Delta f)_{opt}$$

The term 'channel capacity' is used in connection with a digital system or a link. It is the highest data rate (in bps) a system can handle. It is related to the bandwidth Δf and is limited by the S/N ratio of the received signal. Employing information theory, it can be shown that the channel capacity or the maximum bit rate B of a communication channel which is able to carry analog signals covering a bandwidth Δf (in Hz) is given by Shannon's formula:

$$B \text{ (bps)} = \Delta f \log_2 [1 + (S/N)^2]$$

where S is the average signal power and N is the average noise power. If $S/N \gg 1$,

$$B \approx 2\Delta f \log_2(S/N) \approx 6.64\Delta f \log_{10}(S/N)$$

Note In practice, when analog signals are to be transmitted digitally, the bit rate will depend on the sampling rate of the analog signal and the coding scheme. The Nyquist criterion suggests that an analog signal can be transmitted accurately if it is sampled at a rate of at least twice the highest frequency contained in the signal, i.e., if the sampling frequency $f_s \gg 2f_2$. Thus a voice channel of 4 kHz bandwidth would require $f_s = 8$ kHz. A standard coding procedure requires 8 bits to describe the amplitude of each sample, so that the data rate required to transmit a single voice message would be 8×8000 bps = 64 kbps.

Part I: Fiber Optics

2 Ray Propagation in Optical Fibers

After reading this chapter you will be able to understand the following:

- Ray propagation in step-index fibers
- Ray propagation in graded-index fibers
- Effect of multipath time dispersion
- Effect of material dispersion
- Calculation of rms pulse width

2.1 INTRODUCTION

Fiber optics technology uses light as a carrier for communicating signals. Classical wave theory treats light as electromagnetic waves, whereas the quantum theory treats it as photons, i.e., quanta of electromagnetic energy. In fact, this is what is known as the wave-particle duality in modern physics.

Both points of view are valid and valuable in their respective domains. However, it will be easier to understand the propagation of light signals through optical fibers if we think of light as rays that follow a straight-line path in going from one point to another. *Ray optics* employs the geometry of a straight line to explain the phenomena of reflection, refraction, etc. Hence, it is also called *geometrical optics*. Let us review, in brief, certain laws of geometrical optics, which aid the understanding of ray propagation in optical fibers.

2.2 REVIEW OF FUNDAMENTAL LAWS OF OPTICS

The most important optical parameter of any transparent medium is its refractive index n. It is defined as the ratio of the speed of light in vacuum (c) to the speed of light in the medium (v). That is,

$$n = \frac{c}{v} \tag{2.1}$$

As v is always less than c, n is always greater than 1. For air, $n = n_a \approx 1$.

The phenomenon of refraction of light at the interface between two transparent media of uniform indices of refraction is governed by Snell's law. Consider a ray of light passing from a medium of refractive index n_1 into a medium of refractive index n_2 [see Fig. 2.1(a)]. Assume that $n_1 > n_2$ and that the angles of incidence and refraction with respect to the normal to the interface are, respectively, ϕ_1 and ϕ_2. Then, according to Snell's law,

$$n_1 \sin\phi_1 = n_2 \sin\phi_2 \tag{2.2}$$

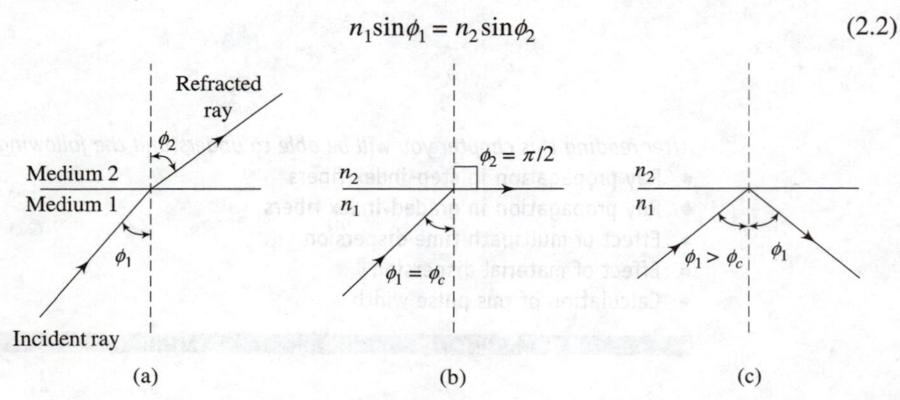

Fig. 2.1 (a) Refraction of a ray of light. (b) Critical ray incident at $\phi_1 = \phi_c$ and refracted at $\phi_2 = \pi/2$. (c) Total internal reflection ($\phi_1 > \phi_c$).

As $n_1 > n_2$, if we increase the angle of incidence ϕ_1, the angle of refraction ϕ_2 will go on increasing until a critical situation is reached, when for a certain value of $\phi_1 = \phi_c$, ϕ_2 becomes $\pi/2$, and the refracted ray passes along the interface. This angle $\phi_1 = \phi_c$ is called the critical angle. If we substitute the values of $\phi_1 = \phi_c$ and $\phi_2 = \pi/2$ in Eq. (2.2), we see that $n_1 \sin\phi_c = n_2 \sin(\pi/2) = n_2$. Thus

$$\sin\phi_c = n_2/n_1 \tag{2.3}$$

If the angle of incidence ϕ_1 is further increased beyond ϕ_c, the ray is no longer refracted but is reflected back into the same medium [see Fig. 2.1(c)]. (This is ideally expected. In practice, however, there is always some tunnelling of optical energy through this interface. The wave carrying away this energy is called the *evanescent wave*. This can be explained in terms of electromagnetic theory, discussed in Chapters 3 and 4.) This is called total internal reflection. It is this phenomenon that is responsible for the propagation of light through optical fibers.

There are several types of optical fibers designed for different applications. We will discuss these in Chapter 4. For the present discussion, we will begin with ray propagation in the simplest kind of optical fiber.

2.3 RAY PROPAGATION IN STEP-INDEX FIBERS

A step-index optical fiber is a thin dielectric waveguide consisting of a solid cylindrical core (diameter, $2a$ = 50–100 μm) of refractive index n_1, surrounded by a coaxial cylindrical cladding (diameter, $2b$ = 120–140 μm) of refractive index n_2 ($n_1 > n_2$), as shown in Fig. 2.2. The refractive index n is a step function of the radial distance r, as shown in Fig. 2.2(a). Hence it is called a *step-index* (SI) fiber.

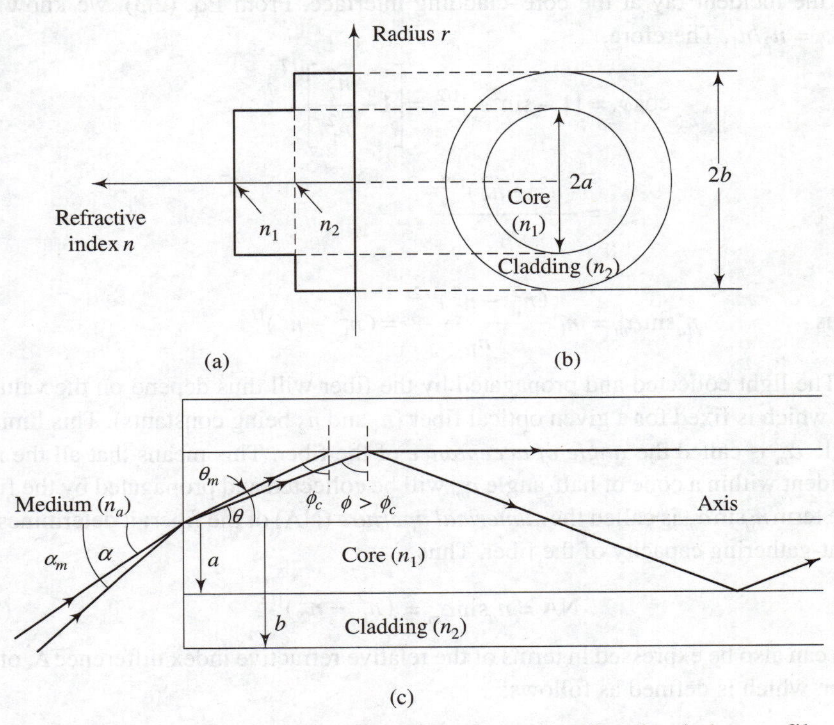

Fig. 2.2 Basic structure of a step-index fiber: (a) refractive index profile, (b) cross section (front view), (c) ray propagation (side view)

The material used for the production of an optical fiber is either silica (glass) or plastic. Assume that such a fiber is placed in a medium (normally air) of refractive index n_a ($n_1 > n_2 > n_a$). If a ray of light enters from the flat end of the fiber at some angle α, from the medium into the core, it will bend towards the normal, making an angle of refraction, θ. This ray then strikes the core–cladding interface at an angle of incidence ϕ. If $\phi > \phi_c$, the ray will undergo total internal reflection. It will suffer multiple reflections at the core–cladding interface and emerge out of the fiber at the other end.

What is the allowed range of α? The propagation of rays by total internal reflection requires ϕ to be greater than ϕ_c and hence θ to be less than $\theta_m = \pi/2 - \phi_c$. Thus, the angle of incidence α should be less than a certain angle α_m. This maximum value of α_m will correspond to the limiting value of $\theta_m = \pi/2 - \phi_c$.

Applying Snell's law at the core–medium (air) interface, we get for $\alpha = \alpha_m$ and corresponding $\theta = \theta_m$,

$$n_a \sin\alpha_m = n_1 \sin\theta_m = n_1 \cos\phi_c$$

for the incident ray at the core–cladding interface. From Eq. (2.3), we know that $\sin\phi_c = n_2/n_1$. Therefore,

$$\cos\phi_c = [1 - \sin^2\phi_c]^{1/2} = \left[1 - \frac{n_2^2}{n_1^2}\right]^{1/2}$$

$$= \frac{(n_1^2 - n_2^2)^{1/2}}{n_1}$$

Thus $$n_a \sin\alpha_m = n_1 \frac{(n_1^2 - n_2^2)^{1/2}}{n_1} = (n_1^2 - n_2^2)^{1/2}$$

The light collected and propagated by the fiber will thus depend on the value of α_m, which is fixed for a given optical fiber (n_1 and n_2 being constants). This limiting angle α_m is called the *angle of acceptance* of the fiber. This means that all the rays incident within a cone of half-angle α_m will be collected and propagated by the fiber. The term $n_a \sin\alpha_m$ is called the *numerical aperture* (NA) of the fiber; it determines the light-gathering capacity of the fiber. Thus

$$\text{NA} = n_a \sin\alpha_m = (n_1^2 - n_2^2)^{1/2} \tag{2.4}$$

NA can also be expressed in terms of the relative refractive index difference Δ, of the fiber, which is defined as follows:

$$\Delta = \frac{n_1^2 - n_2^2}{2n_1^2} \tag{2.5}$$

Hence $$\text{NA} = (n_1^2 - n_2^2)^{1/2} = n_1\sqrt{2\Delta} \tag{2.6}$$

So far, we have considered the propagation of a single ray of light. However, a pulse of light consists of several rays, which may propagate at all values of α, varying from 0 to α_m. The paths traversed by the two extreme rays, one corresponding to $\alpha = 0$ and the other corresponding to α very nearly equal to (but less than) α_m, are shown in Fig. 2.3.

An axial ray travels a distance L inside the core of index n_1 with velocity v in time

$$t_1 = \frac{L}{v} = \frac{Ln_1}{c}$$

since, by definition, $n_1 = c/v$.

The most oblique ray corresponding to $\alpha \approx \alpha_m$ will cover the same length of fiber

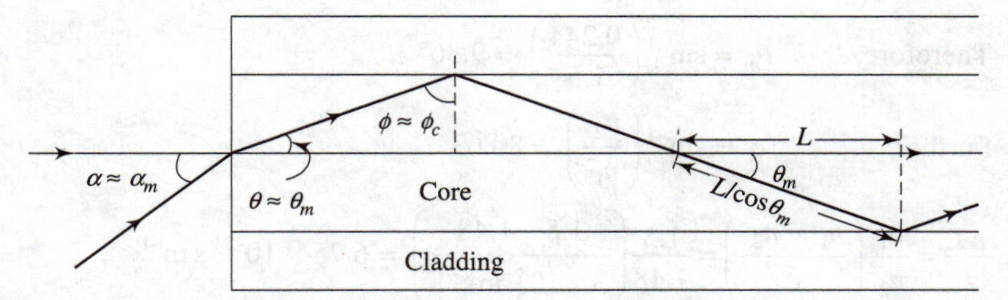

Fig. 2.3 Trajectories of two extreme rays inside the core of a step-index fiber

(axial length L, but actual distance $L/\cos\theta_m$) in time t_2 given by

$$t_2 = \frac{L/\cos\theta_m}{v} = \frac{Ln_1}{c\cos\theta_m} = \frac{Ln_1}{c\sin\phi_c} = \frac{Ln_1}{c(n_2/n_1)} = \frac{Ln_1^2}{cn_2}$$

The two rays are launched at the same time, but will be separated by a time interval ΔT after travelling the length L of the fiber, given by

$$\Delta T = t_2 - t_1 = \frac{Ln_1^2}{cn_2} - \frac{Ln_1}{c} = \frac{Ln_1}{c}\left(\frac{n_1 - n_2}{n_2}\right)$$

Thus a light pulse consisting of rays spread over $\alpha = 0$ to $\alpha = \alpha_m$ will be broadened as it propagates through the fiber, and the pulse broadening per unit length of traversal will be given by

$$\frac{\Delta T}{L} = \frac{n_1}{n_2}\left(\frac{n_1 - n_2}{c}\right) \tag{2.7}$$

This is referred to as *multipath time dispersion* of the fiber.

Example 2.1 If the step-index fiber of Fig. 2.3 has a core of refractive index 1.5, a cladding of refractive index 1.48, and a core diameter of 100 μm, calculate, assuming that the fiber is kept in air, the (a) NA of the fiber; (b) angles α_m, θ_m, and ϕ_c; and (c) pulse broadening per unit length ($\Delta T/L$) due to multipath dispersion.

Solution

$n_1 = 1.5$, $n_2 = 1.48$, $2a = 100$ μm, and $n_a = 1$.

(a) $\text{NA} = \sqrt{n_1^2 - n_2^2} = \sqrt{2.25 - 2.19} = 0.244$

(b) $\text{NA} = n_a \sin\alpha_m = 1 \times \sin\alpha_m = 0.244$

Therefore, $\alpha_m = 14.13°$

Also $n_a \sin\alpha_m = n_1 \sin\theta_m$

or $0.244 = 1.5 \sin\theta_m$

Therefore, $\theta_m = \sin^{-1}\left(\frac{0.244}{1.5}\right) = 9.36°$

Further, $\phi_c = \sin^{-1}\left(\frac{n_2}{n_1}\right) = 80.63°$

(c) $\frac{\Delta T}{L} = \frac{n_1}{n_2}\left(\frac{n_1 - n_2}{c}\right) = \frac{1.5}{1.48}\left(\frac{1.5 - 1.48}{3 \times 10^8 \text{ m s}^{-1}}\right) = 6.75 \times 10^{-11} \text{ s m}^{-1}$

Example 2.2 For the optical fiber of Example 2.1, what are the minimum and maximum number of reflections per metre for the rays guided by it?

Solution

The ray that passes along the axis of the fiber, that is, the one for which $\alpha = 0$, will not be reflected, but the ray that follows the most oblique path, that is, is incident at an angle α very nearly equal to (but less than) α_m, will suffer maximum number of reflections. These can be calculated using the geometry of Fig. 2.3.

$$\tan\theta_m = \frac{a}{L}$$

Hence $L = \frac{a}{\tan\theta_m} = \frac{50 \times 10^{-6} \text{ m}}{\tan(9.36°)} = 3.03 \times 10^{-4} \text{ m}$

Therefore, the maximum number of reflections per metre would be

$$\frac{1}{2L} = 1648 \text{ m}^{-1}$$

All other rays will suffer reflections between these two extremes of 0 and 1648 m^{-1}.

2.4 RAY PROPAGATION IN GRADED-INDEX FIBERS

In a step-index fiber, the refractive index of the core is constant, n_1, and that of the cladding is also constant, n_2; n_1 being greater than n_2. The refractive index n is a step function of the radial distance. A pulse of light launched in such a fiber will get broadened as it propagates through it due to multipath time dispersion. Therefore, such fibers cannot be used for long-haul applications. In order to overcome this problem, another class of fibers is made, in which the core index is not constant but varies with radius r according to the following relation:

$$n(r) = \begin{cases} n_1 = n_0 \left[1 - 2\Delta\left(\frac{r}{a}\right)^{\alpha}\right]^{1/2} & \text{for } r \le a \\ n_2 = n_0\,[1 - 2\Delta]^{1/2} = n_c & \text{for } b \ge r \ge a \end{cases} \tag{2.8}$$

where $n(r)$ is the refractive index at radius r, a is the core radius, b is the radius of the cladding, n_0 is the maximum value of the refractive index along the axis of the core, Δ is the relative refractive index difference, and α is called the *profile parameter*. Such a fiber is called a *graded-index* (GI) fiber. For $\alpha = 1$, the index profile is triangular; for $\alpha = 2$, the profile is parabolic; and for $\alpha = \infty$, the profile is that of a SI fiber, as shown in Fig. 2.4.

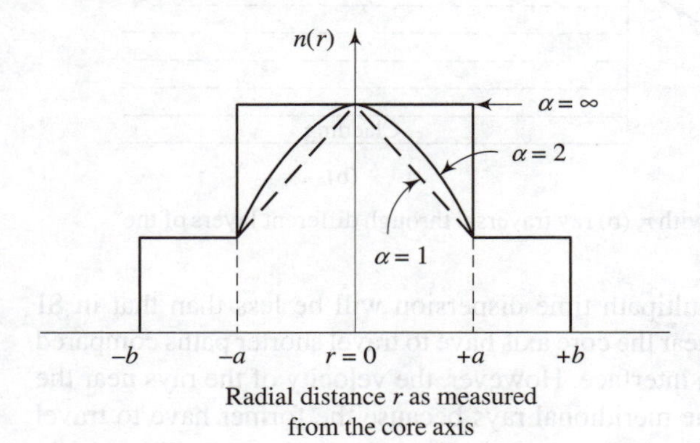

Fig. 2.4 Variation of $n(r)$ with r for different refractive index profiles

For a parabolic profile, which reduces the modal dispersion considerably, as we will see later, the expression for NA can be derived as follows:

$$\text{NA} = (n_1^2 - n_2^2)^{1/2}$$

$$= \left[n_0^2 \left\{ 1 - 2\Delta \left(\frac{r}{a} \right)^2 \right\} - n_0^2 (1 - 2\Delta) \right]^{1/2}$$

$$= n_0 \left[2\Delta \left(1 - \frac{r^2}{a^2} \right) \right]^{1/2} \tag{2.9}$$

Therefore axial NA $= n_0 \sqrt{2\Delta}$; the NA decreases with increasing r and becomes zero at $r = a$.

In order to appreciate ray propagation through a GI fiber, let us first visualize the core of this fiber as having been made up of several coaxial cylindrical layers, as shown in Fig. 2.5.

The refractive index of the central cylinder is the highest, and it goes on decreasing in the successive cylindrical layers. Thus, the meridional ray shown takes on a curved path, as it suffers multiple refractions at the successive interfaces of high to low refractive indices. The angle of incidence for this ray goes on increasing until the condition for total internal reflection is met; the ray then travels back towards the core axis. On the other hand, the axial ray travels uninterrupted.

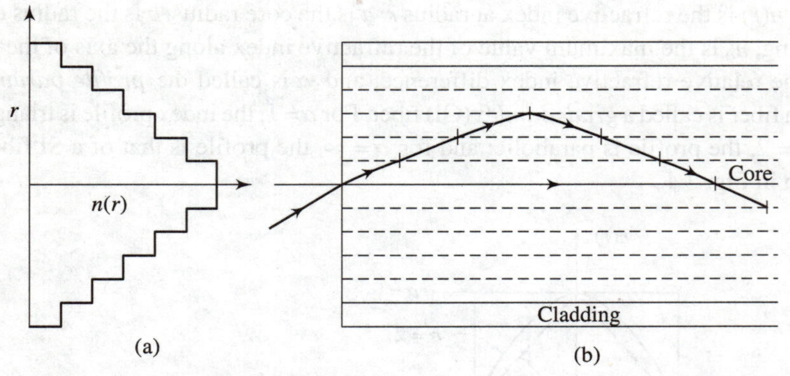

Fig. 2.5 (a) Variation of n with r, (b) ray traversal through different layers of the core

In this configuration, the multipath time dispersion will be less than that in SI fibers. This is because the rays near the core axis have to travel shorter paths compared to those near the core–cladding interface. However, the velocity of the rays near the axis will be less than that of the meridional rays because the former have to travel through a region of high refractive index ($v = c/n$). Thus, both the rays will reach the other end of the fiber almost simultaneously, thereby reducing the multipath dispersion. If the refractive index profile is such that the time taken for the axial and the most oblique ray is same, the multipath dispersion will be zero. In practice, a parabolic profile ($\alpha = 2$), shown in Fig. 2.6, reduces this type of dispersion considerably.

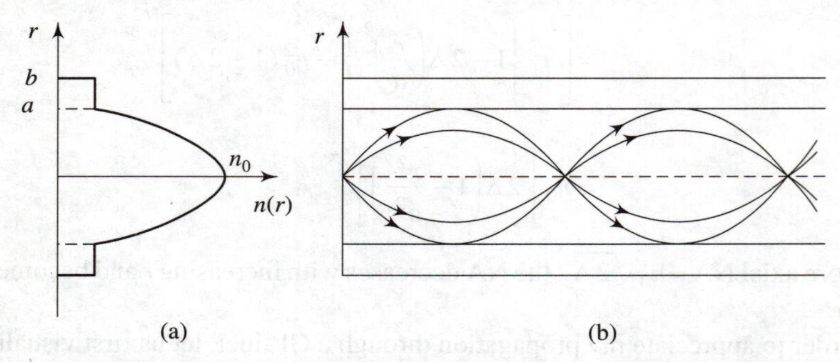

Fig. 2.6 (a) The parabolic profile of a GI fiber, (b) ray path in such a fiber

2.5 EFFECT OF MATERIAL DISPERSION

We know from our earlier discussion that the refractive index n of any transparent medium is given by $n = c/v$, where c is the velocity of light in vacuum and v is its velocity in the medium. In terms of wave theory, v is called the *phase velocity* v_p of the wave in the medium. Thus,

$$v = v_p = c/n \tag{2.10}$$

If $\omega(= 2\pi f)$ is the angular frequency of the wave (in rad/sec), f being the frequency in hertz, and $\beta(= 2\pi/\lambda_m)$ is the propagation constant, λ_m being the wavelength of light in the medium (which is equal to λ/n; λ is the free-space wavelength), then the phase velocity of the wave is also given by

$$v_p = \omega/\beta \tag{2.11}$$

Thus, using Eqs (2.10) and (2.11), we get

$$v_p = c/n = \omega/\beta$$

or

$$n = \frac{c\beta}{\omega} \tag{2.12}$$

At this point, it is important to know that any signal superposed onto a wave does not propagate with the phase velocity v_p of the wave, but travels with a group velocity v_g given by the following expression:

$$v_g = \frac{d\omega}{d\beta} = \frac{1}{d\beta/d\omega} \tag{2.13}$$

In a non-dispersive medium, v_p and v_g are same, as v_p is independent of the frequency ω; but in a dispersive medium, where v_p is a function of ω,

$$v_g = \frac{1}{d\beta/d\omega} = \frac{v_p}{1-(\omega/v_p)(dv_p/d\omega)} \tag{2.14}$$

Thus a signal, which is normally a light pulse, will travel through a dispersive medium (e.g., the core of the optical fiber) with speed v_g. Therefore, for such a pulse, we may define a group index n_g such that

$$n_g = c/v_g \tag{2.15}$$

Let us substitute for v_g from Eq. (2.13) in Eq. (2.15) to get

$$n_g = c\frac{d\beta}{d\omega} = c\frac{d}{d\omega}\left(\frac{\omega n}{c}\right) = \frac{d}{d\omega}(n\omega)$$

$$= n + \omega\frac{dn}{d\omega} \tag{2.16}$$

This is an important expression, relating the group index n_g with the ordinary refractive index or the phase index n.

Since

$$\frac{dn}{d\omega} = \frac{dn}{d\lambda}\frac{d\lambda}{d\omega}$$

and

$$\omega = \frac{2\pi c}{\lambda}$$

$$\frac{d\omega}{d\lambda} = -\frac{2\pi c}{\lambda^2}$$

we have from Eq. (2.16),

$$n_g = n + \frac{2\pi c}{\lambda}\frac{dn}{d\lambda}\left(-\frac{\lambda^2}{2\pi c}\right) = n - \lambda\frac{dn}{d\lambda} \tag{2.17}$$

Thus
$$v_g = \frac{c}{n_g} = \frac{c}{(n - \lambda\, dn/d\lambda)} \tag{2.18}$$

A light pulse, therefore, will travel through the core of an optical fiber of length L in time t given by

$$t = \frac{L}{v_g} = \left[n - \lambda\frac{dn}{d\lambda}\right]\frac{L}{c} \tag{2.19}$$

If the spectrum of the light source has a spread of wavelength $\Delta\lambda$ about λ and if the medium of the core is dispersive, the pulse will spread out as it propagates and will arrive at the other end of length L, over a spread of time Δt. If $\Delta\lambda$ is much smaller than the central wavelength λ, we can write

$$\Delta t = \frac{dt}{d\lambda}\Delta\lambda = \frac{L}{c}\left[\frac{dn}{d\lambda} - \frac{dn}{d\lambda} - \lambda\frac{d^2 n}{d\lambda^2}\right]\Delta\lambda$$

$$= -\frac{L}{c}\lambda\frac{d^2 n}{d\lambda^2}\Delta\lambda \tag{2.20}$$

If $\Delta\lambda$ is the full width at half maximum (FWHM) of the peak spectral power of the optical source at λ, then its relative spectral width γ is given by

$$\gamma = \left|\frac{\Delta\lambda}{\lambda}\right| \tag{2.21}$$

If an impulse of negligible width is launched into the fiber, then $\Delta\lambda$ will be the half power width τ of the output (or the broadened pulse). Thus, the pulse broadening due to material dispersion may be given by

$$\tau = \frac{L}{c}\gamma\left|\lambda^2\frac{d^2 n}{d\lambda^2}\right|$$

or
$$\frac{\tau}{L} = \frac{\gamma}{c}\left|\lambda^2\frac{d^2 n}{d\lambda^2}\right| \tag{2.22}$$

The material dispersion of optical fibers is quoted in terms of the material dispersion parameter D_m given by

$$D_m = \frac{1}{L}\frac{\tau}{\Delta\lambda} = \frac{\lambda}{c}\left|\frac{d^2 n}{d\lambda^2}\right| \tag{2.23}$$

D_m has the units of ps nm^{-1} km^{-1}. Its variation with wavelength for pure silica is shown in Fig. 2.7. The example of pure silica has been chosen because a majority of fibers are manufactured using silica as the host material.

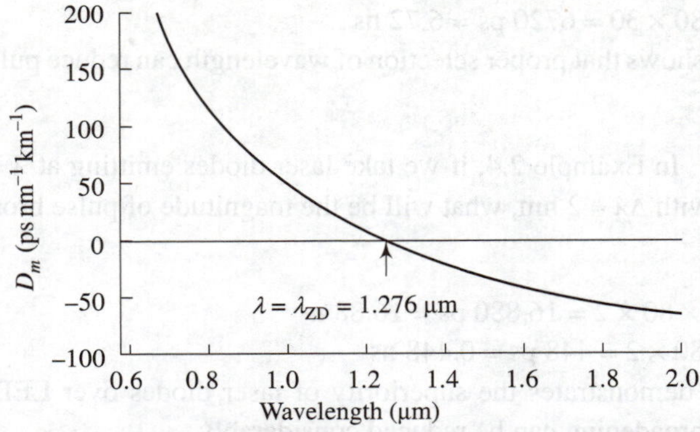

Fig. 2.7 Material dispersion parameter D_m as a function of λ for pure silica (Pyne & Gambling 1975)

Notice that D_m changes sign at $\lambda = \lambda_{ZD} = 1.276$ μm (for pure silica). This point has been frequently referred to as the wavelength of zero material dispersion. This wavelength can be changed slightly by adding other dopants to silica. Thus, the use of an optical source with a narrow spectral width (e.g., injection laser) around λ_{ZD} would substantially reduce the pulse broadening due to material dispersion.

Example 2.3 It is given that for a GaAs LED, the relative spectral width γ at $\lambda = 0.85$ μm is 0.035. This source is coupled to a pure silica fiber (with $|\lambda^2(d^2n/d\lambda^2)| = 0.021$ for $\lambda = 0.85$ μm). Calculate the pulse broadening per kilometre due to material dispersion.

Solution

The pulse broadening per unit length due to material dispersion is given by

$$\frac{\tau}{L} = \frac{\gamma}{c}\left|\lambda^2 \frac{d^2 n}{d\lambda^2}\right| = \frac{0.035}{3 \times 10^8 \text{ m s}^{-1}} \times 0.021 = 2.45 \times 10^{-12} \text{ s m}^{-1}$$

or $$\frac{\tau}{L} = 2.45 \text{ ns km}^{-1}$$

Example 2.4 Calculate the total pulse broadening due to material dispersion for a graded-index fiber of total length 80 km when a LED emitting at (a) $\lambda = 850$ nm and (b) $\lambda = 1300$ nm is coupled to the fiber. In both the cases, assume that $\Delta\lambda \approx 30$ nm. The material dispersion parameters of the fiber for the two wavelengths are −105.5 ps nm^{-1} km^{-1} and −2.8 ps nm^{-1} km^{-1}, respectively.

Solution

(a) Using Eq. (2.23),

$$\tau = D_m L \Delta\lambda = 105.5 \times 80 \times 30 = 253{,}200 \text{ ps} = 253.2 \text{ ns}$$

(b) $\tau = 2.8 \times 80 \times 30 = 6720$ ps = 6.72 ns

This example shows that proper selection of wavelength can reduce pulse broadening considerably.

Example 2.5 In Example 2.4, if we take laser diodes emitting at $\lambda = 850$ nm and $\lambda = 1300$ nm with $\Delta\lambda \approx 2$ nm, what will be the magnitude of pulse broadening?

Solution

(a) $\tau = 105.5 \times 80 \times 2 = 16{,}880$ ps = 16.88 ns

(b) $\tau = 2.8 \times 80 \times 2 = 448$ ps = 0.448 ns

This example demonstrates the superiority of laser diodes over LEDs. With laser diodes, pulse broadening can be reduced considerably.

2.6 THE COMBINED EFFECT OF MULTIPATH AND MATERIAL DISPERSION

In a fiber-optic communication system, the optical power generated by the optical source, e.g., LED, is proportional to the input current to the transmitter. The optical power received by the detector is proportional to the power launched into and propagated by the optical fiber. This, in turn, gives rise to a proportional current at the receiver end, thus giving an overall linearity to the system.

Earlier, we have seen that multipath and material dispersion lead to the broadening of the pulse launched into a fiber. Thus the received pulse represents the impulse response of the fiber. If we assume that the FWHM of the transmitted pulse is τ_0, and the impulse responses due to multipath and material dispersion lead to approximately Gaussian pulses of FWHM τ_1 and τ_2, respectively, as shown in Fig. 2.8, and that the two mechanisms are almost independent of each other, then the received pulse width at half maximum power τ will be given by

$$\tau = [\tau_0^2 + \tau_1^2 + \tau_2^2]^{1/2} \tag{2.24a}$$

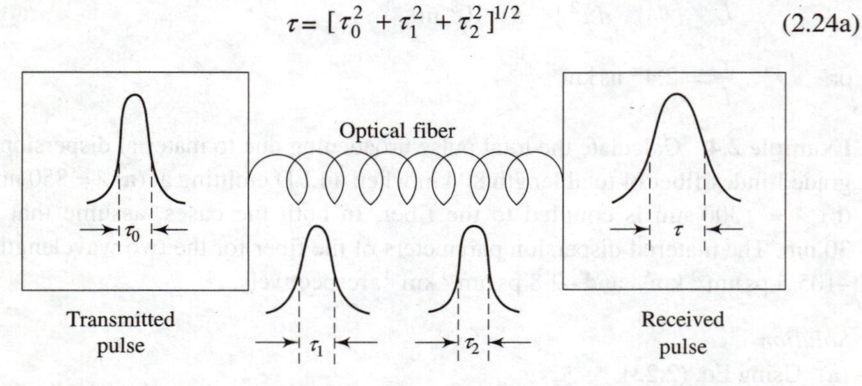

Fig. 2.8 The combined effect of multipath and material dispersion

The temporal pulse broadening per unit length due to both the effects is then given by

$$\frac{\tau}{L} = \left[\left(\frac{\tau_0}{L}\right)^2 + \left(\frac{\tau_1}{L}\right)^2 + \left(\frac{\tau_2}{L}\right)^2\right]^{1/2} \tag{2.24b}$$

where L is the total length of the fiber.

It is also possible to express the same result in terms of the root mean square (rms) pulse widths. The calculation of the rms pulse width is given in the following section. Thus, if the transmitted pulse is Gaussian in shape and has an rms width σ_0 and this pulse is broadened by both multipath and material dispersion leading to nearly Gaussian pulses of rms widths σ_1 and σ_2, respectively, then the rms width σ of the received pulse is given by (Gowar 2001)

$$\frac{\sigma}{L} = \left[\left(\frac{\sigma_0}{L}\right)^2 + \left(\frac{\sigma_1}{L}\right)^2 + \left(\frac{\sigma_2}{L}\right)^2\right]^{1/2} \tag{2.25}$$

2.7 CALCULATION OF RMS PULSE WIDTH

The rms width σ of a pulse is defined as follows. If $p(t)$ is the power distribution in the pulse as a function of time t and the total energy in the pulse is

$$\varepsilon = \int_{-\infty}^{\infty} p(t)\, dt \tag{2.26}$$

then its rms width σ is given by

$$\sigma^2 = \frac{1}{\varepsilon}\int_{-\infty}^{\infty} t^2 p(t)\, dt - \left[\frac{1}{\varepsilon}\int_{-\infty}^{\infty} tp(t)\, dt\right]^2 \tag{2.27}$$

Example 2.6 Calculate the rms pulse width σ of the rectangular pulse shown in Fig. 2.9 in terms of its FWHM τ.

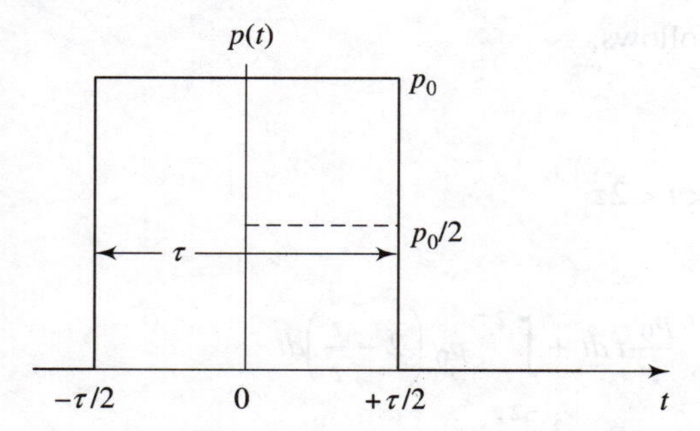

Fig. 2.9 Rectangular pulse

Solution

The total energy in the pulse

$$\varepsilon = \int_{-\infty}^{\infty} p(t)\,dt = \int_{-\tau/2}^{\tau/2} p_0\,dt = p_0\tau$$

$$\sigma^2 = \frac{1}{\varepsilon}\int_{-\tau/2}^{\tau/2} t^2 p_0\,dt - \left[\frac{1}{\varepsilon}\int_{-\tau/2}^{\tau/2} t p_0\,dt\right]^2$$

$$= \frac{1}{p_0\tau}\left[p_0\int_{-\tau/2}^{\tau/2} t^2\,dt\right] - \left[\frac{1}{p_0\tau}\,p_0\int_{-\tau/2}^{\tau/2} t\,dt\right]^2$$

$$= \frac{1}{\tau}\left[\frac{\tau^3}{12}\right] - 0 = \frac{\tau^2}{12}$$

Thus $$\sigma = \frac{\tau}{\sqrt{12}} = \frac{\tau}{2\sqrt{3}}$$

Example 2.7 In a typical fiber-optic communication system, the received power distribution as a function of time t is given by a symmetrical triangular pulse, as shown in Fig. 2.10. Calculate (a) the total energy in the pulse, (b) the mean pulse arrival time, and (c) the rms pulse width.

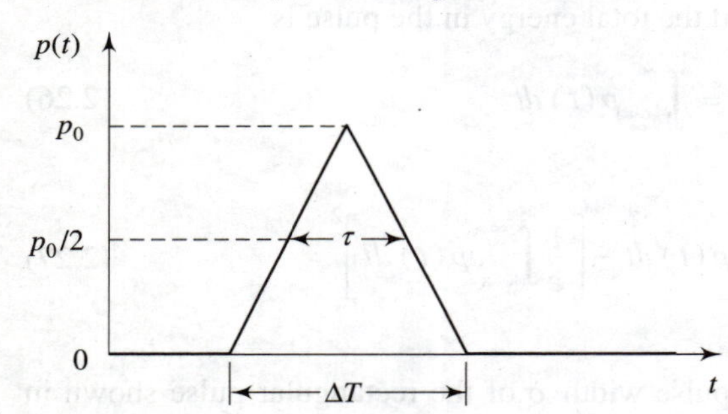

Fig. 2.10 Triangular pulse

Solution

The power $p(t)$ in the pulse varies as follows:

$$p(t) = \frac{p_0}{\tau}t,\quad 0 < t < \tau$$

$$= \frac{-p_0}{\tau}t + 2p_0,\quad \tau < t < 2\tau$$

(a) The total energy in the pulse

$$\varepsilon = \int_{-\infty}^{\infty} p(t)\,dt = \int_0^{\tau} \frac{p_0}{\tau}t\,dt + \int_{\tau}^{2\tau} p_0\left(2 - \frac{t}{\tau}\right)dt$$

$$= \left[\frac{p_0}{\tau}\frac{t^2}{2}\right]_0^{\tau} + \left[2p_0 t - \frac{p_0}{\tau}\frac{t^2}{2}\right]_{\tau}^{2\tau}$$

$$= p_0\tau$$

(b) The mean pulse arrival time $\bar{t}$ is given by

$$\bar{t} = \frac{1}{\varepsilon}\int_{-\infty}^{\infty} p(t)\,t\,dt$$

$$= \frac{1}{p_0\tau}\int_0^{\tau} \frac{p_0}{\tau} t^2 dt + \frac{1}{p_0\tau}\int_{\tau}^{2\tau}\left[2p_0 - \frac{p_0}{\tau}t\right] t\,dt$$

$$= \frac{1}{p_0\tau}\left(\frac{p_0}{\tau}\right)\left[\frac{t^3}{3}\right]_0^{\tau} + \frac{1}{p_0\tau}(2p_0)\left[\frac{t^2}{2}\right]_{\tau}^{2\tau} - \frac{1}{p_0\tau}\left(\frac{p_0}{\tau}\right)\left[\frac{t^3}{3}\right]_{\tau}^{2\tau}$$

$$= \tau$$

(c) The rms pulse width σ is determined as follows:

$$\sigma^2 = \frac{1}{\varepsilon}\int_{-\infty}^{\infty} t^2 p(t)\,dt - [\bar{t}\,]^2$$

$$= \frac{1}{p_0\tau}\int_0^{\tau} \frac{p_0}{\tau} t^3 dt + \frac{1}{p_0\tau}\int_{\tau}^{2\tau}\left[2p_0 - \frac{p_0}{\tau}t\right] t^2 dt - \tau^2$$

$$= \frac{1}{4}\tau^2 + \frac{2}{\tau}\left[\frac{t^3}{3}\right]_{\tau}^{2\tau} - \frac{1}{\tau^2}\left[\frac{t^4}{4}\right]_{\tau}^{2\tau} - \tau^2$$

$$= \frac{\tau^2}{6}$$

or $\quad \sigma = \dfrac{\tau}{\sqrt{6}}$

SUMMARY

- The refractive index (RI) of a transparent medium is the ratio of the speed of light in vacuum to the speed of light in the medium:

$$n = \frac{c}{v}$$

- The phenomenon of refraction of light is governed by Snell's law:

$$n_1 \sin\phi_1 = n_2 \sin\phi_2$$

- An optical fiber is a thin dielectric waveguide made up of a solid cylindrical core of glass (silica) or plastic of RI n_1 surrounded by a coaxial cylindrical cladding of RI n_2. When both n_1 and n_2 are constant, the fiber is called a step-index fiber.

- Within an optical fiber, light is guided by total internal reflection. For light guidance, two conditions must be met: (i) n_1 should be greater than n_2 and (ii) at the core–cladding interface, the light ray must strike at an angle greater than the critical angle ϕ_c, where

$$\phi_c = \sin^{-1}(n_2/n_1)$$

This requires that the ray of light must enter the core of the fiber at an angle less than the acceptance angle α_m.

$$n_a \sin\alpha_m = n_1 \cos\phi_c$$

- The light-gathering capacity of a fiber is expressed in terms of the numerical aperture (NA), which is given by

$$\text{NA} = n_a \sin\alpha_m = (n_1^2 - n_2^2)^{1/2} = n_1\sqrt{2\Delta}$$

- When a pulse of light propagates through the fiber, it gets broadened. Pulse broadening is caused by two mechanisms:
  (i) Multipath time dispersion, which is given by

$$\frac{\Delta T}{L} = \frac{n_1}{n_2}\left(\frac{n_1 - n_2}{c}\right)$$

  To some extent, this type of dispersion can be reduced by index grading, i.e., by varying the RI of the core in a specific manner. The parabolic profile gives best results.
  (ii) Material dispersion, caused by change in the RI of the fiber material with wavelength. This is given by

$$\frac{\tau}{L} = \frac{\gamma}{c}\left|\lambda^2 \frac{d^2 n}{d\lambda^2}\right|$$

- If the power distribution in the pulse as a function of time t is $p(t)$, the total energy in the pulse is given by

$$\varepsilon = \int_{-\infty}^{\infty} p(t)\,dt$$

and its rms width σ is given by

$$\sigma = \left[\frac{1}{\varepsilon}\int_{-\infty}^{\infty} t^2(pt)\,dt - \left\{\frac{1}{\varepsilon}\int_{-\infty}^{\infty} tp(t)\,dt\right\}^2\right]^{1/2}$$

MULTIPLE CHOICE QUESTIONS

2.1 A ray of light is passing from a silica glass of refractive index 1.48 to another silica glass of refractive index 1.46. What is the range of angles (measured with

respect to the normal to the interface) for which this ray will undergo total internal reflection?

(a) 0°–80° (b) 81°–90° (c) 90°–180° (d) 180°–360°

2.2 Light is guided within the core of a step-index fiber by

(a) refraction at the core–air interface.
(b) total internal reflection at the core–cladding interface.
(c) total internal reflection at the outer surface of the cladding.
(d) change in the speed of light within the core.

2.3 A step-index fiber has a core with a refractive index of 1.50 and a cladding with a refractive index of 1.46. Its numerical aperture is

(a) 0.156 (b) 0.244 (c) 0.344 (d) 0.486

2.4 The optical fiber of Question 2.3 is placed in water (refractive index 1.33). The acceptance angle of the fiber will be approximately

(a) 10° (b) 15° (c) 20° (d) 25°

2.5 The axial refractive index of the core, n_0, of a graded-index fiber is 1.50 and the maximum relative refractive index difference Δ is 1%. What is the refractive index of the cladding?

(a) 1.485 (b) 1.50
(c) It will depend on the profile parameter.
(d) It will depend on the radius of the core.

2.6 An impulse is launched into one end of a 30-km optical fiber with a rated total dispersion of 20 ns/km. What will be the width of the pulse at the other end?

(a) 20 ns (b) 100 ns (c) 300 ns (d) 600 ns

2.7 For a typical LED, the relative spectral width γ is 0.030. This source is coupled to a pure silica fiber with $|\lambda^2(d^2n/d\lambda^2)| = 0.020$ at the operating wavelength. What is the pulse broadening per km due to material dispersion?

(a) 2 ps km^{-1} (b) 2 ns km^{-1} (c) 2 μs km^{-1} (d) 2 ms km^{-1}

2.8 A pulse of 100 ns half-width is transmitted through an optical fiber of length 20 km. The fiber has a rated multipath time dispersion of 10 ns km^{-1} and a material dispersion of 2 ns km^{-1}. What will be the half-width of the received pulse?

(a) 100 ns (b) 225 ns (c) 240 ns (d) 340 ns

2.9 A LED is emitting a mean wavelength of $\lambda = 0.90$ μm and its spectral half-width $\Delta\lambda = 18$ nm. What is its relative spectral width?

(a) 0.02 (b) 0.05 (c) 0.90 (d) 18

2.10 A laser diode has a relative spectral width of 2×10^{-3} and is emitting a mean wavelength of 1 μm. What is its spectral half-width?

(a) 1 μm (b) 0.2 μm (c) 20 nm (d) 2 nm

Answers

2.1	(b)	2.2	(b)	2.3	(c)	2.4	(b)	2.5	(a)
2.6	(d)	2.7	(b)	2.8	(b)	2.9	(a)	2.10	(d)

REVIEW QUESTIONS

2.1 What are the functions of the core and cladding in an optical fiber? Why should their refractive indices be different? Would it be possible for the light to be guided without cladding?

2.2 What is total internal reflection? Why is it necessary to meet the condition of total internal reflection at the core–cladding interface?

2.3 What is acceptance angle? If the optical fiber of Example 2.1 is kept in water (RI = 1.33), which of the parameters calculated in parts (a), (b), and (c) will change? What is its new value?
Ans: Only α_m will change and its new value will be 10.57°.

2.4 A step-index fiber has an acceptance angle of 20° in air and a relative refractive index difference of 3%. Estimate the NA and the critical angle at the core–cladding interface.
Ans: 0.34, 76°

2.5 Define numerical aperture (NA) of a fiber. On what factors does it depend?

2.6 A SI fiber has a numerical aperture of 0.17 and a cladding refractive index of 1.46. Determine (a) the acceptance angle of the fiber when it is placed in water (the refractive index of water may be taken as 1.33) and (b) the critical angle at the core–cladding interface.
Ans: (a) 7.34° (b) 83.35°

2.7 The speed of light in vacuum and in the core of a SI fiber is 3×10^8 m s^{-1} and 2×10^8 m s^{-1}, respectively. When the fiber is placed in air, the critical angle at the core–cladding interface is 75°. Calculate the (a) NA of the fiber and (b) multipath time dispersion per unit length.
Ans: (a) 0.388 (b) 1.7×10^{-10} s m^{-1}

2.8 Explain multipath time dispersion and material dispersion. How can these be minimized?

2.9 An impulse is launched into an optical fiber and the received pulse is triangular, as shown in Fig, 2.11. Calculate the (a) total energy in the pulse ε, (b) mean pulse arrival time $\bar{t}$, and (c) rms pulse width σ.
Ans: (a) $p_0\tau$, (b) $4/3\,\tau$ (c) $\tau\sqrt{2/3}$

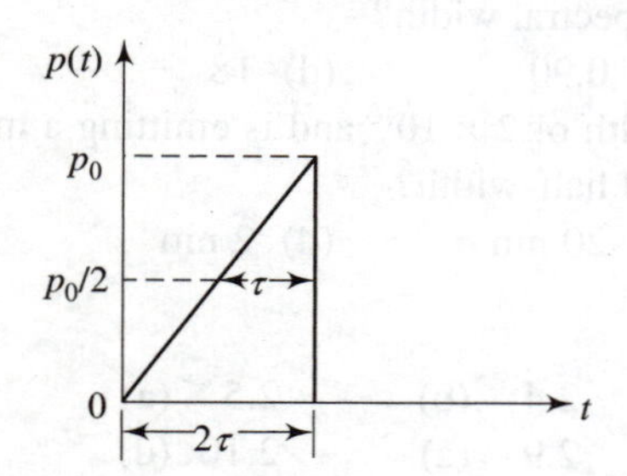

Fig. 2.11

2.10 A plastic fiber of 1 mm diameter has $n_1 = 1.496$ and $n_2 = 1.40$. Calculate its NA and α_m when it is placed in (a) air and (b) water.
Ans: (a) 0.5272, 31.81° (b) 0.5272, 23.35°

3 Wave Propagation in Planar Waveguides

After reading this chapter you will be able to understand the following:

- Light is an electromagnetic wave
- Maxwell's equations
- Solution in an inhomogeneous medium
- Planar optical waveguides
- TE modes of a symmetric step-index waveguide
- Power distribution and confinement factor

3.1 INTRODUCTION

In Chapter 2, we have studied the propagation of light signals through optical fibers using the ray model. However, it has to be realized that light is an electromagnetic wave, that is, a wave consisting of time-varying, mutually perpendicular electric and magnetic fields. To simplify things let us consider plane transverse electromagnetic (TEM) waves travelling in free space (Fig. 3.1). Here the term 'plane' signifies that the waves are polarized in one plane. Thus, referring to Fig. 3.1, if the electric-field vector **E** changes its magnitude in the x-direction but does not change its orientation, i.e., remains in the x-z plane, we say that **E** is x-polarized. Similarly, the magnetic-field vector **H** is y-polarized. In general, the **E** and **H** fields need not be polarized in the x and y directions. The term 'transverse' means that the vectors **E** and **H** are always perpendicular to the direction of propagation, which is the z-axis in the present case.

Now, a ray is considered to be a pencil made of a plane TEM wave, whose wavelength λ tends to zero. In general, this is not true, and hence the ray model should be used with caution. Therefore, this chapter and the next one have been devoted to the discussion of electromagnetic wave propagation through dielectric waveguides. The basis for such a discussion is provided by Maxwell's equations, so let us begin with these.

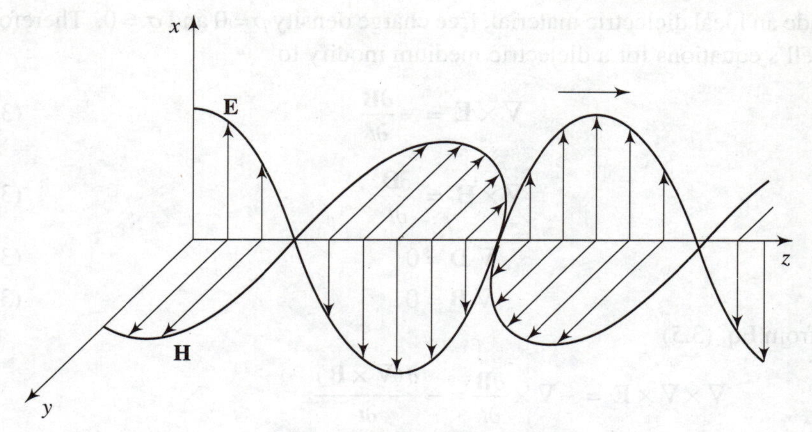

Fig. 3.1 Plane TEM wave

3.2 MAXWELL'S EQUATIONS

These are a set of four equations given below:

$$\nabla \times \mathbf{E} = -\frac{\partial \mathbf{B}}{\partial t} \tag{3.1}$$

$$\nabla \times \mathbf{H} = \mathbf{J} + \frac{\partial \mathbf{D}}{\partial t} \tag{3.2}$$

$$\nabla \cdot \mathbf{D} = \rho \tag{3.3}$$

$$\nabla \cdot \mathbf{B} = 0 \tag{3.4}$$

where the different terms are as follows: (Boldface characters denote vectors and regular fonts denote scalars.) **E** is the electric-field vector in V/m. **D** is the electric displacement vector in C/m^2, it is related to **E** by the formula $\mathbf{D} = \varepsilon\mathbf{E}$, where ε is the dielectric permittivity of the medium. **H** is the magnetic-field vector in A/m. **B** is the magnetic induction vector in H/m, it is related to **H** by the formula $\mathbf{B} = \mu\mathbf{H}$, where μ is the magnetic permeability of the medium. ρ is the charge density of the medium in C/m^3. **J** is the current density in A/m^2; it is related to **E** by the formula $\mathbf{J} = \sigma\mathbf{E}$, where σ is the conductivity of the medium in A/V m. ∇ is the nabla operator defined as

$$\nabla = \hat{\mathbf{i}}\frac{\partial}{\partial x} + \hat{\mathbf{j}}\frac{\partial}{\partial y} + \hat{\mathbf{k}}\frac{\partial}{\partial z}$$

$\hat{\mathbf{i}}$, $\hat{\mathbf{j}}$, and $\hat{\mathbf{k}}$ being unit vectors along the *x*-, *y*-, and *z*-axis, respectively. $\mu = \mu_0\mu_r$, $\mu_0 = 4\pi \times 10^{-7}$ N/A^2 (or H/m) is the permeability of free space, and μ_r is the relative permeability of the material and is very close to unity for most dielectrics. $\varepsilon = \varepsilon_0\varepsilon_r$, $\varepsilon_0 = 8.854 \times 10^{-12}$ C^2/N m^2 is the permittivity of free space, and ε_r is the relative permittivity of the material.

Inside an ideal dielectric material, free charge density $\rho = 0$ and $\sigma = 0$. Therefore, Maxwell's equations for a dielectric medium modify to

$$\nabla \times \mathbf{E} = -\frac{\partial \mathbf{B}}{\partial t} \tag{3.5}$$

$$\nabla \times \mathbf{H} = \frac{\partial \mathbf{D}}{\partial t} \tag{3.6}$$

$$\nabla \cdot \mathbf{D} = 0 \tag{3.7}$$

$$\nabla \cdot \mathbf{B} = 0 \tag{3.8}$$

Now, from Eq. (3.5)

$$\nabla \times \nabla \times \mathbf{E} = -\nabla \times \frac{\partial \mathbf{B}}{\partial t} = -\frac{\partial (\nabla \times \mathbf{B})}{\partial t}$$

$$= -\mu \frac{\partial (\nabla \times \mathbf{H})}{\partial t} = -\mu \frac{\partial}{\partial t}\left(\frac{\partial \mathbf{D}}{\partial t}\right) \text{ [using Eq. (3.6)]}$$

$$= -\mu \frac{\partial^2 \mathbf{D}}{\partial t^2} \tag{3.9}$$

But
$$\nabla \times \nabla \times \mathbf{E} = \nabla(\nabla \cdot \mathbf{E}) - \nabla^2 \mathbf{E} \tag{3.10}$$

where ∇^2 is the Laplacian operator,

$$\nabla^2 \equiv \frac{\partial^2}{\partial x^2} + \frac{\partial^2}{\partial y^2} + \frac{\partial^2}{\partial z^2}$$

and
$$\nabla \cdot \mathbf{E} = \nabla \cdot (\mathbf{D}/\varepsilon) = 0 \text{ [using Eq. (3.7)]} \tag{3.11}$$

Therefore, from Eqs. (3.9)–(3.11), we have

$$-\nabla^2 \mathbf{E} = -\mu \frac{\partial^2 \mathbf{D}}{\partial t^2} = -\mu \frac{\partial^2 (\varepsilon \mathbf{E})}{\partial t^2} = -\mu\varepsilon \frac{\partial^2 \mathbf{E}}{\partial t^2}$$

or
$$\nabla^2 \mathbf{E} - \varepsilon\mu \frac{\partial^2 \mathbf{E}}{\partial t^2} = 0 \tag{3.12}$$

Similarly
$$\nabla^2 \mathbf{H} - \varepsilon\mu \frac{\partial^2 \mathbf{H}}{\partial t^2} = 0 \tag{3.13}$$

We write ψ to represent any Cartesian component of **E** (i.e., E_x, E_y, E_z) or **H** (i.e., H_x, H_y, H_z). Then the vector wave equations (3.12) and (3.13) become a set of scalar wave equations

$$\nabla^2 \psi - \varepsilon\mu \frac{\partial^2 \psi}{\partial t^2} = 0 \tag{3.14}$$

These are linear wave equations, which, in the absence of boundary conditions, have solutions that take the form of plane waves that propagate with a phase velocity given by

$$v_p = \frac{1}{\sqrt{\varepsilon\mu}} \tag{3.15}$$

In vacuum or free space, $\varepsilon_r = \mu_r = 1$, so that $v_p = 1/\sqrt{\varepsilon_0\,\mu_0} = c \approx 3 \times 10^8\ \mathrm{m\,s^{-1}}$.

For an isotropic medium, the refractive index n is related to ε by the relation $n = \sqrt{\varepsilon/\varepsilon_0} \approx \sqrt{\varepsilon_r}$ with $\mu_r \approx 1$. Therefore, $v_p = c/n$. Equation (3.14) then becomes

$$\nabla^2\psi - \frac{1}{v_p^2}\frac{\partial^2\psi}{\partial t^2} = 0 \tag{3.16a}$$

or

$$\nabla^2\psi - \frac{n^2}{c^2}\frac{\partial^2\psi}{\partial t^2} = 0 \tag{3.16b}$$

The solution of this equation, as can be verified by substituting, is of the form

$$\psi = \psi_0 \exp\{i(\omega t - \beta z)\}$$

This represents a plane wave propagating in the positive z-direction with a phase velocity $v_p = \omega/\beta$. Here ω is the angular frequency, β is the propagation phase constant, and $i = \sqrt{-1}$.

3.3 SOLUTION IN AN INHOMOGENEOUS MEDIUM

For an isotropic, linear, non-conducting, non-magnetic, but inhomogeneous medium, the divergence of the electric displacement vector becomes

$$\begin{aligned}\boldsymbol{\nabla}\cdot\mathbf{D} &= \boldsymbol{\nabla}\cdot(\varepsilon\mathbf{E}) = \varepsilon_0\,\boldsymbol{\nabla}\cdot(\varepsilon_r\mathbf{E}) = 0\\ &= \varepsilon_0[\boldsymbol{\nabla}(\varepsilon_r)\cdot\mathbf{E} + \varepsilon_r\,\boldsymbol{\nabla}\cdot\mathbf{E}] = 0\end{aligned} \tag{3.17}$$

which gives

$$\boldsymbol{\nabla}\cdot\mathbf{E} = \left(-\frac{1}{\varepsilon_r}\right)\boldsymbol{\nabla}(\varepsilon_r)\cdot\mathbf{E} \tag{3.18}$$

From Eq. (3.9), we have

$$\nabla\times\nabla\times\mathbf{E} = -\mu\frac{\partial^2\mathbf{D}}{\partial t^2} = -\mu_0\,\varepsilon_0\,\varepsilon_r\,\frac{\partial^2\mathbf{E}}{\partial t^2} \tag{3.19}$$

since $\mu = \mu_0\,\mu_r = \mu_0$ (μ_r being equal to 1) and $\mathbf{D} = \varepsilon\mathbf{E} = \varepsilon_0\,\varepsilon_r\,\mathbf{E}$.

From Eqs (3.10) and (3.19), we get

$$\nabla(\nabla \cdot \mathbf{E}) - \nabla^2 \mathbf{E} = -\mu_0 \varepsilon_0 \varepsilon_r \frac{\partial^2 \mathbf{E}}{\partial t^2}$$

Rearranging, we get

$$\nabla^2 \mathbf{E} - \nabla(\nabla \cdot \mathbf{E}) - \mu_0 \varepsilon_0 \varepsilon_r \frac{\partial^2 \mathbf{E}}{\partial t^2} = 0 \tag{3.20}$$

Substituting for $\nabla \cdot \mathbf{E}$ from Eq. (3.18) in Eq. (3.20), we get

$$\nabla^2 \mathbf{E} + \nabla\left\{\frac{1}{\varepsilon_r} \nabla(\varepsilon_r) \cdot \mathbf{E}\right\} - \mu_0 \varepsilon_0 \varepsilon_r \frac{\partial^2 \mathbf{E}}{\partial t^2} = 0 \tag{3.21}$$

This shows that for an inhomogeneous medium, the equations for the Cartesian components of **E**, i.e., E_x, E_y, and E_z, are coupled. For a homogeneous medium, the second term on the left-hand side of Eq. (3.21) will become zero (as $\nabla \cdot \mathbf{E} = 0$ for a homogeneous medium). In that case, the Cartesian components of **E** will satisfy the scalar wave equation given by Eq. (3.16).

An equation similar to Eq. (3.21) for **H** can be derived as follows. Taking the curl of Eq. (3.6), we have

$$\nabla \times \nabla \times \mathbf{H} = \nabla \times \left(\frac{\partial \mathbf{D}}{\partial t}\right) = \frac{\partial}{\partial t}(\nabla \times \mathbf{D}) = \frac{\partial}{\partial t}(\nabla \times \varepsilon \mathbf{E}) = \varepsilon_0 \frac{\partial}{\partial t}(\nabla \times \varepsilon_r \mathbf{E}) \tag{3.22}$$

But
$$\nabla \times \nabla \times \mathbf{H} = \nabla(\nabla \cdot \mathbf{H}) - \nabla^2 \mathbf{H}$$

From Eq. (3.8),
$$\nabla \cdot \mathbf{B} = \nabla \cdot (\mu_0 \mathbf{H}) = 0$$

Therefore,
$$\nabla \times \nabla \times \mathbf{H} = -\nabla^2 \mathbf{H} \tag{3.23}$$

Equating Eqs (3.22) and (3.23), we get

$$\nabla^2 \mathbf{H} + \varepsilon_0 \frac{\partial}{\partial t}(\nabla \times \varepsilon_r \mathbf{E}) = 0$$

or
$$\nabla^2 \mathbf{H} + \varepsilon_0 \frac{\partial}{\partial t}[(\nabla \varepsilon_r) \times \mathbf{E} + \varepsilon_r (\nabla \times \mathbf{E})] = 0$$

or
$$\nabla^2 \mathbf{H} + \varepsilon_0 (\nabla \varepsilon_r) \times \frac{\partial \mathbf{E}}{\partial t} + \varepsilon_0 \varepsilon_r \frac{\partial}{\partial t}(\nabla \times \mathbf{E}) = 0$$

Employing Eqs (3.5) and (3.6) and rearranging, we get

$$\nabla^2 \mathbf{H} + \frac{1}{\varepsilon_r}(\nabla \varepsilon_r) \times (\nabla \times \mathbf{H}) - \mu_0 \varepsilon_0 \varepsilon_r \frac{\partial^2 \mathbf{H}}{\partial t^2} = 0 \tag{3.24}$$

Again we see that the equations for Cartesian components of **H**, i.e., H_x, H_y, and H_z, are coupled. In order to simplify Eqs (3.21) and (3.24), we set the *z*-coordinate along the direction of propagation of the wave represented by Eqs (3.21) and (3.24). Assume that the refractive index $n = \sqrt{\varepsilon_r}$ does not vary with *y* and *z*. Then

$$\varepsilon_r = n^2 = n^2(x) \tag{3.25}$$

This means that the *y*- and *z*-dependence of the fields, in general, will be of the form $e^{-i(\gamma y+\beta z)}$. However, we may put $\gamma = 0$ without any loss of generality. Thus the equations governing the modes of propagation of E_j or H_j, where $j = x, y$, or z, may be written as follows:

$$\mathrm{E}_j = E_j(x)\, e^{i(\omega t-\beta z)} \tag{3.26}$$

and

$$\mathrm{H}_j = H_j(x)\, e^{i(\omega t-\beta z)} \tag{3.27}$$

where $E_j(x)$ and $H_j(x)$ are the transverse field distributions that do not change as the field propagates through the medium, ω is the angular frequency, and β is the propagation constant.

Now let us write Maxwell's equations (3.5) and (3.6) in Cartesian coordinates. From Eq. (3.5), i.e.,

$$\nabla \times \mathbf{E} = -\frac{\partial \mathbf{B}}{\partial t} = -\mu_0 \frac{\partial \mathbf{H}}{\partial t}$$

we get

$$\hat{\mathbf{i}}\left(\frac{\partial \mathrm{E}_z}{\partial y} - \frac{\partial \mathrm{E}_y}{\partial z}\right) + \hat{\mathbf{j}}\left(\frac{\partial \mathrm{E}_x}{\partial z} - \frac{\partial \mathrm{E}_z}{\partial x}\right) + \hat{\mathbf{k}}\left(\frac{\partial \mathrm{E}_y}{\partial x} - \frac{\partial \mathrm{E}_x}{\partial y}\right)$$

$$= -\mu_0 \left[\frac{\partial}{\partial t}\{\hat{\mathbf{i}}\mathrm{H}_x + \hat{\mathbf{j}}\mathrm{H}_y + \hat{\mathbf{k}}\mathrm{H}_z\}\right] \tag{3.28}$$

and from Eq. (3.6), i.e.,

$$\nabla \times \mathbf{H} = \frac{\partial \mathbf{D}}{\partial t} = \varepsilon_0 \varepsilon_r \frac{\partial \mathbf{E}}{\partial t} = \varepsilon_0 n^2 \frac{\partial \mathbf{E}}{\partial t}$$

we get

$$\hat{\mathbf{i}}\left(\frac{\partial \mathrm{H}_z}{\partial y} - \frac{\partial \mathrm{H}_y}{\partial z}\right) + \hat{\mathbf{j}}\left(\frac{\partial \mathrm{H}_x}{\partial z} - \frac{\partial \mathrm{H}_z}{\partial x}\right) + \hat{\mathbf{k}}\left(\frac{\partial \mathrm{H}_y}{\partial x} - \frac{\partial \mathrm{H}_x}{\partial y}\right)$$

$$= \varepsilon_0 n^2 \frac{\partial}{\partial t}\{\hat{\mathbf{i}}\mathrm{E}_x + \hat{\mathbf{j}}\mathrm{E}_y + \hat{\mathbf{k}}\mathrm{E}_z\} \tag{3.29}$$

As the fields E and H do not vary with y, all the $\partial/\partial y$ terms may be set equal to zero. Substituting the values of E_j and H_j from Eqs (3.26) and (3.27) into Eq. (3.28), we get

$$-\hat{\mathbf{i}}\frac{\partial}{\partial z}\{E_y e^{i(\omega t-\beta z)}\} + \hat{\mathbf{j}}\frac{\partial}{\partial z}\{E_x e^{i(\omega t-\beta z)}\} - \hat{\mathbf{j}}\frac{\partial}{\partial x}\{E_z e^{i(\omega t-\beta z)}\} + \hat{\mathbf{k}}\frac{\partial}{\partial x}\{E_y e^{i(\omega t-\beta z)}\}$$

$$= -\hat{\mathbf{i}}\mu_0 \frac{\partial}{\partial t} H_x e^{i(\omega t-\beta z)} - \hat{\mathbf{j}}\mu_0 \frac{\partial}{\partial t} H_y e^{i(\omega t-\beta z)} - \hat{\mathbf{k}}\mu_0 \frac{\partial}{\partial t} H_z e^{i(\omega t-\beta z)}$$

or

$$\hat{\mathbf{i}}\,(i\beta)\,E_y - \hat{\mathbf{j}}\left[i\beta E_x + \frac{\partial E_z}{\partial x}\right] + \hat{\mathbf{k}}\frac{\partial E_y}{\partial x}$$

$$= -\hat{\mathbf{i}}\mu_0\,(i\omega)\,H_x - \hat{\mathbf{j}}\mu_0\,(i\omega)\,H_y - \hat{\mathbf{k}}\mu_0\,(i\omega)\,H_z$$

Comparing the corresponding components on both sides, we get the following set of equations:

$$i\beta E_y = -i\omega\mu_0 H_x$$

or
$$\beta E_y = -\mu_0 \omega H_x \tag{3.30a}$$

$$i\beta E_x + \frac{\partial E_z}{\partial x} = i\mu_0 \omega H_y \tag{3.30b}$$

$$\frac{\partial E_y}{\partial x} = -i\mu_0 \omega H_z \tag{3.30c}$$

Now substituting the values of E_j and H_j from Eqs (3.26) and (3.27) into Eq. (3.29) and setting to zero all terms containing $\partial/\partial y$, we get

$$-\hat{\mathbf{i}}\frac{\partial}{\partial z}\{H_y e^{i(\omega t-\beta z)}\} + \hat{\mathbf{j}}\left[\frac{\partial}{\partial z}\{H_x e^{i(\omega t-\beta z)}\} - \frac{\partial}{\partial x}\{H_z e^{i(\omega t-\beta z)}\}\right]$$

$$+\hat{\mathbf{k}}\frac{\partial}{\partial x}\{H_y e^{i(\omega t-\beta z)}\}$$

$$= \varepsilon_0 n^2 \hat{\mathbf{i}}\frac{\partial}{\partial t}\{E_x e^{i(\omega t-\beta z)}\} + \varepsilon_0 n^2 \hat{\mathbf{j}}\frac{\partial}{\partial t}\{E_y e^{i(\omega t-\beta z)}\} + \varepsilon_0 n^2 \hat{\mathbf{k}}\frac{\partial}{\partial t}\{E_z e^{i(\omega t-\beta z)}\}$$

or
$$\hat{\mathbf{i}}(i\beta)H_y + \hat{\mathbf{j}}\left\{-i\beta H_x - \frac{\partial H_z}{\partial x}\right\} + \hat{\mathbf{k}}\frac{\partial H_y}{\partial x}$$

$$= \hat{\mathbf{i}}(i\omega\varepsilon_0 n^2)E_x + \hat{\mathbf{j}}(i\omega\varepsilon_0 n^2)E_y + \hat{\mathbf{k}}(i\omega\varepsilon_0 n^2)E_z$$

Comparing the respective components, we get

$$i\beta H_y = -i\omega\varepsilon_0 n^2 E_x$$

or
$$\beta H_y = -\varepsilon_0 n^2 \omega E_x \tag{3.31a}$$

$$-i\beta H_x - \frac{\partial H_z}{\partial x} = i\varepsilon_0 n^2 \omega E_y \tag{3.31b}$$

$$\frac{\partial H_y}{\partial x} = i\varepsilon_0 n^2 \omega E_z \tag{3.31c}$$

where $n^2 = n^2(x)$.

The six equations (3.30a)–(3.30c) and (3.31a)–(3.31c) form two independent sets. Thus, Eqs (3.30a), (3.30c), and (3.31b) involve E_y, H_x, and H_z. Herein the field components E_x, E_z, and H_y are zero. The modes described by these equations are called transverse electric (TE) modes because the electric field has only the transverse component E_y.

We renumber these equations as follows:

$$\left.\begin{aligned} \beta E_y &= -\mu_0 \omega H_x && (3.32) \\ \frac{\partial E_y}{\partial x} &= -i\mu_0 \omega H_z && (3.33) \\ -i\beta H_x - \frac{\partial H_z}{\partial x} &= i\varepsilon_0 \omega n^2(x) E_y && (3.34) \end{aligned}\right\} \text{TE modes}$$

The second set of equations is formed by Eqs (3.31a), (3.31c), and (3.30b), which involve only H_y, E_x, and E_z, and the field components E_y, H_x, and H_z are zero. These are called transverse magnetic (TM) modes because the magnetic field herein has only a transverse component H_y. The second set of equations is also renumbered as follows:

$$\left.\begin{aligned} \beta H_y &= -\varepsilon_0 \omega n^2(x) E_x && (3.35) \\ \frac{\partial H_y}{\partial x} &= i\varepsilon_0 \omega n^2(x) E_z && (3.36) \\ i\beta E_x + \frac{\partial E_z}{\partial x} &= i\mu_0 \omega H_y && (3.37) \end{aligned}\right\} \text{TM modes}$$

3.4 PLANAR OPTICAL WAVEGUIDE

Planar optical waveguides are important components in integrated optical devices. The modal analysis of such waveguides is easier to understand. Hence we will take up their analysis first.

The simplest optical waveguide may have the geometric configuration shown in Fig. 3.2. It consists of a thin dielectric slab of refractive index n_1 and thickness $2a$ sandwiched between two symmetrical dielectric slabs of refractive index n_2 and infinite thickness ($n_2 < n_1$). The waveguide is oriented such that the wave propagates along the z-direction. The y and z dimensions of the guide are assumed to extend to infinity. The thickness of the slabs is along the x-direction (as shown in Fig. 3.2). A ray of light launched into the guide slab or layer would progress by multiple reflections as shown in Fig. 3.3. We may assume that such a ray represents a plane TEM wave travelling at an angle θ with the z-axis. As the refractive index within the guide layer is n_1, the wavelength of light in the layer is reduced to $\lambda_m = \lambda / n_1$, where λ is the wavelength of light in vacuum and the propagation constant is increased to

$$\beta_1 = \frac{2\pi}{\lambda_m} = \frac{2\pi n_1}{\lambda} = k n_1$$

where $k = 2\pi/\lambda$ is the vacuum propagation constant or the propagation vector.

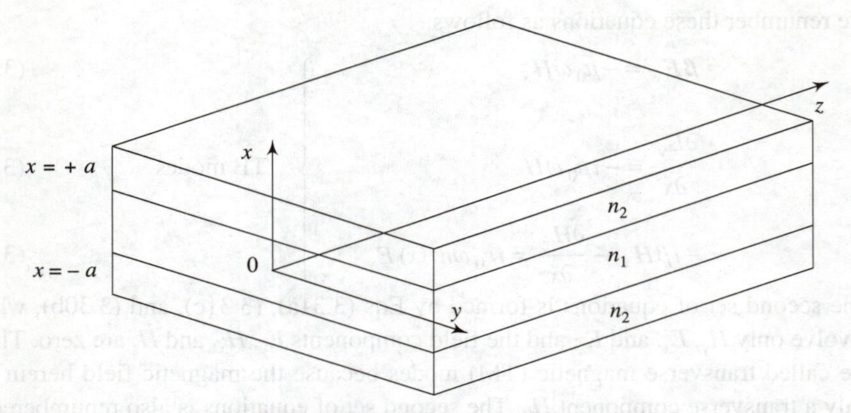

Fig. 3.2 Structure of a planar optical waveguide

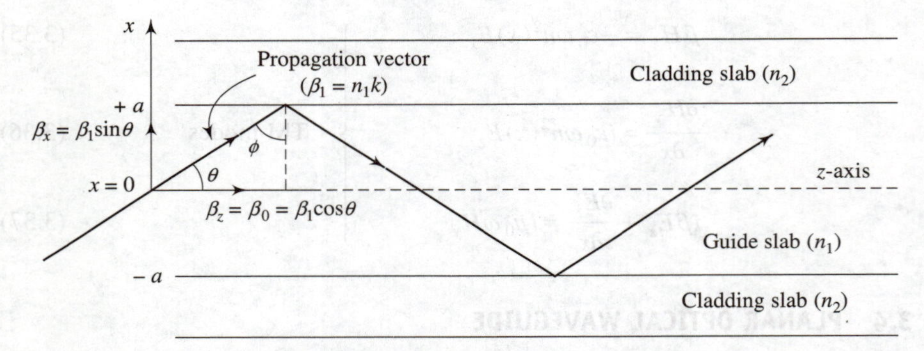

Fig. 3.3 Ray propagation in a planar waveguide

If the propagation vector β_1 makes an angle θ with the z-axis (which is the same as the guide axis), the plane wave may be resolved into two component plane waves propagating in the z and x directions as shown in Fig. 3.3. The component of the propagation vector in the z-direction or, in other words, the effective propagation constant of the guided wave (along the z-direction) will be

$$\beta = \beta_z = \beta_1 \cos\theta \tag{3.38}$$

The limiting value of θ, i.e., θ_m is related to the critical angle ϕ_c at the interface of the guide layer and the cladding slabs, and is given by

$$\sin\phi_c = \cos\theta_m = n_2/n_1 \tag{3.39}$$

Thus the minimum value of β in the z-direction, i.e., $\beta_{\min}$, will be determined by the maximum value of θ, i.e., θ_m or

$$\beta_{\min} = \beta_1 \cos\theta_{\mathrm{m}} = \beta_1 \frac{n_2}{n_1} = \beta_2 \tag{3.40}$$

The maximum value that β can have is β_1, which corresponds to $\theta = 0$, i.e., the plane TEM waves travelling parallel to the guide axis: $\beta_{max} \approx \beta_1$. We, therefore, expect β to lie between β_1 and β_2, or $\beta_2 < \beta < \beta_1$. The component of the propagation vector β_1 in the x-direction is

$$\beta_x = \beta_1 \sin\theta = n_1 k \sin\theta$$

where

$$k = 2\pi/\lambda$$

or

$$\beta_x = \frac{2\pi}{\lambda_m}\sin\theta \tag{3.41}$$

This component of the plane wave is reflected at the interface between the guide layer and the cladding slabs. When the total phase change after two successive reflections at the upper and lower interfaces is equal to $2i\pi$ radians, where i is an integer (0,1, 2, 3, ...), constructive interference will occur and a standing-wave pattern will be formed in the x-direction. This stable field pattern in the x-direction with only a periodic z-dependence is known as a mode. Thus, only a finite number of discrete modes which satisfy the above condition will propagate through the guide, i.e., $4a\beta_x = 2i\pi$

or

$$4a \sin\theta_i = i\lambda_m \tag{3.42}$$

Each value of θ_i corresponds to a particular mode with its own characteristic field pattern and its own propagation constant β_i in the z-direction. Obviously, β_i lies between β_1 and β_2. Since the maximum value that θ_i can take is θ_m, the number of guided modes is limited to

$$M = i_{max} = \frac{4a\sin\theta_m}{\lambda_m} = \frac{4an_1\sin\theta_m}{\lambda} = \frac{4a}{\lambda}(n_1^2 - n_2^2)^{1/2} \tag{3.43}$$

It should be noted that the requirement for the ith mode to be propagated is that $i \le (4a/\lambda)(n_1^2 - n_2^2)^{1/2}$. The mode corresponding to the highest value of i, i.e., i_{max}, does not meet the condition for total internal reflection, as the value of θ_m corresponds exactly to the critical angle ϕ_c, and is refracted at the interface. However, it can propagate freely in the cladding slabs and is said to be a radiation mode.

Example 3.1 A symmetric step-index (SI) planar waveguide is made of glass with $n_1 = 1.5$ and $n_2 = 1.49$. The thickness of the guide layer is 9.83 μm and the guide is excited by a source of wavelength $\lambda = 0.85$ μm. What is the range of the propagation constants? What is the maximum number of modes supported by the guide?

Solution

The phase propagation constant β lies between β_1 and β_2. Here,

$$\beta_1 = kn_1 = \frac{2\pi n_1}{\lambda} = \frac{2\pi}{(0.85 \times 10^{-6})} \times 1.50 = 11.0082 \times 10^6 \text{ m}^{-1}$$

and $$\beta_2 = kn_2 = \frac{2\pi n_2}{\lambda} = \frac{2\pi}{(0.85\times10^{-6})}\times 1.49 = 11.008\times10^6\ \text{m}^{-1}$$

The maximum number of modes that the guide can support is given by Eq. (3.43), i.e.,

$$M = \frac{4a}{\lambda}(n_1^2 - n_2^2)^{1/2} = \frac{2\times(9.83)}{0.85}[(1.5)^2 - (1.49)^2]^{1/2} \approx 4$$

3.5 TE MODES OF A SYMMETRIC STEP-INDEX PLANAR WAVEGUIDE

For the symmetric waveguide structure shown in Fig. 3.2, $n^2(-x) = n^2(x)$. Further, the structure is step-index type, as the guide layer has a refractive index n_1 and the cladding slabs have the refractive index n_2. Both n_1 and n_2 are constant and $n_1 > n_2$. Let us first take up the discussion of TE modes.

Substituting the values of H_x and H_z from Eqs (3.32) and (3.33) into Eq. (3.34), we get

$$-i\beta\left(-\frac{\beta}{\mu_0\omega}\right)E_y - \frac{\partial}{\partial x}\left(\frac{1}{-i\mu_0\omega}\right)\frac{\partial E_y}{\partial x} = i\varepsilon_0\omega n^2(x)E_y$$

Since $E_y = E_y(x)$, the partial derivative involving E_y may as well be written as a full derivative. Thus, rearranging the above equation, we may write

$$\frac{d^2E_y}{dx^2} - \beta^2E_y + \mu_0\varepsilon_0\omega^2n^2(x)E_y = 0$$

or $$\frac{d^2E_y}{dx^2} + [k^2n^2(x) - \beta^2]E_y = 0 \tag{3.44}$$

as $$\mu_0\omega_0 = 1/c^2 \text{ and } \omega/c = k$$

In the waveguide of Fig. 3.2,

$$n(x) = \begin{cases} n_1 & \text{for } |x| < a \\ n_2 & \text{for } |x| > a \end{cases} \tag{3.45}$$

Further, E_y and H_z (and hence $\partial E_y/\partial x$) are continuous at $x = \pm a$ because E_y and H_z are tangential components to the planes represented by $x = \pm a$, and H_z is proportional to $\partial E_y/\partial x$.

Substituting for $n(x)$ from Eq. (3.45) in Eq. (3.44), we get in the guide layer

$$\frac{d^2E_y}{dx^2} + [k^2n_1^2 - \beta^2]E_y = 0 \quad (|x| < a) \tag{3.46}$$

and in the cladding layer

$$\frac{d^2E_y}{dx^2}+[k^2n_2^2-\beta^2]E_y=0\ (|x|>a) \tag{3.47}$$

Let us put

$$u^2=k^2n_1^2-\beta^2=\beta_1^2-\beta^2 \tag{3.48}$$

and

$$w^2=\beta^2-k^2n_2^2=\beta^2-\beta_2^2 \tag{3.49}$$

Thus Eqs (3.46) and (3.47) take the forms

$$\frac{d^2E_y}{dx^2}+u^2E_y=0\ (|x|<a) \tag{3.50}$$

and

$$\frac{d^2E_y}{dx^2}-w^2E_y=0\ (|x|>a) \tag{3.51}$$

For the wave to be guided through the layer, both parameters u and w must be real. This implies that

$$\beta_1^2\ (=k^2n_1^2)>\beta^2>\beta_2^2\ (=k^2n_2^2) \tag{3.52}$$

With these conditions, the solutions in the guide layer are oscillatory, while those in the cladding layers decay exponentially. This is what is exactly required. Thus, for a guided-wave solution, the propagation constant β must lie between β_1 and β_2. The same inference was drawn from ray analysis.

Since the refractive index $n(x)$ is symmetrically distributed about $x=0$, the solutions are either symmetric or antisymmetric functions of x. Therefore, we must have

$$E_y(-x)=E_y(x)\ \text{symmetric modes} \tag{3.53}$$

$$E_y(-x)=-E_y(x)\ \text{antisymmetric modes} \tag{3.54}$$

For the symmetric mode, the electric-field distribution takes the form

$$E_y(x)=\begin{cases}A\cos ux, & |x|<a \quad (3.55)\\ C\exp(-w|x|), & |x|>a \quad (3.56)\end{cases}$$

where A and C are constants. The continuity of $E_y(x)$ and dE_y/dx at $x=\pm a$ gives the following equations:

$$A\cos(ua)=Ce^{-wa} \tag{3.57}$$

and

$$-uA\sin(ua)=-wCe^{-wa} \tag{3.58}$$

Dividing Eq. (3.58) by Eq. (3.57), one gets

$$u\tan(ua)=w$$

or

$$ua\tan(ua)=wa \tag{3.59}$$

Now, we define a new dimensionless waveguide parameter called the normalized frequency parameter V. From Eqs (3.48) and (3.49), we have

$$u^2 + w^2 = k^2 n_1^2 - \beta^2 + \beta^2 - k^2 n_2^2 = k^2 (n_1^2 - n_2^2)$$

or $$(ua)^2 + (wa)^2 = k^2 a^2 (n_1^2 - n_2^2) = \left(\frac{2\pi}{\lambda}\right)^2 a^2 (n_1^2 - n_2^2)$$

Thus, V is defined as

$$V = \{(ua)^2 + (wa)^2\}^{1/2} = \frac{2\pi a}{\lambda}(n_1^2 - n_2^2)^{1/2} \tag{3.60}$$

In terms of V, Eq. (3.59) may be written as

$$ua \tan(ua) = \{V^2 - (ua)^2\}^{1/2} \tag{3.61}$$

For antisymmetric modes, the solutions take the form

$$E_y(x) = \begin{cases} B \sin ux, & |x| < a \quad (3.62) \\ \dfrac{x}{|x|} D \exp(-w|x|), & |x| > a \quad (3.63) \end{cases}$$

where B and D are constants. Following exactly the above procedure, we get

$$-ua \cot(ua) = wa \tag{3.64}$$

or, in terms of the parameter V, we have

$$-ua \cot(ua) = \{V^2 - (ua)^2\}^{1/2} \tag{3.65}$$

In order to solve the transcendental equations (3.59) and (3.64), in Fig. 3.4, we plot ua as a function of wa for $ua \tan(ua) = wa$ (bold solid lines) and $-ua \cot(ua) = wa$ (dashed lines). Equation (3.60), i.e., $V^2 = \{(ua)^2 + (wa)^2\}$= constant, is plotted as arcs of circles for constant V-values (lightface solid lines). The points of intersection of the bold solid and the dashed lines with the arc of a circle of radius V determine the propagation constants for the waveguide corresponding to the symmetric and antisymmetric modes, respectively. The values of ua and wa for different V-values corresponding to different modes are shown in Fig. 3.4. (For a clear understanding of the calculation of u and w for different modes, see Example 3.3.)

From Fig. 3.4 we can derive the following information:

(i) For $0 \le V \le \pi/2$ (i.e., for an arc of a circle of radius corresponding to $V < \pi/2$), there is only one intersection with the bold solid curve marked $m = 0$. This is the only solution for the guided TE mode. That is, the waveguide supports only one discrete TE mode and this mode is symmetric in x.

(ii) For $\pi/2 \le V \le \pi$ (i.e., for an arc of circle of radius corresponding to $\pi/2 < V < \pi$), the arc intersects at two points; one on the bold solid line $m = 0$ and the other on the dashed line $m = 1$. Thus we have two TE modes, one symmetric and the other antisymmetric. In general, if

$$(2m)\frac{\pi}{2} \le V \le (2m+1)\frac{\pi}{2} \tag{3.66}$$

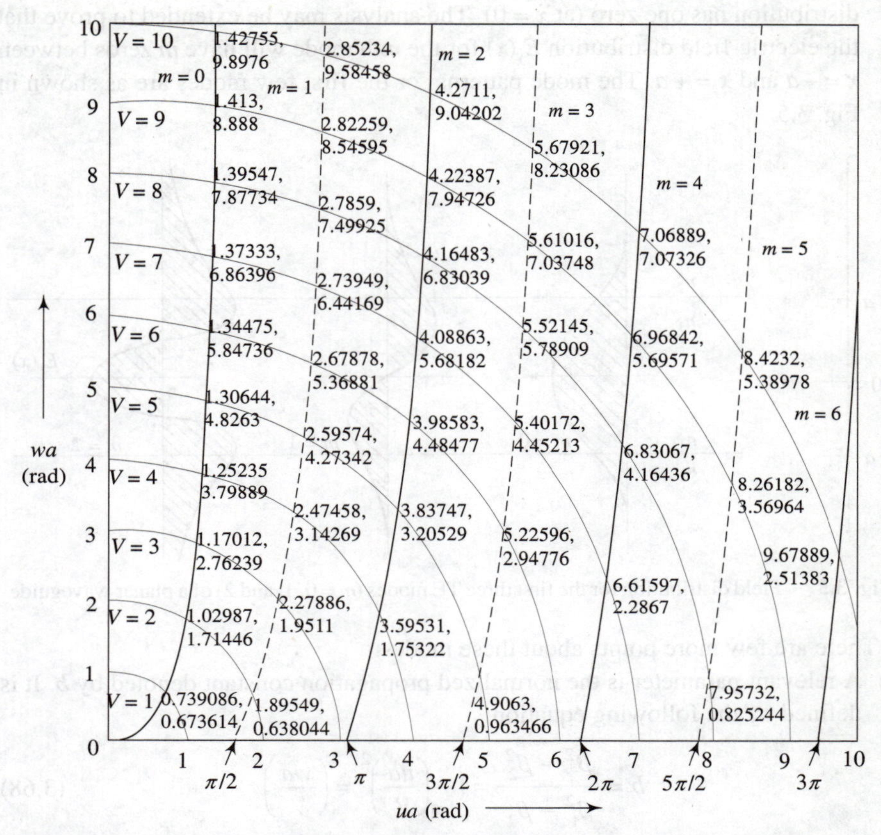

Fig. 3.4 Graphical solution of Eqs (3.59) and (3.64) for obtaining the propagation parameters *ua* and *wa* of TE modes for constant values of *V* in a planar waveguide

we will have $m + 1$ symmetric and m antisymmetric modes, and if

$$(2m+1)\frac{\pi}{2} \leq V \leq (2m+2)\frac{\pi}{2} \tag{3.67}$$

we will have $m + 1$ symmetric and $m + 1$ antisymmetric modes, where $m = 0, 1, 2, \ldots$; $m = 0, 2, 4, \ldots$ correspond to symmetric modes and $m = 1, 3, 5, \ldots$ correspond to antisymmetric modes. The maximum number of TE modes, M, supported by the guide would be an integer close to or greater than $2V/\pi$. This is in agreement with Eq. (3.43), which was obtained on the basis of the ray model.

(iii) An interesting point to be noted in Fig. 3.4 is that for the fundamental mode ($m = 0$), ua always lies between 0 and $\pi/2$ and the corresponding electric field $E_y(x)$ for $|x| < a$ will have no zeros. For the next mode ($m = 1$), which is antisymmetric in x, ua lies between $\pi/2$ and π and the corresponding field distribution has one zero (at $x = 0$). The analysis may be extended to prove that the electric-field distribution $E_y(x)$ for the mth mode will have m zeros between $x = -a$ and $x = +a$. The mode patterns for the first few modes are as shown in Fig. 3.5.

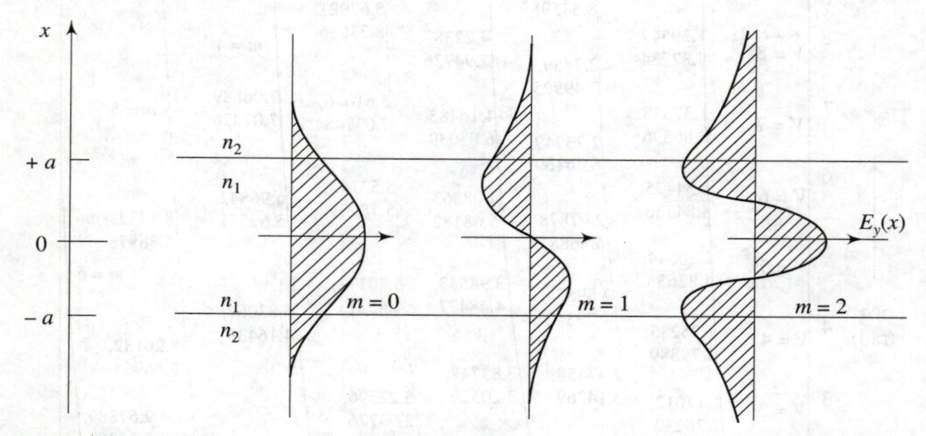

Fig. 3.5 Field distribution for the first three TE modes ($m = 0$, 1, and 2) of a planar waveguide

There are few more points about these modes:

(i) A relevant parameter is the normalized propagation constant denoted by b. It is defined by the following equation:

$$b = \frac{\beta^2 - \beta_2^2}{\beta_1^2 - \beta_2^2} = 1 - \left(\frac{ua}{V}\right)^2 = \left(\frac{wa}{V}\right)^2 \tag{3.68}$$

For a given guided mode, the value of b lies between 0 and 1. The variation of b with V for first few modes is shown in Fig. 3.6. When β becomes equal to β_2, $b = 0$ and the mode is said to have reached the 'cut-off'. Thus, at cut-off, $ua = V = V_c$ and $wa = 0$. This occurs at

$$ua = V_c = \frac{m\pi}{2} \tag{3.69}$$

Since $\quad V = \dfrac{2\pi a}{\lambda}(n_1^2 - n_2^2)^{1/2}$ [from Eq. (3.60)]

equating Eqs (3.60) and (3.69), we can obtain the thickness of the guide layer necessary to support m modes:

$$\frac{m\pi}{2} = \frac{2\pi a}{\lambda}(n_1^2 - n_2^2)^{1/2}$$

or

$$2a = \frac{m\lambda}{2(n_1^2 - n_2^2)^{1/2}} \tag{3.70}$$

where $m = 0, 1, 2, \ldots$. Note that fundamental mode has no cut-off frequency.

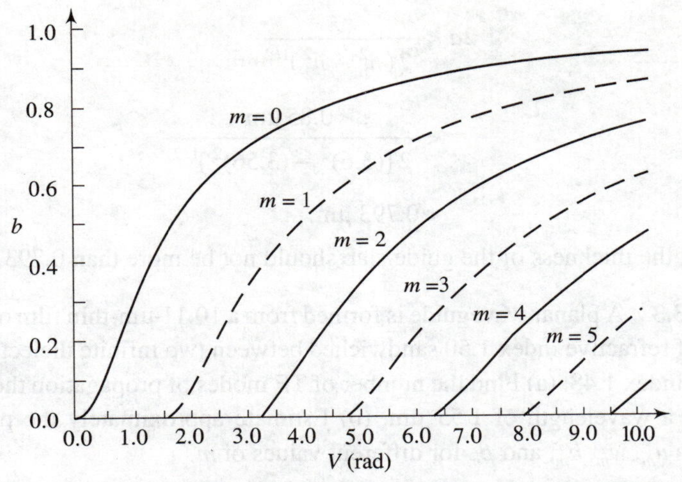

Fig. 3.6 Variation of the normalized propagation parameter b for TE modes with the normalized frequency V for a planar waveguide

(ii) The modes for which $\beta^2 < \beta_2^2$ are called radiation modes. These are continuous. In terms of the ray model, these correspond to the rays for which total internal reflection does not occur, and get refracted into the cladding.

(iii) All the modes are orthogonal. If the field $E_y(x)$ corresponding to a guided mode with a propagation constant β_m is represented by $\psi_m(x)$ and its complex conjugate by $\psi_m^*(x)$, then it can be shown that

$$\int_{-\infty}^{\infty} \psi_m^*(x)\, \psi_m(x)\, dx = 0 \quad \text{for } m \neq n$$

This is known as the condition of orthogonality.

Having determined the electric-field distribution $E_y(x)$, it is possible to calculate the H_x and H_z components of the magnetic field by substituting the value of E_y in Eq. (3.32) and $\partial E_y / \partial x$ in Eq. (3.33). A similar analysis may be done for the TM modes of a planar waveguide (see Review Question 3.7).

Example 3.2 What should be the maximum thickness of the guide slab of a symmetrical SI planar waveguide so that it supports only the fundamental TE mode? Take $n_1 = 3.6$, $n_2 = 3.56$, and $\lambda = 0.85$ µm.

Solution

For the waveguide to support only the fundamental mode, V should be less than $\pi/2$ (see Fig. 3.4), i.e.,

$$\frac{2\pi a}{\lambda}(n_1^2 - n_2^2)^{1/2} < \pi/2$$

or
$$2a < \frac{\lambda}{2(n_1^2 - n_2^2)^{1/2}}$$
$$< \frac{0.85\ \mu\text{m}}{2\{(3.6)^2 - (3.56)^2\}^{1/2}}$$
$$< 0.793\ \mu\text{m}$$

Therefore, the thickness of the guide slab should not be more than 0.793 μm.

Example 3.3 A planar waveguide is formed from a 10.11-μm-thin film of dielectric material of refractive index 1.50 sandwiched between two infinite dielectric slabs of refractive index 1.48. (a) Find the number of TE modes of propagation the guide can support at a wavelength of 1.55 μm. (b) Estimate approximately the propagation parameters u_m, w_m, b_m, and β_m for different values of m.

Solution

(a) $$V = \frac{2\pi a}{\lambda}(n_1^2 - n_2^2)^{1/2} = \frac{\pi \times 10.11 \times 10^{-6}\ \text{m}}{1.55 \times 10^{-6}\ \text{m}}[(1.50)^2 - (1.48)^2]^{1/2}$$
$$\approx 5$$

From Fig. 3.4, we can see that for this value of V, the number of TE modes of propagation is four, corresponding to m = 0, 1, 2, and 3. Thus there are two symmetric modes corresponding to $m = 0$ and $m = 2$ and two antisymmetric modes corresponding to $m = 1$ and $m = 3$.

(b) The abscissae of the points of intersection of the *ua* versus *wa* curves (see Fig. 3.4) for m = 0, 1, 2, and 3 with the quadrant of a circle of radius $V = 5$ give the values of $u_m a$, and the corresponding ordinates give the values of $w_m a$. Dividing these values of $u_m a$ and $w_m a$ by a gives the values of u_m and w_m. In the present case $a = 5.055 \times 10^{-6}$ m. The values of b_m can be obtained using Eq. (3.68). Here, the subscript m simply denotes the value of u, w, or b for the mth mode. The results are given in Table 3.1.

Table 3.1

m	$u_m a$ (rad)	u_m (m^{-1})	$w_m a$ (rad)	w_m (m^{-1})	$b_m = \left(\frac{w_m a}{V}\right)^2$
0	1.30644	2.58445×10^5	4.8263	9.54757×10^5	0.93172
1	2.59574	5.13499×10^5	4.27342	8.45384×10^5	0.73048
2	3.83747	7.59143×10^5	3.20529	6.34083×10^5	0.41095
3	4.9063	9.70583×10^5	0.963466	1.90596×10^5	0.03713

The values of β_m may be obtained using the relation

$$\beta_m = [b_m (\beta_1^2 - \beta_2^2) + \beta_2^2]^{1/2}$$

where $\beta_1 = 2\pi n_1/\lambda = 6.08 \times 10^6$ m^{-1} and $\beta_2 = 2\pi n_2/\lambda = 5.99 \times 10^6$ m^{-1}. Therefore the phase propagation constant for the mth mode will be given by

$$\begin{aligned}\beta_m &= [b_m (0.9764 \times 10^{12}) + 35.99 \times 10^{12}]^{1/2} \text{ m}^{-1}\\ &= [0.9764 b_m + 35.99]^{1/2} \times 10^6 \text{ m}^{-1}\end{aligned}$$

3.6 POWER DISTRIBUTION AND CONFINEMENT FACTOR

In Fig. 3.5, we have seen that the electric-field distribution of the guided modes extends beyond the boundary of the guide layer, and the extent of penetration into the cladding layer depends on the thickness and the mode number. It is important to know, in many situations, the fraction of guided optical power confined within the guide layer. This fraction is called the confinement factor.

The power flow is given by Poynting vector defined by

$$\mathbf{S} = \mathbf{E} \times \mathbf{H}$$

where **E** and **H** are normally expressed in the complex form. However, the actual fields are the real part of the complex form. Thus, taking the time average of the Poynting vector, we get

$$\begin{aligned}\langle \mathbf{S} \rangle &= \langle \text{Re}\,\mathbf{E} \times \text{Re}\,\mathbf{H} \rangle\\ &= \frac{1}{2}\text{Re}\,\langle \mathbf{E} \times \mathbf{H}^* \rangle\end{aligned}$$

where **H*** is the complex conjugate of **H**. Thus the time average of **S** along the z-direction will be given by

$$\langle S_z \rangle = \frac{1}{2}\langle (E_x H_y - H_x E_y) \rangle \tag{3.71}$$

We have seen that, for TE modes, the field components E_x and E_z vanish and

$$H_x = -\frac{\beta}{\mu_0 \omega} E_y$$

Therefore,

$$\langle S_z \rangle = \frac{1}{2}\frac{\beta}{\omega \mu_0} |E_y^2| \tag{3.72}$$

For a particular mode, the power associated per unit area per unit length in the y-direction will thus be given by

$$P = \frac{1}{2}\frac{\beta}{\omega \mu_0}\int_{x=-\infty}^{\infty} |E_y^2|\, dx \tag{3.73}$$

The power inside the guide layer will be given by

$$P_{\text{in}} = \frac{1}{2}\frac{\beta}{\omega\mu_0}\int_{-a}^{a} E_y^2\,dx \tag{3.74}$$

and the power outside the guide layer will be given by

$$P_{\text{out}} = \frac{1}{2}\frac{\beta}{\omega\mu_0}\left[\int_{-\infty}^{-a} E_y^2\,dx + \int_{a}^{\infty} E_y^2\,dx\right] \tag{3.75}$$

For symmetric TE modes, substituting for E_y from Eq. (3.55), we get

$$P_{\text{in}} = \frac{1}{2}\frac{\beta}{\omega\mu_0}2\int_0^a (A\cos ux)^2\,dx = \frac{\beta}{\omega\mu_0}A^2\int_0^a \frac{1}{2}(1+\cos 2ux)\,dx$$

or
$$P_{\text{in}} = \frac{\beta}{2\omega\mu_0}A^2\left[x + \frac{1}{2u}\sin 2ux\right]_0^a = \frac{\beta}{2\omega\mu_0}A^2\left[a + \frac{1}{2u}\sin 2ua\right] \tag{3.76}$$

Substituting the value of E_y from Eq. (3.56), we obtain

$$P_{\text{out}} = \frac{1}{2}\frac{\beta}{\omega\mu_0}2\int_a^{\infty} (Ce^{-wx})^2\,dx = \frac{\beta}{\omega\mu_0}C^2\left[-\frac{1}{2w}e^{-2wx}\right]_a^{\infty}$$

or
$$P_{\text{out}} = \frac{\beta}{\omega\mu_0}C^2\left[\frac{1}{2w}e^{-2wa}\right] \tag{3.77}$$

From Eqs (3.76) and (3.77), we can derive the final expression for P as follows:

$$P = P_{\text{in}} + P_{\text{out}}$$

$$= \frac{\beta}{2\omega\mu_0}A^2\left[a + \frac{1}{2u}\sin 2ua\right] + \frac{\beta}{\omega\mu_0}C^2\left[\frac{1}{2w}e^{-2wa}\right]$$

$$= \frac{\beta A^2}{2\omega\mu_0}\left[a + \frac{1}{2u}\sin 2ua + \left(\frac{C}{A}\right)^2\frac{1}{w}e^{-2wa}\right]$$

Putting the value of $C/A = e^{wa}\cos xa$ from Eq. (3.57), we get

$$P = \frac{\beta A^2}{2\omega\mu_0}\left[a + \frac{1}{2u}\sin 2ua + \frac{1}{w}\cos^2 ua\right]$$

$$= \frac{\beta A^2}{4\omega\mu_0}\left[2a + \frac{2}{w} + \frac{2\sin(ua)\cos(ua)}{u} - \frac{2\sin^2 ua}{w}\right]$$

$$= \frac{\beta A^2}{4\omega\mu_0}\left[\left(2a + \frac{2}{w}\right) + \frac{2\sin(ua)\cos(ua)}{uwa}\{wa - ua\tan ua\}\right]$$

From Eq. (3.59), the term within the braces becomes zero. This gives the following expression for P:

$$P = \frac{\beta A^2}{4\omega\mu_0}\left(a + \frac{1}{w}\right) \tag{3.78}$$

A similar expression may be derived for power carried by antisymmetric modes.

The fraction of power confined within the guide layer, called the confinement factor G, may be calculated as follows:

$$G = \frac{P_{\text{in}}}{P_{\text{in}} + P_{\text{out}}} \tag{3.79}$$

$$= \frac{a + \frac{1}{2u}\sin 2ua}{\left[a + \frac{1}{2u}\sin 2ua\right] + \left[\frac{1}{w}e^{-2wa}\right]\left(\frac{C}{A}\right)^2} \tag{3.80}$$

Putting the value of $C/A = e^{wa}\cos ua$, from Eq. (3.57), in Eq. (3.80), we get

$$G = \frac{\left[a + \frac{1}{2u}\sin 2ua\right]}{\left[a + \frac{1}{2u}\sin 2ua\right] + \left[\frac{1}{w}e^{-2wa}(e^{wa}\cos ua)^2\right]}$$

$$= \left[\frac{a + \frac{1}{2u}\sin 2ua + \frac{1}{w}\cos^2 ua}{a + \frac{1}{2u}\sin 2ua}\right]^{-1}$$

$$= \left[1 + \frac{\cos^2 ua}{wa\{1 + [\sin(ua)\cos(ua)]/ua\}}\right]^{-1} \tag{3.81}$$

A similar expression may be derived for antisymmetric modes. Some important points may be noted from Eq. (3.81).

(i) As $a \to 0$, $G \to 0$; that is, for a very thin guide layer, there is no guidance of light. Rather, almost all the power is lost in the cladding.

(ii) As the G-factor depends on u and w and both these parameters in turn depend on m, G will vary with the mode number.

(iii) It can be shown that G increases, first rapidly and then slowly, with increase in the thickness of the guide layer for each mode.

Example 3.4 Calculate the G-factor for the fundamental TE mode supported by the waveguide of Example 3.2. Take $2a = 0.793$ μm.

Solution

For thickness $2a = 0.793$ μm, $V = \pi/2$. The point of intersection of an arc of $V = \pi/2$ with the curve $ua \tan ua = wa$ for $m = 0$ gives

$$ua = 0.934 \text{ and } wa = 1.262$$

Using Eq. (3.80), the confinement factor G may be calculated as follows:

$$G = \left[1 + \frac{\cos^2(0.934)}{1.262\{1 + (\sin\ 0.934)\ (\cos 0.934)/0.934\}}\right]^{-1}$$

$$= 0.8436$$

This means that only 84.36% of the total power carried by the fundamental mode is confined within the guide layer; the remaining 15.64% extends into the cladding slabs.

SUMMARY

- Light is an electromagnetic wave.
- Maxwell's theory is based on a set of four equations. These are used to derive a wave equation describing the propagation of electromagnetic waves in any medium.
- In an isotropic and homogeneous dielectric medium, the wave equation takes the form

$$\nabla^2 \psi - \frac{n^2}{c^2}\frac{\partial^2 \psi}{\partial t^2} = 0$$

  where ψ represents the scalar field components of **E** or **H**. Its solution is of the form

$$\psi = \psi_0 \exp\{i(\omega t - \beta z)\}$$

  where the symbols have their usual meaning.
- A mode is a stable (electric or magnetic) field pattern in the transverse direction (e.g., the x-direction) with only a periodic z-dependence. In an isotropic and inhomogeneous dielectric medium, the modes of wave propagation are described by the following sets of equations.

$$\left.\begin{aligned} \beta E_y &= \mu_0 \omega H_x \\ \frac{\partial E_y}{\partial x} &= -\, i\ \mu_0 \omega H_z \\ -\, i\beta H_x - \frac{\partial H_z}{\partial x} &= i\varepsilon_0\, \omega n^2(x)\, E_y \end{aligned}\right\} \text{TE modes}$$

$$\left.\begin{aligned} \beta H_y &= -\varepsilon_0 \omega n^2(x) E_x \\ \frac{\partial H_y}{\partial x} &= i\varepsilon_0 \omega n^2(x) E_z \\ i\beta E_x + \frac{\partial E_z}{\partial x} &= i\mu_0 \omega H_y \end{aligned}\right\} \text{TM modes}$$

- **The maximum number of TE modes, M, supported by a symmetrical step-index planar waveguide is an integer close to or greater than $2V/\pi$, where**

$$V = \frac{2\pi}{\lambda} a \sqrt{n_1^2 - n_2^2}$$

- **The normalized propagation constant is given by**

$$b = \frac{\beta^2 - \beta_1^2}{\beta_1^2 - \beta_2^2} = 1 - \left(\frac{ua}{V}\right)^2 = \left(\frac{wa}{V}\right)^2$$

- **For a given guided mode, b lies between 0 and 1, and β^2 lies between β_1^2 and β_2^2.**
- **The cut-off of a mode occurs at $b = 0$, $ua = V_c = m\pi/2$, and $wa = 0$. The fundamental mode has no cut-off frequency.**
- **The thickness of a guide layer that can support m modes is given by**

$$2a = \frac{m\lambda}{2(n_1^2 - n_2^2)^{1/2}}$$

- **The thickness of a guide layer required to support only the fundamental mode is given by**

$$2a \le \frac{\lambda}{2(n_1^2 - n_2^2)^{1/2}}$$

- **The fraction of guided optical power that is confined within a guide layer is called the confinement factor. It depends on the thickness of the guide layer and the mode number.**

MULTIPLE CHOICE QUESTIONS

3.1 In a symmetrical SI planar waveguide, the refractive index of the guide layer is $n_1 = 1.5$ and $\Delta = 0.001$. If the thickness of the guide layer is 8.5 μm, for what wavelength will the guide support only a fundamental mode?
(a) 0.85 μm (b) 1.14 μm (c) 1.30 μm (d) 1.55 μm

3.2 A symmetrical SI planar waveguide is to be excited by a source of central wavelength $\lambda = 0.85$ μm. Assume that $n_1 = 1.5$ and $\Delta = 0.01$. What should be the thickness of the guide layer so that it supports one symmetric and one antisymmetric TE mode?

(a) 4.00 μm (b) 6.00 μm (c) 8.00 μm (d) 10.00 μm

3.3 For a given guided mode, the normalized propagation constant lies between

(a) $-\infty$ and $+\infty$. (b) 0 and ∞. (c) 0 and 1. (d) -1 and $+1$.

3.4 The range of the phase propagation constant ($\beta_2 < \beta < \beta_1$) for a wave guided by a planar waveguide can be obtained if the following is known:

(a) Wavelength of the excitation source.
(b) Refractive index of the guide layer.
(c) Refractive index of the cladding slabs.
(d) All of the above.

3.5 Inside an ideal dielectric medium,

(a) the free charge density ρ is zero but the conductivity σ is nonzero.
(b) ρ is nonzero and σ is zero.
(c) both ρ and σ are zero.
(d) both ρ and σ are nonzero.

3.6 For symmetric TE modes supported by a symmetrical SI planar waveguide, the stable electric-field distribution inside the guide layer takes

(a) a sinusoidal form.
(b) a cosinusoidal form.
(c) an exponential form.
(d) a tangential form.

3.7 The maximum number of TE modes supported by a symmetrical SI planar waveguide is an integer greater than

(a) $2V/\pi$. (b) V/π. (c) V. (d) zero.

3.8 For a typical symmetrical SI planar waveguide, the value of the V-parameter is 5. The guide supports the following TE modes.

(a) Two symmetric and three antisymmetric.
(b) Three symmetric and two antisymmetric.
(c) Two symmetric and two antisymmetric.
(d) One symmetric and one antisymmetric.

3.9 The electric-field distribution $E_y(x)$ in a planar waveguide corresponding to the $m = 3$ TE mode will have the following number of zeros between $x = -a$ and $x = +a$.

(a) Three (b) Two (c) One (d) None

3.10 In a planar waveguide, for a typical mode, 70% of its total power remains in the guide layer and the remaining 30% extends into the cladding slabs. Its confinement factor for the mode is

(a) 0.30. (b) 0.40. (c) 0.70. (d) 1.00.

Answers

3.1	(b)	3.2	(a)	3.3	(c)	3.4	(d)	3.5	(c)
3.6	(b)	3.7	(a)	3.8	(c)	3.9	(a)	3.10	(c)

REVIEW QUESTIONS

3.1 What is a plane TEM wave? How will the wavelength of this wave change with the medium through which it is propagating?

3.2 What is the significance of Maxwell's equations? Write these equations for the metallic and dielectric media.

3.3 What is the difference between phase velocity and group velocity? Write expressions for these.

3.4 Distinguish between the propagation parameters k, β, and b. How are they interrelated?

3.5 Prepare a list of assumptions that have been made while deriving Eqs (3.50) and (3.51).

3.6 What are modes? How does one distinguish between symmetric and antisymmetric modes of a planar SI waveguide?

3.7 Perform a detailed analysis of the TM modes (which are characterized by the field components E_x, E_z, and H_y) of a symmetrical SI planar wave guide.

[**Hint:**

Step I: Substituting for E_x and E_z from Eqs (3.35) and (3.36), respectively, in Eq. (3.37), we obtain

$$\frac{d^2 H_y}{dx^2} - \left[\frac{1}{n^2(x)} \frac{dn^2(x)}{dx}\right] \frac{dH_y}{dx} + [k^2 n^2(x) - \beta^2] H_y = 0$$

Step II: Substitute for $n(x)$ in this equation in the region $|x| < a$ and $|x| > a$, to get two differential equations for the two regions.

Step III: Apply the boundary conditions. Here H_y and E_z are tangential components to the planes $x = \pm a$. Thus H_y and $[1/n^2(x)]\,(dH_y/dx)$ must be continuous at $x = \pm a$.

Step IV: Write the solutions for symmetric and antisymmetric modes and arrive at the following transcendental equations:

$$ua \tan ua = \left(\frac{n_1}{n_2}\right)^2 wa \quad \text{for symmetric modes}$$

$$-\,ua \cot ua = \left(\frac{n_1}{n_2}\right)^2 wa \quad \text{for antisymmetric modes}$$

where the symbols have their usual meaning.

Step V: Next, solve these equations graphically and discuss the results.]

3.8 A planar waveguide is formed from a 14-μm-thick film of dielectric material of refractive index 1.46 sandwiched between two infinite dielectric slabs of refractive index 1.455. (a) Calculate the number of TE modes of propagation that the guide supports at a wavelength of 1.3 μm. (b) Estimate the propagation parameters u_m and w_m for different values of m and hence estimate b_m and β_m for each.

Ans: (a) Three modes corresponding to $m = 0$, 1, and 2.

(b) Since $V = 4$, the abscissae of the intersecting points of the ua versus wa curves (see Fig. 3.4) for $m = 0$, 1, and 2 with the quadrant of a circle of radius $V = 4$ give the values of $u_m a$, and the corresponding ordinates give the values of $w_m a$. Similarly, from Fig. 3.6, one can get the values of b_m corresponding to $V = 4$. β_m can be obtained from $\beta_m = \{\beta_2^2 + b_m(\beta_1^2 - \beta_2^2)\}^{1/2}$, where $\beta_1 = 2\pi n_1/\lambda$ and $\beta_2 = 2\pi n_2/\lambda$.

3.9 Calculate the maximum thickness of the guide slab of a symmetrical planar waveguide so that it supports the first 10 modes. Take $n_1 = 3.6$, $n_2 = 3.58$, and $\lambda = 0.90$ μm. Calculate also the maximum and minimum values of the propagation constant β.

Ans: $2a = 11.875$ μm, $\beta_1 = 2.513 \times 10^5$ m^{-1}, $\beta_2 = 2.499 \times 10^5$ m^{-1}

3.10 A planar waveguide is formed from a 10-μm-thick core film of dielectric material of refractive index 1.5 sandwiched between the cladding slabs of a similar material for which the relative refractive index difference (with respect to the core) is 0.02. (a) Calculate the V-parameter and the number of TE modes of propagation that the guide supports at a wavelength of 1.55 μm. (b) Estimate the propagation parameters u_m and w_m for different values of m and hence estimate b_m and β_m for each.

Ans: (a) $V = 6$; four modes corresponding to $m = 0$, 1, 2, and 3.

(b) The values of ua and wa for different values of m may be obtained from Fig. 3.4.

3.11 (a) Derive an expression for the confinement factor G for the antisymmetric TE modes of a symmetrical SI planar waveguide.

(b) Calculate the G-factors for the modes supported by a planar waveguide whose guide layer has a thickness 6.523 μm, $n_1 = 1.50$, and $n_2 = 1.48$. The guide is excited by a light of wavelength $\lambda = 1.0$ μm.

Ans: (a)
$$G = \left[1 + \frac{\sin^2(ua)}{wa\left\langle 1 - \dfrac{\sin(ua)\cos(ua)}{ua}\right\rangle}\right]^{-1}$$

(b) $G_0 = 0.98828$, $G_1 = 0.94889$, $G_2 = 0.85992$, $G_3 = 0.50960$

4 Wave Propagation in Cylindrical Waveguides

After reading this chapter you will be able to understand the following:

- Modal analysis of an ideal step-index fiber
- Modal power distribution
- Analysis of graded-index fiber
- Calculation of number of modes
- Limitations of multimode fibers

4.1 INTRODUCTION

In the previous chapter, we have discussed how an electromagnetic wave propagates through an ideal step-index (SI) planar waveguide. Using graphical techniques for solving wave equations, we could arrive at the propagation parameters for different modes. In this chapter, we will study wave propagation in cylindrical waveguides. The method of analysis is almost similar except that the mathematics involved gets more complicated.

4.2 MODAL ANALYSIS OF AN IDEAL SI OPTICAL FIBER

Consider an ideal step-index fiber consisting of a uniform cylindrical dielectric core of radius a and refractive index n_1, surrounded by an infinitely thick uniform dielectric cladding of refractive index n_2. We attempt a solution of the wave equation for electric and magnetic fields for the modes propagated by such a fiber.

Earlier, in Chapter 3 [Eq. 3.21], we have seen that the equation for an electric field in an isotropic, linear, non-conducting, and non-magnetic, but inhomogeneous medium is given by

$$\nabla^2 \mathbf{E} + \nabla\left\{\frac{1}{\varepsilon_r}\nabla(\varepsilon_r)\cdot\mathbf{E}\right\} - \mu_0\,\varepsilon_0\,\varepsilon_r\,\frac{\partial^2 \mathbf{E}}{\partial t^2} = 0 \tag{4.1}$$

Similarly, the equation for a magnetic field is given by

$$\nabla^2 \mathbf{H} + \frac{1}{\varepsilon_r}\{\nabla(\varepsilon_r) \times (\nabla \times \mathbf{H})\} - \mu_0 \varepsilon_0 \varepsilon_r \frac{\partial^2 \mathbf{H}}{\partial t^2} = 0 \tag{4.2}$$

From Eqs (4.1) and (4.2) it is evident that the different components of **E** as well as **H** are coupled.

In the present case, the refractive index $n = \sqrt{\varepsilon_r}$ is constant ($= n_1$) up to $r \le a$ and equal to n_2 for points $r > a$, but there is a discontinuity at $r = a$. Let us assume that this discontinuity is small (i.e., $n_1 \approx n_2$). This is called a weakly guiding approximation. With this approximation, the second term on the left-hand side (LHS) of Eqs (4.1) and (4.2) may be neglected and each Cartesian component of **E** and **H** satisfies the scalar wave equation; putting $\varepsilon_r = n^2$,

$$\nabla^2 \Psi = \varepsilon_0 \mu_0 n^2 \frac{\partial^2 \Psi}{\partial t^2} \tag{4.3}$$

where Ψ represents the scalar E or H field. The fiber boundary conditions have cylindrical symmetry and we assume that the direction of propagation of the electromagnetic waves is along the axis of the fiber, which we take to be the z-axis. In the scalar wave approximation, the modes may be assumed to be nearly transverse and they may possess an arbitrary state of polarization. These linearly polarized modes are referred to as LP modes. The propagation constants of the TE and TM modes are nearly equal.

In cylindrical coordinates (r, ϕ, z), Eq. (4.3) may be written as

$$\nabla^2 \Psi - \varepsilon_0 \mu_0 n^2 \frac{\partial^2 \Psi}{\partial t^2} = \frac{\partial^2 \Psi}{\partial r^2} + \frac{1}{r}\frac{\partial \Psi}{\partial r} + \frac{1}{r^2}\frac{\partial^2 \Psi}{\partial \phi^2}$$

$$+ \frac{\partial^2 \Psi}{\partial z^2} - \varepsilon_0 \mu_0 n^2 \frac{\partial^2 \Psi}{\partial t^2} = 0 \tag{4.4}$$

Since n may depend on the transverse coordinates (r, ϕ), though it usually depends only on r, and the wave is propagating along the z-direction, we may write the solution of Eq. (4.4) as

$$\Psi(r, \phi, z, t) = \psi(r, \phi) e^{i(\omega t - \beta z)} \tag{4.5}$$

Substituting the value of Ψ from Eq. (4.5) in Eq. (4.4), we get

$$e^{i(\omega t - \beta z)} \frac{\partial^2 \psi}{\partial r^2} + \frac{1}{r} e^{i(\omega t - \beta z)} \frac{\partial \psi}{\partial r} + \frac{1}{r^2} e^{i(\omega t - \beta z)} \frac{\partial^2 \psi}{\partial \phi^2}$$

$$+ \psi(-\beta^2) e^{i(\omega t - \beta z)} - \varepsilon_0 \mu_0 n^2 (-\omega^2) \psi e^{i(\omega t - \beta z)} = 0$$

or

$$\frac{\partial^2 \psi}{\partial r^2} + \frac{1}{r}\frac{\partial \psi}{\partial r} + \frac{1}{r^2}\frac{\partial^2 \psi}{\partial \phi^2} + [\varepsilon_0 \mu_0 \omega^2 n^2 - \beta^2]\psi = 0$$

Putting $\varepsilon_0\mu_0 = 1/c^2$ and $\omega/c = k$, the free-space wave number, in the above equation, we get

$$\frac{\partial^2\psi}{\partial r^2}+\frac{1}{r}\frac{\partial\psi}{\partial r}+\frac{1}{r^2}\frac{\partial^2\psi}{\partial\phi^2}+[n^2k^2-\beta^2]\psi=0 \tag{4.6}$$

Since the fiber under consideration has cylindrical symmetry, the variables can be separated:

$$\psi(r,\phi)=R(r)\,\Phi(\phi) \tag{4.7}$$

where R is a function of only r and Φ is a function of only ϕ. Substituting ψ from Eq. (4.7) in Eq. (4.6), we get

$$\Phi\frac{\partial^2 R}{\partial r^2}+\frac{1}{r}\Phi\frac{\partial R}{\partial r}+\frac{R}{r^2}\frac{\partial^2\Phi}{\partial\phi^2}+[n^2k^2-\beta^2]R\Phi=0$$

Since the derivatives involved are dependent either on r or ϕ only, the partial derivatives may be replaced by full derivatives. Further, dividing the entire LHS by $R\Phi$, we get

$$\frac{1}{R}\frac{d^2R}{dr^2}+\frac{1}{Rr}\frac{dR}{dr}+\frac{1}{r^2}\frac{1}{\Phi}\frac{d^2\Phi}{d\phi^2}+[n^2k^2-\beta^2]=0$$

or
$$\frac{r^2}{R}\left(\frac{d^2R}{dr^2}+\frac{1}{r}\frac{dR}{dr}\right)+r^2[n^2k^2-\beta^2]=-\frac{1}{\Phi}\frac{d^2\Phi}{d\phi^2}=l^2 \quad \text{(say)} \tag{4.8}$$

where l is a constant, known as an azimuthal eigenvalue .

The dependence of Φ on ϕ will be of the form $e^{il\phi}$. For the function to be single-valued, i.e., $\Phi(\phi+2\pi)=\Phi(\phi)$, the constant l is required to be an integer, that is,

$$l=0,1,2,3,\ldots \tag{4.9}$$

Therefore the complete transverse field will be given by

$$\Psi(r,\phi,z,t)=R(r)e^{il\phi}e^{i(\omega t-\beta z)} \tag{4.10}$$

The radial part of Eq. (4.8) may be written as

$$\frac{r^2}{R}\left(\frac{d^2R}{dr^2}+\frac{1}{r}\frac{dR}{dr}\right)+r^2(n^2k^2-\beta^2)=l^2$$

which can be rearranged to give

$$r^2\frac{d^2R}{dr^2}+r\frac{dR}{dr}+[r^2(n^2k^2-\beta^2)-l^2]R=0 \tag{4.11}$$

We know that $n=n_1$ for $r\le a$ and $n=n_2$ for $r>a$. Thus, substituting the value of n in Eq. (4.11), we obtain for the case of a step-index fiber,

$$r^2\frac{d^2R}{dr^2}+r\frac{dR}{dr}+[r^2(k^2n_1^2-\beta^2)-l^2]R=0,\qquad r\le a \tag{4.12}$$

and
$$r^2\frac{d^2R}{dr^2}+r\frac{dR}{dr}+[r^2(k^2n_2^2-\beta^2)-l^2]R=0,\qquad r>a \tag{4.13}$$

In order to simplify the above equations, we put

$$u^2 \equiv (k^2 n_1^2 - \beta^2)\, a^2 \tag{4.14}$$

and
$$w^2 \equiv (\beta^2 - k^2 n_2^2)\, a^2 \tag{4.15}$$

The normalized waveguide parameter V for the fiber is defined by

$$V = (u^2 + w^2)^{1/2} = ka(n_1^2 - n_2^2)^{1/2} = \frac{2\pi a}{\lambda}(n_1^2 - n_2^2)^{1/2} \tag{4.16}$$

Substituting the values of u and w in Eqs (4.12) and (4.13), we get

$$r^2 \frac{d^2R}{dr^2} + r\frac{dR}{dr} + \left(\frac{u^2 r^2}{a^2} - l^2\right) R = 0, \quad r \le a \tag{4.17}$$

and
$$r^2 \frac{d^2R}{dr^2} + r\frac{dR}{dr} - \left(\frac{w^2 r^2}{a^2} + l^2\right) R = 0, \quad r > a \tag{4.18}$$

Equations (4.17) and (4.18) are second-order equations and hence should possess two independent solutions. The solutions corresponding to Eq. (4.17) are the Bessel function of the first kind and the modified Bessel function of the first kind. The solutions corresponding to Eq. (4.18) are the Bessel function of the second kind and the modified Bessel function of the second kind. The modified Bessel function of the first kind has a discontinuity at the origin and the Bessel function of the second kind has an asymptotic form. Hence these are discarded in the solutions for fiber modes. For the solutions to be well behaved, that is, be finite at $r = 0$ and tend to zero as $r \to \infty$, it is essential that both u and w are real. Therefore a valid solution of Eq. (4.17) would be given by the first kind of Bessel function of order l and that of Eq. (4.18) would be given by the second kind of modified Bessel function of order l. Thus

$$R(r) = AJ_l\left(\frac{ur}{a}\right), \quad r < a \tag{4.19}$$

and
$$R(r) = BK_l\left(\frac{wr}{a}\right), \quad r > a \tag{4.20}$$

The Bessel function of the first kind of order l and argument x, denoted by $J_l(x)$, is defined in terms of an infinite series as follows:

$$J_l(x) = \sum_{n=0}^{\infty} \frac{(-1)^n}{n!\,\Gamma(n + l + 1)}\left(\frac{x}{2}\right)^{2n+l} \tag{4.21}$$

where $x = ur/a$ and the gamma function $\Gamma(n + l + 1) = (n + l)!$. For $l = 0$,

$$J_0(x) = 1 - \frac{\frac{1}{4}x^2}{(1!)^2} + \frac{\left(\frac{1}{4}x^2\right)^2}{(2!)^2} - \frac{\left(\frac{1}{4}x^2\right)^3}{(3!)^2} + \cdots$$

and for $l = 1$,

$$J_1(x) = \frac{1}{2}x - \frac{\left(\frac{1}{2}x\right)^3}{2!} - \frac{\left(\frac{1}{2}x\right)^5}{2!3!} - \cdots$$

and so on for higher values of l.

The second kind of modified Bessel function of order l is given by

$$K_l(\tilde{x}) = (\pi/2)\, i^{-(l+1)}\, H_l(-i\tilde{x}), \quad \tilde{x} = \frac{wr}{a} \tag{4.22}$$

where $H_l(-i\tilde{x})$ is a Hankel function, which is a linear combination of Bessel functions of the first (J_l) and second (Y_l) kind. These functions have been chosen because for $x \ll 1$,

$$J_l(x) = \frac{1}{l!}\left(\frac{x}{2}\right)^l, \quad l = 0, 1, \ldots \tag{4.23}$$

$$K_l(\tilde{x}) = (l-1)!\,2^{l-1}\,\tilde{x}^{-l}, \quad l \geq 1 \tag{4.24}$$

and for $x \gg 1$,

$$J_l(x) = \sqrt{\frac{2}{\pi x}}\cos\left[x - \frac{\pi(2l+1)}{4}\right] \tag{4.25}$$

and

$$K_l(\tilde{x}) = \sqrt{\frac{2}{\pi x}}\,e^{-\tilde{x}}\left[1 + \frac{4l^2 - 1}{8\tilde{x}}\right] \tag{4.26}$$

Thus $J_l(x)$ is a well behaved function for $r < a$ and $K_l(\tilde{x})$ is well behaved for $r > a$.

For further analysis, some recurrence relations for these functions (with argument x) and some asymptotic forms are given as follows:

$$J_{-l} = (-1)^l J_l \tag{4.27}$$

$$J_l' = \frac{1}{2}(J_{l-1} - J_{l+1}) = \pm J_{l\mp 1} \mp \frac{lJ_l}{x} \tag{4.28}$$

$$J_{l\mp 1} = \frac{2lJ_l}{x} - J_{l\pm 1} \tag{4.29}$$

$$J_{l\mp 2} = \frac{2(l\mp 1)J_{l\mp 1}}{x} - J_l \tag{4.30}$$

$$K_l = K_{-l} \tag{4.31}$$

$$K_l' = -\frac{1}{2}(K_{l-1} + K_{l+1}) = \mp\frac{lK_l}{x} - K_{l\mp 1} \tag{4.32}$$

$$K_{l\mp 1} = \mp\frac{2lK_l}{x} + K_{l\pm 1} \tag{4.33}$$

$$K_{l \mp 2} = \mp \frac{2\,(l \mp 1)\,K_{l \mp 1}}{x} + K_l \tag{4.34}$$

Here, prime denotes the first derivative. For $\tilde{x} \ll 1$

$$\frac{K_0}{K_1} = \tilde{x} ln \frac{2}{1.782\,\tilde{x}} \tag{4.35}$$

$$\frac{K_{l-1}}{K_l} = \frac{\tilde{x}}{2\,(l-1)}, \quad l \geq 2 \tag{4.36}$$

$$\frac{K_{l+1}}{K_l} = \frac{2l}{\tilde{x}}, \quad l \geq 1 \tag{4.37}$$

For $\tilde{x} \gg 1$

$$\frac{K_{l \mp 1}}{K_l} = 1 + \frac{1 \mp 2l}{2\,\tilde{x}} \tag{4.38}$$

Since ψ is continuous at $r = a$, $R(r)$ must be continuous at $r = a$. Imposing this condition, we can get the values of constants A and B. Thus from Eqs (4.19) and (4.20) we have

$$A = \frac{R\,(a)}{J_l\,(u)} \tag{4.39}$$

and

$$B = \frac{R\,(a)}{K_l\,(w)} \tag{4.40}$$

Substituting the values of R and Φ in Eq. (4.7), we get the transverse dependence of the modal fields as follows:

$$\psi(r, \phi) = AJ_l\left(\frac{ur}{a}\right)\begin{bmatrix}\cos(l\phi);\, r < a \\ \sin(l\phi)\end{bmatrix} \tag{4.41}$$

and

$$\psi(r, \phi) = BK_l\left(\frac{wr}{a}\right)\begin{bmatrix}\cos(l\phi);\, r > a \\ \sin(l\phi)\end{bmatrix} \tag{4.42}$$

Now $\partial\psi/\partial r$ is also continuous at $r = a$:

$$\left.\frac{\partial \psi}{\partial r}\right|_{r=a} = A\frac{u}{a}J_l{}'\left(\frac{ur}{a}\right)\cos l\phi\Bigg|_{r=a} = A\frac{u}{a}J_l{}'(u)\cos l\phi$$

$$\left.\frac{\partial \psi}{\partial r}\right|_{r=a} = B\frac{w}{a}K_l{}'\left(\frac{wr}{a}\right)\cos l\phi\Bigg|_{r=a} = B\frac{w}{a}K_l{}'(w)\cos l\phi$$

Thus, substituting the values of A and B from Eqs (4.39) and (4.40),

$$\frac{R\,(a)}{J_l\,(u)}\frac{u}{a}J_l{}'(u)\cos l\phi = \frac{R\,(a)}{K_l\,(w)}\frac{w}{a}K_l{}'(w)\cos l\phi$$

or
$$\frac{u\,J_l'(u)}{J_l(u)} = \frac{wK_l'(w)}{K_l(w)}$$

Thus the continuity of ψ and $\partial\psi/\partial r$ at the core–cladding interface ($r = a$) leads to an eigenvalue equation of the form

$$\frac{uJ_l'(u)}{J_l(u)} = \frac{wK_l'(w)}{K_l(w)} \tag{4.43}$$

Substituting the values of J_l' and K_l' from Eqs (4.28) and (4.32), respectively, in Eq. (4.43), we get

$$\frac{u}{J_l(u)}\left[\pm J_{l\mp1}(u) \mp \frac{lJ_l(u)}{u}\right] = \frac{w}{K_l(w)}\left[\mp \frac{lK_l(w)}{w} - K_{l\mp1}(w)\right]$$

or
$$\pm u\frac{J_{l\mp1}(u)}{J_l(u)} \mp l = \mp l - \frac{wK_{l\mp1}(w)}{K_l(w)}$$

This equation may be written in either of the following two forms:

$$\frac{uJ_{l+1}(u)}{J_l(u)} = \frac{wK_{l+1}(w)}{K_l(w)} \tag{4.44}$$

or
$$\frac{uJ_{l-1}(u)}{J_l(u)} = -w\frac{K_{l-1}(w)}{K_l(w)} \tag{4.45}$$

One can obtain, from Eqs (4.44) and (4.45), the values of u and w for various values of l and the corresponding values of the propagation constant β. For β-values lying within the range

$$\beta_2^2\ (= n_2^2 k^2) < \beta^2 < \beta_1^2\ (= n_1^2 k^2) \tag{4.46}$$

the radial part of the field, $R(r)$, in the core is given by the Bessel function $J_l(x)$, $x = ur/a$, which is oscillatory in nature. Hence there exist m allowed solutions of β for each value of l. Therefore, each allowed value of β is characterized by two integers l and m. The first integer l is associated with two circular functions $\cos l\phi$ and $\sin l\phi$ corresponding to the azimuthal part of the solution, and the second integer m is associated with the mth root of the eigenvalue equation corresponding to the radial part of the solution. These are known as guided modes.

From Eq. (4.16), we know that $V^2 = u^2 + w^2$. Thus the solution of the transcendental equations (for given values of l and V) will give universal curves describing the dependence of u or w on V. The value of β can be calculated by substituting the values of u (or w) and V in the defining equations. Alternatively, we can define the normalized propagation constant b as follows:

$$b = \frac{\beta^2 - \beta_2^2}{\beta_1^2 - \beta_2^2} = \frac{\beta^2 - n_2^2 k^2}{n_1^2 k^2 - n_2^2 k^2} = \frac{w^2}{V^2} = 1 - \frac{u^2}{V^2} \tag{4.47}$$

Since β lies between β_1 ($= n_1 k$) and β_2 ($= n_2 k$) for the guided modes, the value of b will lie between 0 (for $\beta = \beta_2$) and 1 (for $\beta = \beta_1$). The plots of b as a function of V, shown in Fig. 4.1, form universal curves.

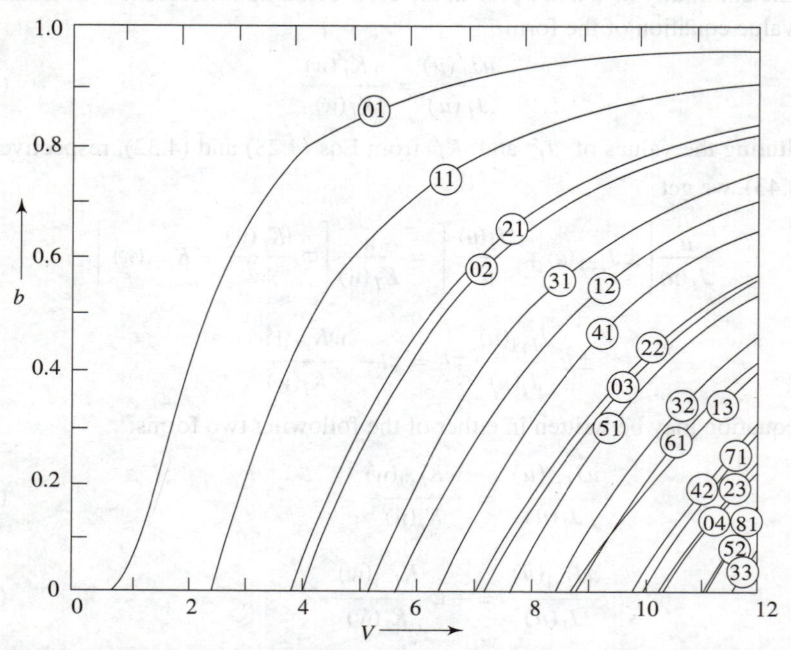

Fig. 4.1 Plots of the normalized propagation constant b as a function of the normalized frequency parameter V for a SI fiber (Gloge 1971)

A mode ceases to be guided when $\beta^2 < \beta_2^2$. Such modes are called radiation modes. In terms of the ray model, these modes correspond to rays that undergo refraction, rather than total internal reflection, at the core–cladding interface. When such modes are excited, they quickly leak away from the core. The condition $\beta = \beta_2$ corresponds to what is known as the cut-off of a mode. Thus, at $\beta = \beta_2$; $b = 0$, $w = 0$, and $u = V = V_c$, where V_c is the cut-off frequency. It is important to mention here that

$$\lim_{w \to 0} w \frac{K_{l-1}(w)}{K_l(w)} \to 0, \quad l = 0, 1, 2, 3, \ldots$$

Therefore, the RHS of Eq. (4.45) vanishes as $w \to 0$. Thus, Eq. (4.45) may be used for finding the cut-off frequencies of different modes.

For $l = 0$, we get from Eq. (4.45)

$$u \frac{J_{-1}(u)}{J_0(u)} = -w \frac{K_{-1}(w)}{K_0(w)}$$

Using Eqs (4.27) and (4.31), we get

$$J_{-1}(u) = -J_1(u) \quad \text{and} \quad K_{-1}(w) = K_1(w)$$

Substituting the value of $J_{-1}(u)$ and $K_{-1}(w)$ in the above equation, we obtain

$$\frac{uJ_1(u)}{J_0(u)} = w\frac{K_1(w)}{K_0(w)} \tag{4.48}$$

At the cut-off frequency, $w = 0, u = V = V_c$, the RHS = 0, and Eq. (4.48) is transformed into

$$\frac{V_c J_1(V_c)}{J_0(V_c)} = 0$$

or

$$J_1(V_c) = 0 \tag{4.49}$$

The roots of Eq. (4.49) give the values of the cut-off frequency for $l = 0$ and $m = 1, 2, 3, \ldots$.

Similarly, for $l = 1$, we get from Eq. (4.45)

$$\frac{uJ_0(u)}{J_1(u)} = -w\frac{K_0(w)}{K_1(w)} \tag{4.50}$$

At the cut-off frequency, $w = 0$, $u = V = V_c$, the RHS = 0, and we obtain

$$J_0(V_c) = 0 \tag{4.51}$$

The roots of Eq. (4.51) give V_c for $l = 1$ and $m = 1, 2, 3, \ldots$.

For $l \geq 2$ modes, the following equation gives the values of V_c:

$$J_{l-1}(V_c) = 0, \quad V_c \neq 0$$

It should be mentioned here that, for $l \geq 2$, the root $V_c = 0$ must not be included because

$$\lim_{V\to 0} V\frac{J_{l-1}(V)}{J_l(V)} \neq 0 \quad \text{for } l \geq 2$$

The values of the cut-off frequencies for the first few LP modes are given in Table 4.1. Figure 4.2 shows the oscillatory nature of Bessel functions J_0 and J_1 and their roots.

Table 4.1 Cut-off frequencies of the first few lower order LP_{lm} modes in a SI fiber

l	m			
	1	2	3	4
0	0	3.832	7.106	10.173
1	2.405	5.520	8.654	11.790
2	3.832	7.016	10.173	13.324
3	5.136	8.417	11.620	14.796

Before proceeding further, a few important points must be mentioned about the modes:

(i) The $l = 0$ modes have twofold degeneracy corresponding to two orthogonal linearly polarized states.

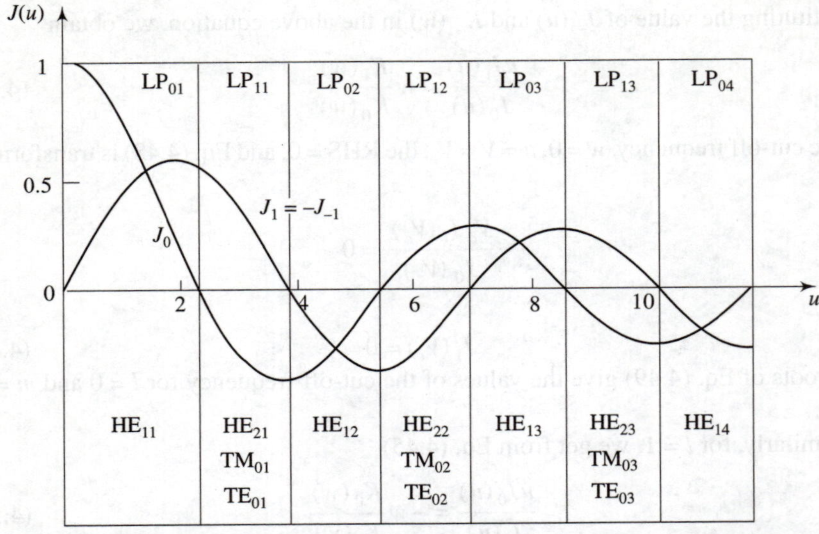

Fig. 4.2 Plot of Bessel functions J_0 and J_1, indicating the range of allowed values of u for lower order modes (Gloge 1971)

(ii) The $l \geq 1$ modes have fourfold degeneracy as each polarization state may have ϕ-dependence of the $\cos l\phi$ type or the $\sin l\phi$ type.

(iii) We will see in the next section that the total number of modes (when $V \gg 1$) guided along a step-index fiber is given approximately by $M = V^2/2$. As we know from Eq. (4.16) that the V-value of the fiber is dependent on the dimensions, the numerical aperture of the fiber, and the wavelength of the light signal; the total number of guided modes in a particular fiber at a specific wavelength is fixed.

Example 4.1 A step-index fiber has a core diameter of 7.2 μm, a core index of 1.46, and a relative refractive index difference of 1%. A light of wavelength 1.55 μm is used to excite the modes in the fiber. Find (a) the normalized frequency parameter V, (b) the propagation constants β for these modes, and (c) the phase velocity of each mode.

Solution

(a) $$V = \frac{2\pi a}{\lambda} n_1 \sqrt{2\Delta}$$

$$= 2\frac{\pi \times 7.2 \times 10^{-6}}{1.55 \times 10^{-6}} \times 1.46\sqrt{2 \times 0.01}$$

$$= 6.02$$

(b) Draw a vertical line on the graph of b versus V (the plots of Fig. 4.1) at $V \approx 3$. The line intersects the curves for only two modes, namely, the LP_{01} and LP_{11} modes. This implies that the fiber is supporting only these two modes.

(c) The β-values can be obtained using Eq. (4.47), i.e.,

$$b_{lm} = \frac{\beta_{lm}^2 - \beta_2^2}{\beta_1^2 - \beta_2^2}$$

or
$$\beta_{lm} = [\beta_2^2 + b_{lm}(\beta_1^2 - \beta_2^2)]^{1/2}$$

$$\beta_1 = kn_1 = \frac{2\pi}{\lambda} n_1 = \frac{2\pi}{1.55 \times 10^6} \times 1.46 = 5.915 \times 10^6 \text{ m}^{-1}$$

$$\beta_2 = kn_2 = \frac{2\pi}{1.55 \times 10^{-6}} \times 1.4454 = 5.855 \times 10^6 \text{ m}^{-1}$$

[since $n_2 = n_1(1 - \Delta) = 1.46(1 - 0.01) = 1.4454$].

$$\beta_{lm} = [34.281 \times 10^{12} + b_{lm}(0.7062 \times 10^{12})]^{1/2} \text{ m}^{-1}$$

From Fig. 4.1, we get approximately $b_{01} = 0.62$ and $b_{11} = 0.18$. Therefore,

$$\beta_{01} = (34.281 + 0.62 \times 0.7062)^{1/2} \times 10^6 \text{ m}^{-1}$$
$$= 5.8922 \times 10^6 \text{ m}^{-1}$$

and
$$\beta_{11} = (34.281 + 0.18 \times 0.7062)^{1/2} \times 10^6 \text{ m}^{-1}$$
$$= 5.8658 \times 10^6 \text{ m}^{-1}$$

(d) Phase velocity

$$V_p = \frac{\omega}{\beta} = \frac{2\pi c}{\lambda\beta}$$

Thus
$$(V_p)_{01} = \frac{2\pi \times 3 \times 10^8}{1.55 \times 10^{-6} \times 5.8922 \times 10^6} = 2.06 \times 10^8 \text{ m s}^{-1}$$

and
$$(V_p)_{02} = \frac{2\pi \times 3 \times 10^8}{1.55 \times 10^{-6} \times 5.8658 \times 10^{-6}} = 2.07 \times 10^8 \text{ m s}^{-1}$$

4.3 FRACTIONAL MODAL POWER DISTRIBUTION

Once again, using the scalar approximation, let us calculate the fractional modal power distribution in the core and the cladding of a SI fiber. The power in the core is given by the integral

$$P_{\text{core}} = (\text{constant}) \int_{r=0}^{a} \int_{\phi=0}^{2\pi} |\psi(r, \phi)|^2 \, r \, dr \, d\phi$$

$$= \text{constant}\{R(a)\}^2 \int_{r=0}^{a} \frac{J_l^2\,(ur/a)\, r\, dr}{J_l^2(u)} \int_{\phi=0}^{2\pi} \cos^2 l\phi\, d\phi$$

where we have substituted the value of $\psi(r, \phi)$ from Eq. (4.41) and taken the ϕ-dependence to be of the form $\cos(l\phi)$. The solution of the above integral can be shown to be

$$P_{\text{core}} = C\pi a^2 \left[1 - \frac{J_{l-1}(u)\, J_{l+1}\,(u)}{J_l^2(u)} \right] \tag{4.52}$$

where C is a constant.

Similarly, the power distribution in the cladding may be obtained by solving the integral

$$P_{\text{cladding}} = (\text{constant}) \int_{r=a}^{\infty} \int_{\phi=0}^{2\pi} |\psi(r, \phi)|^2\, r\, dr\, d\phi$$

Substituting the value of $\psi(r, \phi)$ from Eq. (4.42) in the above relation and again taking the $\cos(l\phi)$ form, we get

$$P_{\text{cladding}} = (\text{constant})\, \{R(a)\}^2 \int_{r=a}^{\infty} \frac{K_l^2\,(wr/a)}{K_l^2\,(w)}\, r\, dr \int_{\phi=0}^{2\pi} \cos^2 l\phi\, d\phi$$

$$= C\pi a^2 \left[\frac{K_{l-1}(w)\, K_{l+1}\,(w)}{K_l^2\,(w)} - 1 \right] \tag{4.53}$$

Adding Eqs (4.52) and (4.53), we get an expression for the total power P_T as follows:

$$P_T = P_{\text{core}} + P_{\text{cladding}}$$

$$= C\pi a^2 \left[1 - \frac{J_{l-1}(u)\, J_{l+1}\,(u)}{J_l^2\,(u)} \right] + C\pi a^2 \left[\frac{K_{l-1}\,(w) K_{l+1}\,(w)}{K_l^2\,(w)} - 1 \right]$$

$$= C\pi a^2 \left[\frac{K_{l-1}\,(w)\, K_{l+1}\,(w)}{K_l^2\,(w)} - \frac{J_{l-1}\,(u)\, J_{l+1}\,(u)}{J_l^2\,(u)} \right] \tag{4.54}$$

Multiplying the eigenvalue equations (4.45) and (4.44), we get

$$\frac{u^2\, J_{l-1}\,(u)\, J_{l+1}\,(u)}{J_l^2\,(u)} = \frac{w^2\, K_{l-1}(u)\, K_{l+1}\,(u)}{K_l^2\,(w)} \tag{4.55}$$

Using Eq. (4.55), Eq. (4.54) may be written as

$$P_T = C\pi a^2 \left(\frac{K_{l-1}\,(w)\, K_{l+1}\,(w)}{K_l^2\,(w)} \right) \left(1 + \frac{w^2}{u^2} \right)$$

$$= C\pi a^2 \frac{V^2}{u^2}\left[\frac{K_{l-1}(w)K_{l+1}(w)}{K_l^2(w)}\right] \tag{4.56}$$

Again using Eq. (4.55), Eq. (4.52) may be written as

$$P_{\text{core}} = C\pi a^2\left[1+\frac{w^2}{u^2}\frac{K_{l-1}(w)K_{l+1}(w)}{K_l^2(w)}\right] \tag{4.57}$$

Dividing Eq. (4.57) by Eq. (4.56), we obtain the fractional power propagating in the core as

$$\frac{P_{\text{core}}}{P_T} = \frac{C\pi a^2\left[1+\dfrac{w^2}{u^2}\dfrac{K_{l-1}(w)K_{l+1}(w)}{K_l^2(w)}\right]}{C\pi a^2\dfrac{V^2}{u^2}\left[\dfrac{K_{l-1}(w)K_{l+1}(w)}{K_l^2(w)}\right]}$$

$$= \left[\frac{u^2}{V^2}\frac{K_l^2(w)}{K_{l-1}(w)K_{l+1}(w)}+\frac{w^2}{V^2}\right] \tag{4.58}$$

The fractional power propagating in the cladding is given by

$$\frac{P_{\text{cladding}}}{P_T} = 1-\frac{P_{\text{core}}}{P_T} = \frac{u^2}{V^2}\left[1-\frac{K_l^2(w)}{K_{l-1}(w)K_{l+1}(w)}\right] \tag{4.59}$$

Figure 4.3 shows the plots of fractional power propagating in the core and the cladding for some lower order LP modes. It is interesting to note that for the first two lower order modes, the power flow is mostly in the cladding near cut-off. Using Eqs (4.35)–(4.37), it can be shown that as $V \to V_c$, $w \to 0$, and $u \to V_c$,

$$\frac{P_{\text{core}}}{P_T} \to \begin{cases} 0 & \text{for } l = 0 \text{ and } 1 \\ \dfrac{(l-1)}{l} & \text{for } l \ge 2 \end{cases} \tag{4.60}$$

However, for larger values of l, the power remains in the core even at or just beyond cut-off. Another point to be mentioned is that the power associated with a particular mode is mostly confined in the core for large values of V.

Example 4.2 For the step-index fiber of Example 4.1, what is the fractional power propagating in the cladding for the two modes?

Solution

The step-index fiber of Example 4.1 has a V-parameter approximately equal to 3. For this value of V, from Fig. 4.3, we get for the modes LP_{01} and LP_{11}

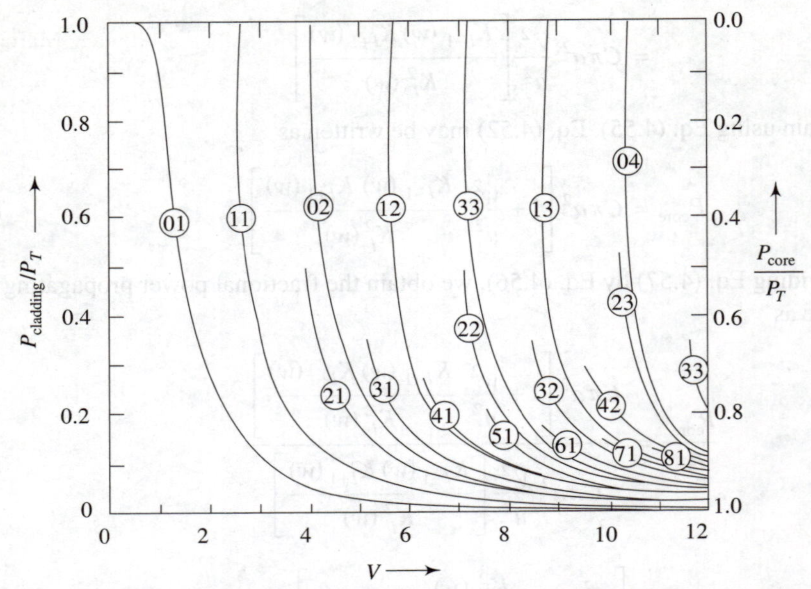

Fig. 4.3 Plots of fractional power contained in the core and cladding of a SI fiber as a function of V (Gloge 1971)

$$\left(\frac{P_{\text{cladding}}}{P_T}\right)_{01} = 0.11$$

and

$$\left(\frac{P_{\text{cladding}}}{P_T}\right)_{11} = 0.347$$

that is, for the LP_{01} and LP_{11} modes, respectively, 11% and ≈ 35% power flows in the cladding.

4.4 GRADED-INDEX FIBERS

We know that the refractive index profile for a multimode graded-index (GI) fiber is given by

$$n(r) = n_0\{1 - 2\Delta\,(r/a)^{\alpha}\}^{1/2} \quad r \le a \tag{4.61}$$

$$= n_0(1 - 2\Delta)^{1/2} = n_c \quad r \ge a \tag{4.62}$$

where $\Delta = (n_0^2 - n_c^2)/2n_0^2 \approx \dfrac{(n_0 - n_c)}{n_0}$

when $\Delta \ll 1$. n_0 is the refractive index at $r = 0$, n_c is the refractive index of the cladding, and α is called the profile parameter.

In general, for a cylindrical dielectric waveguide, e.g., an optical fiber, the electric and magnetic fields are governed by Eqs (4.1) and (4.2), respectively. The exact solutions of these equations for graded-index fibers are difficult to obtain. However, using the WKB (Wentzel-Kramers-Brillouin) approximation (Gloge & Marcatili 1973) it is possible to study the propagation characteristics of these fibers. The propagation constant β_p of the pth mode in a GI fiber with an α-profile may be given by

$$\beta_p \approx \beta_0 \left[1 - 2\Delta\left(\frac{p}{M_g}\right)\right]^{1/2} \tag{4.63}$$

where $p = 1, 2, 3, \ldots, M_g$ and $\beta_0 = kn_0$. Here, M_g represents the total number of guided modes given by

$$M_g = k^2 n_0^2 a^2 \Delta \left(\frac{\alpha}{\alpha + 2}\right) \tag{4.64}$$

Substituting the values of Δ and k, we get

$$M_g = \frac{1}{2}\left(\frac{\alpha}{\alpha + 2}\right)\left[\frac{2\pi a}{\lambda}(n_0^2 - n_c^2)^{1/2}\right]^2 \tag{4.65}$$

Equation (4.65) gives the approximate modal volume or the number of modes guided by the α-profile graded-index fiber. For a step-index fiber ($\alpha = \infty$), $n = n_1$ in the core and $n = n_2$ in the cladding. Therefore, the modal volume M_s for such a fiber will be given by

$$M_s \approx \frac{1}{2}\left[\frac{2\pi a}{\lambda}(n_1^2 - n_2^2)^{1/2}\right]^2 \tag{4.66}$$

We know that the normalized frequency parameter is given by

$$V = \frac{2\pi a}{\lambda}(n_1^2 - n_2^2)^{1/2}$$

Therefore

$$M_s = \frac{V^2}{2} \tag{4.67}$$

For a graded-index fiber, the numerical aperture (NA) for the guided rays varies with r as follows:

$$\text{NA}(r) = \begin{cases} \{n^2(r) - n_c^2\}^{1/2} & \text{for } r < a \\ 0 \quad \text{for } r \geq a \end{cases}$$

However, for small variation of $n(r)$ with r,

$$\text{NA} \approx \{n_0^2 - n_c^2\}^{1/2}$$

and

$$V \approx \frac{2\pi a}{\lambda}(n_0^2 - n_c^2)^{1/2}$$

With this approximation then

$$M_g \cong \frac{\alpha}{\alpha + 2}\frac{V^2}{2} \tag{4.68}$$

Care must be taken when using Eqs (4.60)–(4.68) because the WKB analysis is valid for highly multimoded fibers with V much greater than 1.

For a parabolic profile, $\alpha = 2$ and

$$M_g \approx \frac{V^2}{4}$$

It can be shown (Okamoto & Okoshi 1976) that the cut-off value of the normalized frequency, V_c, to support a single mode in a graded-index fiber is given by

$$V_c = 2.405\left(1 + \frac{2}{\alpha}\right)^{1/2} \tag{4.69}$$

Thus it is possible to determine the structural and/or operational parameters of the fiber which give single-mode operation. We have obtained a similar result for a SI fiber ($\alpha = \infty$) in our earlier discussion.

Example 4.3 A step-index fiber has a numerical aperture of 0.17 and a core diameter of 100 μm. Determine the normalized frequency parameter of the fiber when light of wavelength 0.85 μm is transmitted through it. Also estimate the number of guided modes propagating in the fiber.

Solution

From Eq. (4.16), we have

$$V = \frac{2\pi a}{\lambda}(n_1^2 - n_2^2)^{1/2} = \frac{2\pi a(\mathrm{NA})}{\lambda} = \frac{\lambda(100\ \mu\mathrm{m})(0.17)}{(0.85\ \mu\mathrm{m})}$$

$$= 62.83$$

Therefore, $$M_s = \frac{V^2}{2} = 1974$$

Example 4.4 A multimode step-index fiber has a relative refractive index difference of 2% and a core refractive index of 1.5. The number of modes propagating at a wavelength of 1.3 μm is 1000. Calculate the diameter of the fiber core.

Solution

$$M_s = \frac{V^2}{2} = \frac{1}{2}\left[\frac{2\pi a}{\lambda} n_1 \sqrt{2\Delta}\right]^2 = \frac{1}{2}\,\frac{4\pi^2 a^2 n_1^2\, 2\Delta}{\lambda^2}$$

Therefore, $$2a = \frac{\lambda}{\pi n_1}\left(\frac{M_s}{\Delta}\right)^{1/2} = \frac{1.3}{\pi \times 1.5}\left(\frac{1000}{0.02}\right)^{1/2} \approx 62\ \mu\mathrm{m}$$

Example 4.5 A graded-index fiber with a parabolic profile supports the propagation of 700 guided modes. The fiber has a relative refractive index difference of 2%, a core refractive index of 1.45 and a core diameter of 75 μm. Calculate the wavelength of light propagating in the fiber. Further, estimate the maximum diameter of the fiber core which can give single-mode operation at the same wavelength.

Solution

Use Eq. (4.68) for the mode volume to evaluate V. With $\alpha = 2$ for the parabolic profile,

$$V = \sqrt{4M_g} = \sqrt{4 \times 700} = 52.91$$

and

$$\lambda = \frac{2\pi a}{V} n_1 \sqrt{2\Delta} = \frac{\pi \times (75\,\mu\text{m})}{52.91} \times (2 \times 0.02)^{1/2}$$

$$= 1.3\ \mu\text{m}$$

The cut-off value of the normalized frequency, V_c, for single-mode operation in a graded-index fiber is given by Eq (4.69). Thus, with $\alpha = 2$,

$$V_c = 2.405\sqrt{2}$$

The maximum core diameter may be obtained as follows:

$$2a = \frac{V_c \lambda}{\pi n_1 \sqrt{(2\Delta)}} = \frac{2.405\sqrt{2} \times (1.3\,\mu\text{m})}{\pi \times 1.45 \times \sqrt{(2 \times 0.02)}} = 4.85\ \mu\text{m}$$

4.5 LIMITATIONS OF MULTIMODE FIBERS

As we have seen in the previous section, multimode fibers support many modes. The higher order modes (corresponding to oblique rays in terms of the ray model) travel slower and hence arrive at the other end of the fiber later than the lower order modes (corresponding to axial rays). This means that different modes travel with different group velocities. Thus, a light pulse propagating through such a fiber will get broadened. This is called multipath dispersion or intermodal dispersion. In Chapter 2, we have calculated the pulse dispersion per unit length for a step-index fiber ($\alpha = \infty$); it is given by

$$\frac{\Delta T}{L} = \frac{n_1}{n_2}\left(\frac{n_1 - n_2}{c}\right) \approx \frac{n_1 \Delta}{c} \tag{4.70}$$

Similarly, it can be shown that the pulse dispersion per unit length for a graded- index fiber with a parabolic profile ($\alpha = 2$) may be given by

$$\frac{\Delta T}{L} = \frac{n_0}{2c}\Delta^2 \tag{4.71}$$

and that for a GI fiber with an optimum profile ($\alpha = 2 - 2\Delta$) may be given by

$$\frac{\Delta T}{L} \approx \frac{n_0}{8c}\Delta^2 \tag{4.72}$$

Thus, pulse broadening due to intermodal dispersion, varying from about 0.05 nm/km to 80 nm/km, depending on the value of α and the core index n_0, has been observed in multimode fibers. This severely restricts the use of such fibers in long-haul communications.

Further, a light pulse has a number of spectral components (of different frequencies), and the group velocity of a mode varies with frequency. Therefore different spectral components in the pulse propagate with slightly different group velocities, resulting in pulse broadening. This phenomenon is called group velocity dispersion (GVD) or intramodal dispersion. We will study in detail this type of dispersion as well as attenuation in single-mode fibers in Chapter 5.

In spite of these limitations, multimode fibers are used in local area networks and short-haul communication links. However, prior to using them, the link power budget and the rise-time budget (discussed in Chapter 12) must be analysed.

SUMMARY

- The wave equation, together with the boundary conditions for cylindrical wave guides, describes the propagation of electromagnetic waves in step-index and graded-index optical fibers. Solving these equations for an ideal step-index fiber, under the weakly guiding approximation, gives a set of solutions:

$$\Psi(r, \phi, z, t) = R(r)\, e^{il\phi}\, e^{i(\omega t - \beta z)}$$

where
$$R(r) = \begin{cases} AJ_l\left(\dfrac{ur}{a}\right), & r < a \\ BK_l\left(\dfrac{wr}{a}\right), & r > a \end{cases}$$

The Bessel functions $J_l(ur/a)$ are oscillatory in nature, and hence there exist m allowed solutions (corresponding to m roots of J_l) for each value of l. Thus, the propagation phase constant β is characterized by two integers l and m.

- The number of modes for smaller V and the propagation parameters for different modes can be found from the universal b versus V curves. However, for $V \gg 1$, the modal volumes for graded-index and step-index fibers may be calculated using the following approximate relations:

$$M_g = \frac{1}{2}\left(\frac{\alpha}{\alpha+2}\right)\left[\frac{2\pi a}{\lambda}(n_0^2 - n_c^2)^{1/2}\right]^2$$

$$\approx \left(\frac{\alpha}{\alpha+2}\right)\frac{V^2}{2}$$

and
$$M_s = \frac{1}{2}\left[\frac{2\pi a}{\lambda}(n_1^2 - n_2^2)^{1/2}\right]^2$$

$$\simeq \frac{V^2}{2}$$

- The cut-off value of the normalized frequency V_c to support a single mode in a graded-index fiber is given by

$$V_c = 2.405\left(1+\frac{2}{\alpha}\right)^{1/2}$$

- Intermodal dispersion in multimode fibers restricts their use in long-haul communications.

MULTIPLE CHOICE QUESTIONS

4.1 A step-index fiber has a core of refractive index 1.5 and a cladding of refractive index 1.49. The core diameter is 100 μm. How many guided modes are supported by the fiber if the wavelength of light is 0.85 μm?
(a) 180 (b) 570 (c) 1160 (d) 2040

4.2 A graded-index fiber has a triangular profile with $n_0 = 1.48$ and $\Delta = 0.02$. If it is excited by a source of $\lambda = 1.0$ μm, what is the range of phase propagation constants for the modes supported by the fiber?
(a) 2.438–5.327 μm^{-1} (b) 4.289–7.142 μm^{-1}
(c) 6.315–8.342 μm^{-1} (d) 8.620–9.299 μm^{-1}

4.3 If the core diameter of the fiber of Question 4.2 is 50 μm, what is the value of the normalized frequency parameter?
(a) 17.61 (b) 24.53 (c) 31.41 (d) 50.72

4.4 For any multimode optical fiber, what is the range of the normalized propagation parameter?
(a) 0–1 (b) 1–10
(c) Cannot be calculated unless λ is known.
(d) Cannot be calculated unless the profile parameter is known.

4.5 For the optical fiber of Question 4.3, what is the total number of modes supported by the fiber?
(a) 74 (b) 164 (c) 203 (d) 500

4.6 In a step-index fiber, what is the cut-off frequency of the LP_{11} mode?
(a) 0.0 (b) 2.405 (c) 3.832 (d) 5.520

4.7 An unclad fiber with a core refractive index of 1.46 and core diameter of 60 μm is placed in air. What is the normalized frequency for the fiber when light of wavelength 0.85 μm is transmitted through it?
(a) 40.74 (b) 103.23 (c) 221.65 (d) 375.02

4.8 A GI fiber with a parabolic profile has an axial refractive index of 1.46 and Δ of 0.5%. What is the pulse broadening per unit length due to intermodal dispersion?
(a) 30.4 nm/km (b) 60.8 ns/km (c) 60.8 ps/km (d) Zero

4.9 In a multimode SI fiber, the higher order modes propagate within the fiber with
(a) lower group velocity than the lower order modes.
(b) higher group velocity than the lower order modes.
(c) same group velocity as that of lower order modes.
(d) random group velocity.

4.10 Pulse broadening in GI fibers is due to
(a) intermodal dispersion. (b) intramodal dispersion.
(c) both (a) and (b). (d) none of these.

Answers

4.1	(d)	4.2	(d)	4.3	(c)	4.4	(a)	4.5	(b)
4.6	(b)	4.7	(c)	4.8	(c)	4.9	(a)	4.10	(c)

REVIEW QUESTIONS

4.1 State the boundary conditions and the assumptions made while arriving at the solutions of the wave equation for an ideal step-index fiber. How well are these conditions and assumptions satisfied in a real optical fiber?

4.2 (a) Distinguish between multimode step-index and graded-index fibers. What is the difference between multimode and single-mode fibers?
(b) A multimode SI fiber has a core diameter of 50 μm, a core index of 1.46, and a relative refractive index difference of 1%. It is operating at a wavelength of 1.3 μm. Calculate (i) the refractive index of the cladding, (ii) the normalized frequency parameter V, and (iii) the total number of modes guided by the fiber.
Ans: (i) 1.445 (ii) 25 (iii) 312

4.3 (a) What is the difference between the propagation phase constant β and the normalized propagation parameter b? How are they related?
(b) The refractive indices of the core and cladding of a SI fiber are 1.48 and 1.465, respectively. Light of wavelength 0.85 μm is guided through it. Calculate the minimum and maximum values of the propagation phase constant β.
Ans: $10.82 \times 10^6\ \text{m}^{-1}$, $10.93 \times 10^6\ \text{m}^{-1}$

4.4 A graded-index single-mode fiber has a core axis refractive index of 1.5, a triangular index profile ($\alpha = 1$) in the core, and a relative refractive index difference of 1.3%. Calculate the core diameter of the fiber if it has to transmit (i) $\lambda = 1.3$ μm and (ii) $\lambda = 1.55$ μm.
Ans: (i) 7.1 μm (ii) 8.5 μm

4.5 Assuming $\varepsilon = 0$ and $dn_0/d\lambda = 0$, show that the ratio of the rms pulse broadening in step-index fibers (σ_{step}) to the rms pulse broadening in GI fibers (σ_{graded}) (for an optimum profile) is given by $\sigma_{step}/\sigma_{graded} \approx 10/\Delta$.

4.6 Gloge (1972) has shown that the effective number of modes guided by a curved multimode GI fiber of radius a is given by

$$(M_g)_{\text{eff}} = (M_g)\left[1 - \frac{(\alpha+2)}{2\alpha\Delta}\left\{\frac{2a}{R} + \left(\frac{3}{2n_c kR}\right)^{2/3}\right\}\right]$$

where α is the profile parameter, Δ is the relative refractive index difference, n_c is the refractive index of the cladding, $k = 2\pi/\lambda$, and M_g is the total number of

guided modes in a straight fiber given by Eq. (4.77). Find the radius of curvature R such that the effective number of guided modes reduces to half its maximum value. Assume that $\alpha = 2$, $n_c = 1.48$, $\Delta = 0.01$, $a = 50$ µm, and $\lambda = 0.85$ µm.
Ans: $R \approx 0.66$ cm

4.7 Equations (4.41) and (4.42) give $\psi_z(r, \phi)$, the axial components of E or H, i.e., E_z or H_z. Obtain the transverse components E_r, E_ϕ, H_r, and H_ϕ using the following relations:

$$E_r = -\frac{i}{k_r^2}\left(\beta\frac{\partial E_z}{\partial r} + \omega\mu\frac{1}{r}\frac{\partial H_z}{\partial \phi}\right)$$

$$E_\phi = -\frac{i}{k_r^2}\left(\frac{\beta}{r}\frac{\partial E_z}{\partial \phi} - \omega\mu\frac{\partial H_z}{\partial r}\right)$$

$$H_r = -\frac{i}{k_r^2}\left(\beta\frac{\partial H_z}{\partial r} - \omega\varepsilon\frac{1}{r}\frac{\partial E_z}{\partial \phi}\right)$$

$$H_\phi = -\frac{i}{k_r^2}\left(\frac{\beta}{r}\frac{\partial H_z}{\partial \phi} + \omega\varepsilon\frac{\partial E_z}{\partial r}\right)$$

where k_r is the radial component of the propagation vector **k** in the waveguide and has an amplitude given by

$$k_r = k_1\sqrt{\omega^2\,\varepsilon\mu - \beta^2}$$

4.8 The local numerical aperture of a graded-index fiber at a radial distance r is given by

$$\text{NA}(r) = \{n^2(r) - (n_c^2)\}^{1/2} \quad \text{for } r < a$$

Show that the rms value of the NA of the fiber for an α-profile (taken over the core area) is given by

$$(\text{NA})_{\text{rms}} = \left[\left(\frac{\alpha}{\alpha+2}\right)(n_0^2 - n_c^2)\right]^{1/2}$$

4.9 A graded-index fiber has a core diameter of 50 µm, $\alpha = 2$, $n_0 = 1.460$, and $n_c = 1.445$. If it is excited by a source of $\lambda = 1.3$ µm, calculate (a) $(\text{NA})_{\text{rms}}$; (b) β_0, Δ, and V; and (c) the total number of propagating modes.
Ans: (a) 0.1476 (b) 7056 mm^{-1}, 0.0102, 25.23 (c) 159

4.10 A graded-index fiber with a triangular profile supports the propagation of 500 modes. The core axis refractive index is 1.46 and the core diameter is 75 µm. If the wavelength of light propagating through the fiber is 1.3 µm, calculate (a) the relative refractive index difference Δ of the fiber and (b) the maximum diameter of the fiber core which would give single-mode operation at the same wavelength.
Ans: (a) 0.021 (b) 5.76 µm

4.11 A graded-index fiber with a parabolic profile has a core diameter of 70 μm, $n_0 = 1.47$, and $n_c = 1.45$. If it is excited by a source of $\lambda = 1.3$ μm, calculate (a) β_0, β_c, and Δ and (b) V and M_g.

Ans: (a) 7.10×10^6 m^{-1}, 7.00×10^6 m^{-1}, 0.01388 (b) 46.88, 418

5 Single-mode Fibers

After reading this chapter you will be able to understand the following:

- Single-mode fibers (SMFs)
- Characteristic parameters of SMFs
- Dispersion in SMFs
- Attenuation in SMFs
- Design of SMFs

5.1 INTRODUCTION

In the past two decades, single-mode fibers have emerged as a viable means for fiber-optic communication. They have become the most widely used fiber type, especially for long-haul communications. The major reason for this is that they exhibit the largest transmission bandwidth, high-quality transfer of signals because of the absence of modal noise, very low attenuation, compatibility with integrated optics technology, and long expected installation lifetime.

In this chapter, we discuss (i) single-mode fibers, (ii) their characteristic parameters, (iii) the factors/mechanisms which contribute to dispersion and/or attenuation in such fibers, and (iv) how the design of such fibers can be optimized for specific applications.

5.2 SINGLE-MODE FIBERS

An optical fiber that supports only the fundamental mode (LP_{01} mode or HE_{11} mode) is called a single-mode fiber. This type of fiber is designed such that all higher order modes are cut off at the operating wavelength. We have seen in Chapter 4 that the number of modes supported by a fiber is governed by the V-parameter, and the cut-off condition of various modes is also determined by V. The cut-off normalized frequency V_c for the LP_{11} mode in a step-index (SI) fiber (from Table 4.1) is 2.405. Thus single-mode operation (of the fundamental LP_{01} mode) in an SI fiber is possible over a normalized frequency range $0 < V < 2.405$ (since there is no cut-off for the

LP_{01} mode, $V_c = 0$ for this mode). We know from Eq. (4.16) that the normalized frequency parameter V is given by

$$V = \frac{2\pi a}{\lambda}(n_1^2 - n_2^2)^{1/2}$$

In terms of the relative refractive index difference Δ, this relation may also be written as

$$V = \frac{2\pi a n_1}{\lambda}(2\Delta)^{1/2}$$

Thus for a particular value of λ, if the value of V is to be made lower than 2.405, then either the diameter ($2a$) of the fiber or the relative refractive index difference Δ of the fiber has to be made smaller. Both these factors create practical problems. With small core diameter, it becomes difficult to launch the light into the fiber and also to join (splice) the fibers in the field. It is also practically difficult to get very low values of Δ.

An alternate solution for getting higher values of V_c and hence large core diameters is that instead of using an SI fiber, a graded-index (GI) fiber may be used. The cut-off value [see Eq. (4.69)] of the normalized frequency V_c for single-mode operation in a GI fiber is given by

$$V_c = 2.405\left(1 + \frac{2}{\alpha}\right)^{1/2}$$

Thus V_c will increase by a factor of $\sqrt{2}$ for a parabolic profile ($\alpha = 2$) and by a factor of $\sqrt{3}$ for a triangular profile ($\alpha = 1$).

Another problem with low V-values and low Δ for the single-mode operation is that the modal field of the LP_{01} mode extends well into the cladding. It has been shown that for V-values less than 1.4, more than half the modal power propagates through the cladding. Thus the evanescent field, which decays exponentially, may extend to a considerable distance in the cladding. To ensure that this power propagating in the cladding is not lost, the thickness of the cladding should be large (cladding radius greater than 50 μm). Further, the absorption and scattering losses in the cladding should also be minimum.

Before we take up the different types of SMFs, let us discuss the parameters that characterize an SMF and those that are helpful in evaluating the dispersion, attenuation, and the jointing and bending losses.

5.3 CHARACTERISTIC PARAMETERS OF SMFs

5.3.1 Mode Field Diameter

In multimode fibers, the numerical aperture and core diameters are considered important from the point of view of predicting the performance of these fibers. But in single-mode fibers, the radial distribution of the optical power in the propagating fundamental mode plays an important role. Therefore, the *mode field diameter* (MFD) of the propagating mode constitutes a fundamental parameter characteristic of a single-mode fiber. This is also known as *mode spot size*. It may be defined as follows. For step-index and graded-index (parabolic profile) single-mode fibers operating near the cut-off wavelength λ_c, the field distribution of the fundamental mode is approximately Gaussian and may be expressed by the following equation

$$\psi(r) = \psi_0 \exp(-r^2/w^2) \tag{5.1}$$

where $\psi(r)$ is the electric or magnetic field at the radius r, ψ_0 is the axial field (at $r = 0$), and w is the mode field radius, which is the radial distance from the axis at which ψ_0 drops to ψ_0/e. Thus the MFD is $2w$. As the power is proportional to ψ^2, the MFD may also be defined as the radial distance between the $1/e^2$ power points (in the power versus radius graph).

If the power per unit area at a radial distance r in the LP_{01} mode is $P(r)$, the power in the annular ring between r and $r + dr$ will be $2\pi rP(r)dr$. So the MFD may be defined as

$$\text{MFD} = 2w = 2\left[\frac{\int_0^\infty r^2\, P(r)\, 2\pi\, dr}{\int_0^\infty P(r)\, 2\pi r\, dr}\right]^{1/2}$$

$$= 2w = 2\left[\frac{\int_0^\infty r^2 \psi^2(r)\, dr}{\int_0^\infty r\psi^2(r)\, dr}\right]^{1/2} \tag{5.2}$$

Here, $\psi(r)$ is the field distribution in the LP_{01} mode.

For a step-index fiber, the mode field radius w can be approximately given by the following expression (Marcuse 1977):

$$w \approx a\left[0.65 + \frac{1.619}{V^{3/2}} + \frac{2.879}{V^6}\right] \tag{5.3}$$

where a is the core radius. This formula gives the value of w to within about 1% for V-values nearly in the range 0.8–2.5.

Figure 5.1 shows the plot of normalized spot size (w/a) as a function of the V-parameter calculated using Eq. (5.3). It is quite clear that for a fiber of given radius, the normalized spot size increases as V becomes smaller (or as λ becomes longer). However, as the wavelength increases, the modal field is less well confined within the core. Therefore, single-mode fibers are so designed that the cut-off wavelength λ_c is not too far from the wavelength for which the fiber is designed.

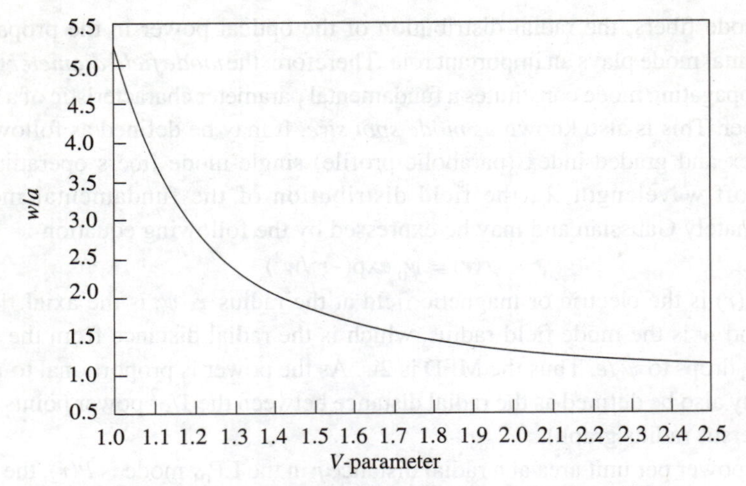

Fig. 5.1 Normalized spot size (w/a) as a function of V

There is another definition of MFD that is more complex and is called the Peterman-2 spot size. This is defined as (Peterman 1983)

$$(\mathrm{MFD})_P = 2w_P = 2\left[\frac{2\int_0^\infty r\psi^2(r)\,dr}{\int_0^\infty r\,(d\psi/dr)^2\,dr}\right]^{1/2} \tag{5.4}$$

where w_P is called the Peterman-2 mode radius. There is an empirical relation for w_P which is accurate to within about 1% for a step-index fiber with V-values between 1.5 and 2.5, and is given by (Hussey & Martinez 1985)

$$w_P \approx w - a\left[0.016 + \frac{1.567}{V^7}\right] \tag{5.5}$$

where w is mode field radius given by Eq. (5.3). This definition has the advantage that the fiber dispersion and loss caused by the offset at a joint can be related to w_P. It can also be related to the far-field pattern of the single-mode fiber.

Example 5.1 A typical step-index fiber has a core refractive index of 1.46, a relative refractive index difference of 0.003, and a core radius of 4 μm. Calculate w and w_P at $\lambda = 1.30$ μm and 1.55 μm.

Solution

The normalized frequency parameter of the fiber is

$$V = \frac{2\pi a}{\lambda} n_1\sqrt{2\Delta} = \frac{2\pi\times 4\times 10^{-6}\times 1.46\sqrt{2\times 0.003}}{\lambda\,(\mathrm{m})} = \frac{2.842\times 10^{-6}}{\lambda\,(\mathrm{m})}$$

Therefore, for $\lambda = 1.30$ μm, $V = 2.186$, and for $\lambda = 1.55$ μm, $V = 1.834$. Using approximate expressions (5.3) and (5.5), we get at $\lambda = 1.30$ μm,

$$w \approx 4 \times 10^{-6} \left[0.65 + \frac{1.619}{(2.186)^{3/2}} + \frac{2.879}{(2.186)^6} \right]$$
$$= 4 \times 10^{-6} [0.65 + 0.5009 + 0.02556]$$
$$= 4.7058 \times 10^{-6} \text{ m}$$
$$= 4.7058 \text{ μm}$$

and
$$w_P = 4.7508 \text{ μm} - (4 \text{ μm}) \left[0.016 + \frac{1.567}{(2.186)^7} \right]$$
$$= 4.6155 \text{ μm}$$

Similarly, at $\lambda = 1.55$ μm,

$$w = (4 \text{ μm}) \left[0.65 + \frac{1.619}{(1.834)^{3/2}} + \frac{2.879}{(1.834)^6} \right] = 5.5098 \text{ μm}$$

and
$$w_P = 5.5098 \text{ μm} - (4 \text{ μm}) \left[0.016 + \frac{1.567}{(1.834)^7} \right]$$
$$= 5.3559 \text{ μm}$$

5.3.2 Fiber Birefringence

We have discussed in Sec. 4.2 that a single-mode fiber actually supports two orthogonal linearly polarized modes that are degenerate. In a single-mode fiber of a perfectly cylindrical core of uniform diameter, these modes have the same mode index n and hence the same propagation constant β. Therefore they travel with the same velocity. However, real fibers are not perfectly cylindrical. They exhibit variation in the diameter due to the presence of non-uniform stress, bends, twists, etc. along the length of the fiber. Hence the propagation constants of the two polarization components become different and the fiber becomes birefringent. Modal birefringence is defined by

$$\delta n = |n_x - n_y| \tag{5.6}$$

assuming arbitrarily that the two modes are polarized in the x- and y-direction and that n_x and n_y are the mode indices for the two polarization components. The corresponding change in the propagation constants may be given by

$$\delta\beta = \frac{2\pi}{\lambda} \delta n \tag{5.7}$$

where λ is the wavelength of light in vacuum. This leads to a periodic exchange of modal power between the two components, which can be understood in a simple manner as follows.

Suppose linearly polarized light is launched into a single-mode fiber such that both the modes are excited. We assume that the two polarization components have the same amplitude and there is no phase difference at the launching end. But as the light propagates down the fiber length, one mode goes out of phase with respect to the other due to different phase propagation constants. Thus at any point along the fiber length (for a random phase difference) the two components will produce elliptically polarized light. At a phase difference of $\pi/2$, circularly polarized light will be produced. In this way the polarization progresses from linear to elliptical to circular to elliptical and back to linear. This sequence of alternating states of polarization continues along the fiber length.

The length L_p of the fiber over which the polarization rotates through an angle of 2π radians (i.e., the length over which the phase difference between the two polarized modes becomes 2π) is called the *beat length* of the fiber. It is given by

$$L_p = \frac{2\pi}{\delta\beta} \tag{5.8}$$

In conventional single-mode fibers, birefringence varies along the length of the fiber randomly in both magnitude as well as direction. Therefore, linearly polarized light launched in such a fiber quickly attains an arbitrary state of polarization. The dispersion arising because of fiber birefringence is discussed in Sec. 5.4.4.

Example 5.2 A beat length of 10 cm is observed in a typical single-mode fiber when light of wavelength 1.0 μm is launched into it. Calculate (a) the difference between the propagation constants for the two orthogonally polarized modes, and (b) the modal birefringence.

Solution

(a) Using Eq. (5.8), we get

$$\delta\beta = \frac{2\pi}{L_p} = \frac{2\pi}{10\times 10^{-2}\ \text{m}} = 62.83\ \text{m}^{-1}$$

(b) Modal birefringence [from Eq. (5.7)]

$$\delta n = \delta\beta\frac{\lambda}{2\pi} = 62.83\ \text{m}^{-1} \times \frac{1\times 10^{-6}\ \text{m}}{2\pi} = 9.999\times 10^{-6}$$

5.4 DISPERSION IN SINGLE-MODE FIBERS

5.4.1 Group Velocity Dispersion

In a multimode fiber, different modes at a single spectral frequency travel with different group velocities, giving rise to pulse dispersion. In Sec. 2.3, using the ray model we have derived the delay difference between the fastest mode (corresponding to the axial ray) and the slowest mode (corresponding to the most oblique ray). Equation (2.7) gives the multipath time dispersion per unit length of the optical fiber. In terms of mode theory, this is called *intermodal dispersion*. As expected, multimode step-index fibers exhibit large intermodal dispersion. By choosing an appropriate graded-index profile, this type of dispersion can be reduced considerably.

In single-mode fibers, intermodal dispersion is absent because the power of the launched pulse is carried by a single mode. However, pulse broadening does not vanish altogether, due to chromatic dispersion. A light pulse, although propagating as a fundamental mode, has a number of spectral components (of different frequencies), and the group velocity of the fundamental mode varies with frequency. Therefore different spectral components in the pulse propagate with slightly different group velocities, resulting in pulse broadening. This phenomenon is called *group velocity dispersion* (GVD) or *intramodal dispersion*. This has two components, namely, material dispersion and waveguide dispersion. Let us see how the GVD limits the performance of single-mode fibers.

Suppose we have a single-mode fiber of length L and a pulse is launched at one end of this fiber. Then a spectral component of a given frequency ω would travel with a group velocity V_g given by

$$V_g = \frac{1}{d\beta / d\omega}$$

After travelling length L, this component would arrive at the other end after time τ_g, known as the *group delay time*. The latter may be given by

$$\tau_g = \frac{L}{V_g} = L\frac{d\beta}{d\omega} \tag{5.9}$$

If the spectral width of the pulse is not too wide, the delay difference per unit frequency along the propagation length may be expressed as $d\tau_g/d\omega$. For the spectral components that are $\Delta\omega$ apart, the total delay difference ΔT over length L may be given by

$$\Delta T = \frac{d\tau_g}{d\omega}\Delta\omega$$

Substituting for τ_g from Eq. (5.9), we get

$$\Delta T = L\left(\frac{d^2\beta}{d\omega^2}\right)\Delta\omega = L\beta_2\Delta\omega \tag{5.10}$$

The factor $\beta_2 \equiv d^2\beta/d\omega^2$ is known as the GVD parameter. It determines the magnitude of pulse broadening as the pulse propagates inside the fiber.

If the spectral spread is measured in terms of the wavelength range $\Delta\lambda$ around the central wavelength λ, then replacing ω by $2\pi c/\lambda$ and $\Delta\omega$ by $(-2\pi c/\lambda^2)\Delta\lambda$, we can write Eq. (5.10) as follows:

$$\Delta T = \frac{d\tau_g}{d\lambda}\delta\lambda = \frac{d}{d\lambda}\left(\frac{L}{V_g}\right)\Delta\lambda = L\frac{d}{d\lambda}\left(\frac{1}{V_g}\right)\Delta\lambda \tag{5.11}$$

The factor

$$D = \frac{1}{L}\frac{d\tau_g}{d\lambda} = \frac{d}{d\lambda}\left(\frac{1}{V_g}\right) = \frac{-2\pi c}{\lambda^2}\beta_2 \tag{5.12}$$

is called the *dispersion parameter*. It is defined as the pulse spread per unit length per unit spectral width of the source and is measured in units of ps nm^{-1} km^{-1}. It is the combined effect of waveguide dispersion D_w and material dispersion D_m present together. However, we can write to a very good approximation

$$D = D_w + D_m \tag{5.13}$$

In the following sections we will discuss each of these separately and obtain separate expressions for D_w and D_m considering that in each case the other one is absent.

5.4.2 Waveguide Dispersion

In Sec. 4.2, we have defined the normalized propagation constant b of a mode as follows [see Eq. (4.47)]:

$$b = \left(1 - \frac{u^2}{V^2}\right) = \frac{\beta^2 - \beta_2^2}{\beta_1^2 - \beta_2^2} = \frac{\beta^2 - k^2 n_2^2}{k^2 n_1^2 - k^2 n_2^2} = \frac{(\beta/k)^2 - n_2^2}{n_1^2 - n_2^2}$$

In this expression, β/k is called the *mode index* n. Rewriting the above expression, we get

$$b = \frac{n^2 - n_2^2}{n_1^2 - n_2^2} = \frac{(n - n_2)(n + n_2)}{(n_1 - n_2)(n_1 + n_2)}$$

Taking $n_1 \approx n_2$, which is true for most fibers, we can write

$$b \approx \frac{n - n_2}{n_1 - n_2}$$

or the mode index $n \approx n_2 + b(n_1 - n_2)$. (5.14)

But β and n are related by the expression

$$\beta = \frac{\omega}{c} n \tag{5.15}$$

Therefore, from Eqs (5.14) and (5.15), we get

$$\beta = \frac{\omega}{c}[n_2 + (n_1 - n_2)b] \tag{5.16}$$

Let us assume that there is no material dispersion, and hence n_1 and n_2 do not vary with λ. But b is a function of V and hence of ω. Therefore, using Eq. (5.16) we can write

$$\frac{1}{V_g} = \frac{d\beta}{d\omega} = \frac{1}{c}[n_2 + (n_1 - n_2)b] + \frac{\omega}{c}(n_1 - n_2)\frac{db}{d\omega}$$

$$= \frac{1}{c}[n_2 + (n_1 - n_2)b] + \frac{\omega}{c}(n_1 - n_2)\frac{db}{dV}\frac{dV}{d\omega} \tag{5.17}$$

We know that for a step-index fiber of core radius a,

$$V = \frac{2\pi a}{\lambda}\sqrt{(n_1^2 - n_2^2)} = \frac{\omega}{c}a\sqrt{(n_1^2 - n_2^2)} \tag{5.18}$$

Therefore, using Eq. (5.18), we get

$$\frac{dV}{d\omega} = \frac{a}{c}\sqrt{(n_1^2 - n_2^2)} = \frac{V}{\omega} \tag{5.19}$$

The same relation can also be expressed as $dV/d\lambda = V/\lambda$. Substituting the value of $dV/d\omega$ from Eq. (5.19) into Eq. (5.17), we get

$$\frac{1}{V_g} = \frac{1}{c}[n_2 + (n_1 - n_2)b] + \frac{1}{c}(n_1 - n_2)V\frac{db}{dV}$$

$$= \frac{n_2}{c} + \frac{(n_1 - n_2)}{c}\left[\frac{d}{dV}(bV)\right]$$

$$= \frac{n_2}{c}\left[1 + \frac{(n_1 - n_2)}{n_2}\left\{\frac{d}{dV}(bV)\right\}\right]$$

Since the relative refractive index difference Δ can be approximately given by $(n_1 - n_2)/n_2$, we have

$$\frac{1}{V_g} = \frac{n_2}{c}\left[1 + \Delta\frac{d}{dV}(bV)\right] \tag{5.20}$$

From this expression, it is clear that the group velocity V_g of the mode varies with V and hence with ω, even in the absence of material dispersion. This is called *waveguide dispersion*. Thus the group delay, that is, the time taken by a mode to travel the length L of the fiber due to waveguide dispersion will be given by

$$\tau_w = \frac{L}{V_g} = \frac{Ln_2}{c}\left[1 + \Delta\frac{d}{dV}(bV)\right] \tag{5.21}$$

For a source of spectral width $\Delta\lambda$, the corresponding pulse spread due to waveguide dispersion may be obtained from the derivative of the group delay with respect to wavelength, that is,

$$\Delta\tau_w = \frac{d\tau_w}{d\lambda}\Delta\lambda \approx \left[\frac{L}{c} n_2 \Delta \frac{d^2}{dV^2}(bV)\frac{dV}{d\lambda}\right]\Delta\lambda$$

Since $$V = \frac{2\pi a n_1 \sqrt{2\Delta}}{\lambda}$$

$$\frac{dV}{d\lambda} = \frac{-2\pi a n_1 \sqrt{2\Delta}}{\lambda^2} = -\frac{V}{\lambda}$$

Therefore, $$\Delta\tau_w \approx -\frac{L}{c} n_2 \Delta \frac{1}{\lambda}\left[V\frac{d^2}{dV^2}(bV)\right]\Delta\lambda$$

or $$\frac{\Delta\tau_w}{L} = -\frac{n_2\Delta}{c\lambda}\left[V\frac{d^2}{dV^2}(bV)\right]\Delta\lambda$$

$$= D_w\lambda \tag{5.22}$$

where $$D_w = -\frac{n_2\Delta}{c\lambda}\left[V\frac{d^2}{dV^2}(bV)\right] \tag{5.23}$$

is called the waveguide dispersion parameter or simply the waveguide dispersion.

Calculation of actual values of the term within square brackets is not a simple task, and hence the following easy-to-use empirical formula (Marcuse 1979) may be employed:

$$V\frac{d^2}{dV^2}(bV) \approx 0.080 + 0.549\,(2.834 - V)^2 \tag{5.24}$$

From Eq. (5.3), we see that D_w depends on n_2, Δ, λ, the V-parameter, and $V(d^2/dV^2)(bV)$. The value of the last term, i.e., $V(d^2/dV^2)(bV)$ is positive for V-values in the range 0.5–3, which means that in the normal case, for single-mode fibers, D_w will remain negative for wavelengths in the range of interest, e.g., 1.0–1.7 μm.

Example 5.3 A step-index single-mode fiber has a core index of 1.45, a relative refractive index difference of 0.3%, and a core diameter of 8.2 μm. Calculate the waveguide dispersion parameter for this fiber at $\lambda = 1.30$ μm and 1.55 μm.

Solution

It is given that $n_1 = 1.45$ and $\Delta = 0.003$. Hence

$$n_2 \approx n_1 (1 - \Delta) = 1.45\,(1 - 0.003) = 1.4456$$

Since $$V = \frac{2\pi a}{\lambda} n_1 \sqrt{2\Delta}$$

at $\lambda_1 = 1.30$ μm,

$$V_1 = \frac{\pi \times 8.2}{1.3} \times 1.45 \sqrt{2 \times 0.003}$$
$$= 2.2256$$

and at $\lambda_2 = 1.55$ μm,

$$V_2 = \frac{\pi \times 8.2 \times 1.45 \sqrt{2 \times 0.003}}{1.55}$$
$$= 1.8667$$

Using Eq. (5.24), we get at $V = V_1$,

$$V \frac{d^2}{dV^2}(bV) \approx 0.080 + 0.549\,(2.834 - 2.2256)^2$$
$$= 0.2832$$

and at $V = V_2$,

$$V \frac{d^2}{dV^2}(bV) \approx 0.080 + 0.549\,(2.834 - 21.8667)^2$$
$$= 0.5936$$

Thus, using Eq. (5.23), the waveguide dispersion parameter D_w at $\lambda_1 = 1.30$ μm may be given by

$$(D_w)_{\lambda = 1.3\,\mu\text{m}} = -\frac{1.4456 \times 0.003}{(3 \times 10^8 \text{ m s}^{-1})\,(1.30\,\mu\text{m})}\,[0.2832]$$
$$= -3.149 \times 10^{-12} \text{ s}\,\mu\text{m}^{-1}\,\text{m}^{-1}$$
$$= -3.149 \text{ ps nm}^{-1}\,\text{km}^{-1}$$

and D_w at $\lambda_2 = 1.55$ μm may be given by

$$(D_w)_{\lambda = 1.55\,\mu\text{m}} = -\frac{1.4456 \times 0.003}{(3 \times 10^8 \text{ m s}^{-1})(1.55\,\mu\text{m})}\,[0.5936]$$
$$= -5.536 \times 10^{-12} \text{ s}\,\mu\text{m}^{-1}\,\text{m}^{-1}$$
$$= -5.536 \text{ ps nm}^{-1}\,\text{km}^{-1}$$

5.4.3 Material Dispersion

Material dispersion is the result of variation of the refractive index of the material of the fiber (i.e., silica) with the wavelength of light propagating through the fiber. Since the group velocity V_g of a mode is a function of the refractive index, the various spectral components of a given mode will propagate with different speeds. Therefore, if the spectral width $\Delta\lambda$ of the source, e.g., LED, is larger, the pulse broadening due to this effect may be significant. Material dispersion has already been discussed in Sec. 2.5. To know the resultant effect in the case of a single-mode fiber, let us rewrite Eq. (2.23) for the material dispersion parameter D_m as follows:

$$D_m = -\frac{\lambda}{c}\frac{d^2 n}{d\lambda^2} \tag{5.25}$$

(We have removed the modulus sign here and retained the minus sign.) D_m is zero at $\lambda = 1.276$ μm for pure silica, and hence this wavelength is referred to as the zero-dispersion wavelength λ_{ZD}.

D_m can also be expressed by an approximate empirical relation (Agarwal 2002) (in a limited range of wavelengths 1.25–1.66 μm):

$$D_m \approx 122\left(1 - \frac{\lambda_{ZD}}{\lambda}\right) \tag{5.26}$$

Figure 5.2 shows the variation of D_m, D_w, and their sum $D = D_w + D_m$ for a typical conventional single-mode fiber. The effect of waveguide dispersion is to shift λ_{ZD} by about 30–40 μm so that the total dispersion D is zero near 1.32 μm. However, material dispersion is dominant above λ_{ZD}, say, around 1.55 μm, giving rise to high values of D. The window around 1.55 μm is of considerable interest for fiber-optic communication systems, as silica fibers exhibit least attenuation around 1.55 μm.

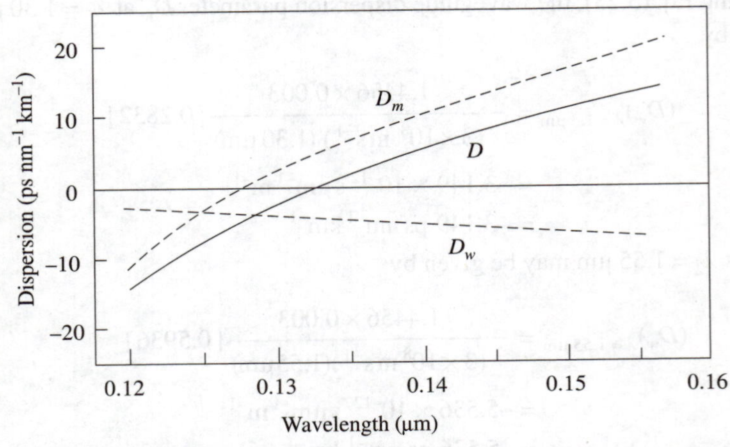

Fig. 5.2 Variation of D_w, D_m, and their sum $D = D_w + D_m$ for a typical single-mode fiber (Keck 1985)

We have seen in Eq. (5.23) that D_w depends on fiber parameters such as Δ, n_2, V, the core diameter, etc. Therefore, it is possible to design the fiber such that λ_{ZD} is shifted to 1.55 μm. Such fibers are called dispersion-shifted fibers (DSFs). It is also possible to modify D_w such that the total dispersion D is small over a wide range of wavelengths (typically from 1.3 to 1.6 μm). Such fibers are called dispersion-flattened fibers (DFFs). Figure 5.3 shows the variation of D for a typical standard (conventional) single-mode fiber, DSF, and DFF. We will take up the design aspects of different types of SMFs in Sec. 5.6.

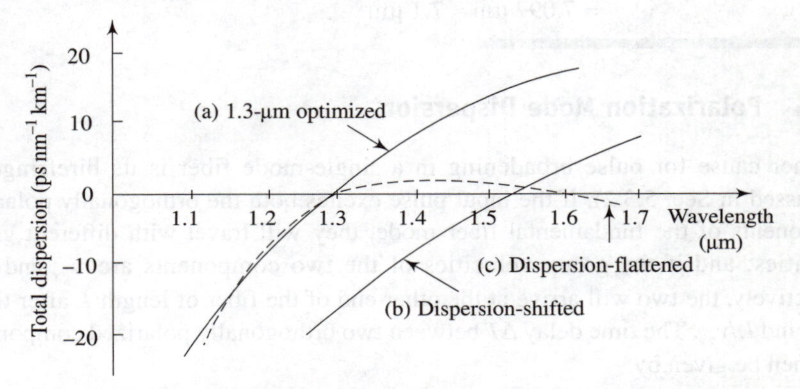

Fig. 5.3 Variation of total dispersion D for standard (conventional) SMF, DSF, and DFF

Example 5.4 A step-index single-mode fiber exhibits material dispersion of 6 ps nm^{-1} km^{-1} at an operating wavelength of 1.55 μm. Assume that n_1 = 1.45 and $\Delta = 0.5\%$. Calculate the diameter of the core needed to make the total dispersion of the fiber zero at this wavelength.

Solution

The total dispersion $D = D_m + D_w = 0$. Since $D_m = 6$ ps nm^{-1} km^{-1}, $D_w = -6$ ps nm^{-1} km^{-1}. Combining Eqs (5.23) and (5.24), we get

$$D_w = -\frac{n_2 \Delta}{c\lambda}[0.080 + 0.549\,(2.834 - V)^2]$$

$$= -\frac{1.4427 \times 0.005}{3 \times 10^8\ \mathrm{m\,s^{-1}}}\frac{1}{1.55\,\mu\mathrm{m}}[0.080 + 0.549(2.834 - V^2)]$$

$$= -15.51\ (\mathrm{ps\,nm^{-1}\,km^{-1}})\ [0.080 + 0.549\ (2.834 - V)^2]$$

As the required value of D_w is -6 ps nm^{-1} km^{-1}, we have

$$-6 = -15.51\ [0.080 + 0.549\ (2.834 - V)^2]$$

This gives $V = 2.0863$

Since $$V = \frac{2\pi a n_1 \sqrt{2\Delta}}{\lambda}$$

the core diameter needed to make total dispersion zero will be given by

$$2a = \frac{V\lambda}{\pi n_1 \sqrt{2\Delta}} = \frac{2.0863 \times (1.55\,\mu\text{m})}{\pi \times 1.45 \sqrt{2 \times 0.005}}$$

$$= 7.097\ \mu\text{m} \approx 7.1\ \mu\text{m}$$

5.4.4 Polarization Mode Dispersion

Another cause for pulse broadening in a single-mode fiber is its birefringence (discussed in Sec. 5.3.2). If the input pulse excites both the orthogonally polarized components of the fundamental fiber mode, they will travel with different group velocities; and if the group velocities of the two components are v_{gx} and v_{gy}, respectively, the two will arrive at the other end of the fiber of length L after times L/v_{gx} and L/v_{gy}. The time delay ΔT between two orthogonally polarized components will then be given by

$$\Delta T = \left| \frac{L}{v_{gx}} - \frac{L}{v_{gy}} \right| \tag{5.27}$$

This difference in the propagation times gives rise to pulse broadening. This is called *polarization mode dispersion* (PMD). This can be a limiting factor, particularly in long-haul fiber-optic communication systems operating at high bit rates. In writing Eq. (5.27), we have assumed that the fiber has constant birefringence. This is true only in polarization-maintaining fibers. In conventional single-mode fibers, birefringence varies randomly, and hence Eq. (5.27) is not valid in such fibers. However, an approximate estimate of PMD-induced pulse broadening can be made using the following relation:

$$\Delta T \approx D_{\text{PMD}} \sqrt{L} \tag{5.28}$$

where D_{PMD}, which is measured in units of $\text{ps}/\sqrt{\text{km}}$, is the average PMD parameter and L is the length of the fiber. Values of D_{PMD} normally vary from 0.01 to 10 $\text{ps}/\sqrt{\text{km}}$.

5.5 ATTENUATION IN SINGLE-MODE FIBERS

Signal attenuation (or fiber loss) is another major factor limiting the performance of a fiber-optic communication system, as it reduces the optical power reaching the receiver. Every receiver needs a certain minimum incident power for accurate signal recovery. Hence the fiber loss plays a major role in determining the maximum repeaterless transmission distance between the transmitter and the receiver. In this section, we discuss, in general, various mechanisms that give rise to signal attenuation in all types of fibers, and single-mode fibers in particular.

In general, the attenuation of optical power P with distance z inside an optical fiber is governed by Beer's law:

$$\frac{dP}{dz} = -\alpha P \tag{5.29}$$

where α is the coefficient of attenuation. Thus, for a particular wavelength, if P_{in} is the (transmitted) optical power at the input end of the optical fiber of length L and P_{out} the (received) optical power at the other end of the fiber, then

$$P_{out} = P_{in} \exp(-\alpha L) \tag{5.30}$$

It should be noted here that Beer's law refers only to the losses due to absorption. However, there are other mechanisms such as scattering and bending that also contribute to the losses. Further, α varies with the wavelength and the material of the fiber. Therefore, relation (5.30) should be treated as an approximation.

The attenuation coefficient α is usually expressed in terms of decibels per unit length (dB km^{-1}) following the relation

$$\alpha\,(\text{dB km}^{-1}) = \frac{10}{L}\log_{10}\left(\frac{P_{in}}{P_{out}}\right) \tag{5.31}$$

where the length L is measured in kilometres.

5.5.1 Loss Due to Material Absorption

Optical fibers are normally made of silica-based glass. As light passes through the fiber, it may be absorbed by one or more major components of glass. This is called *intrinsic absorption*. Impurities within the fiber material may also absorb light. This is called *extrinsic absorption*.

Intrinsic absorption results from the electronic and vibrational resonances associated with specific molecules of glass. For pure silica, electronic resonances in the form of absorption bands have been observed in the ultraviolet range, whereas vibrational resonances have been observed in the infrared range. Empirical relationships have been found for different glass compositions.

Extrinsic absorption results from the presence of impurities in the glass. Transition-metal ions such as Cr^{3+}, Cu^{2+}, Fe^{2+}, Fe^{3+}, Ni^{2+}, Mn^{2+}, etc. absorb strongly in the wavelength range 0.6–1.6 μm. Thus the impurity content of these metal ions should be reduced to below 1 ppb (part per billion) in order to obtain loss below 1 dB/km. Another source of extrinsic absorption is the presence of OH^- ions that are incorporated in the fiber material during the manufacturing process. Though the vibrational resonance of OH^- ions peaks at 2.73 μm, its overtones produce strong absorption at 1.38, 0.95, and 0.72 μm. Typically, the OH^- ion concentration should be less than 10 ppb to obtain a loss below 10 dB/km at 1.38 μm. The attenuation versus wavelength curve for a typical fiber is shown in Fig. 5.4.

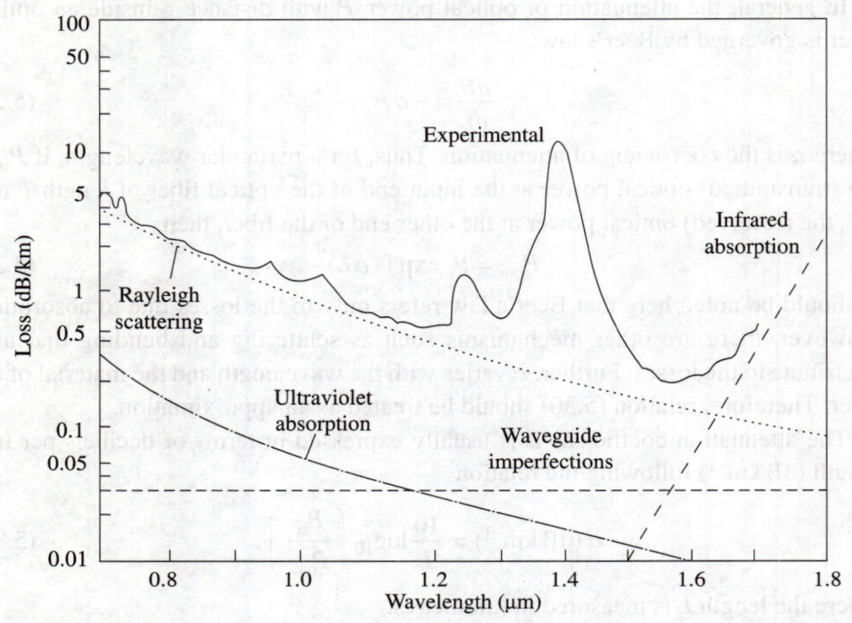

Fig. 5.4 Attenuation versus wavelength curve for a low-loss optical fiber (Miya et al. 1979)

5.5.2 Loss Due to Scattering

There are two scattering mechanisms which cause the linear transfer of some or all of the optical power contained within one guided mode (proportional to the mode power) into a different mode. This transfer process results in the attenuation of power. The mechanisms involved are *Rayleigh scattering* and *Mie scattering*.

Rayleigh scattering is a loss mechanism arising from the microscopic variations in the density of the fiber material. Such a variation in density is a result of the composition of glass. In fact, glass is a randomly connected network of molecules. Therefore, such a configuration naturally has some regions in which the density is either lower or higher than its average value. For multicomponent glass, compositional variation is another factor causing density fluctuation. These variations in the density lead to the random fluctuation of the refractive index on a scale comparable to the wavelength of light. Consequently, the propagating light is scattered in almost all the directions. This is known as Rayleigh scattering. The attenuation caused by such scattering is proportional to $1/\lambda^4$. The contribution to attenuation due to Rayleigh scattering is shown in Fig. 5.4. Scattering may also be caused by waveguide imperfections. For example, irregularities at the core–cladding interface, refractive index difference along the fiber, fluctuation in the core diameter, etc. may lead to additional scattering losses. These are known as losses due to Mie scattering.

5.5.3 Bending Losses

An optical fiber tends to radiate propagating power whenever it is bent. Two types of bends may be encountered: (i) macrobends with radii much larger than the fiber diameter and (ii) random microbends of the fiber axis that may arise because of faulty cabling.

The loss due to macrobending may be explained as follows. We know that every guided core mode has a modal electric-field distribution that has a tail extending into the cladding. This evanescent field decays exponentially with distance from the core. Thus, a part of the the energy of the propagating mode travels in the cladding. At the macrobend, as shown in Fig. 5.5, the evanescent field tail on the far side of the centre of curvature must move faster to keep up with the field in the core. At a critical distance r_c from the fiber axis, the field tail would have to move faster than the speed of light in the cladding (i.e., c/n_2, where c is the speed of light in vacuum and n_2 is the refractive index of the cladding) in order to keep up with the core field. As this is impossible, the optical energy in the tail beyond r_c is lost through radiation.

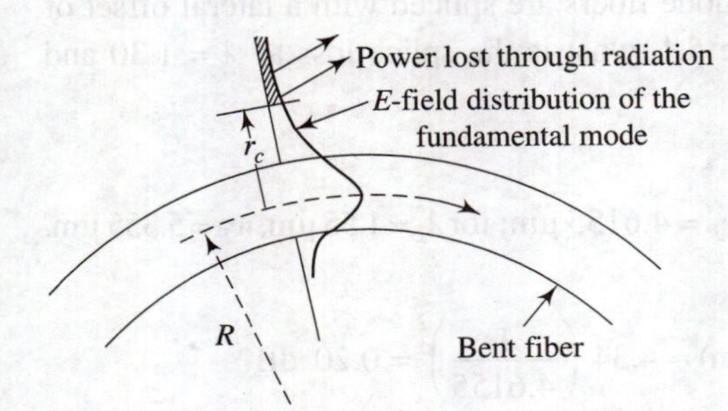

Fig. 5.5 Radiation loss at the macrobending (R is the radius of macrobend)

Further, radiation loss in optical fibers is also caused by mode coupling resulting from microbends along the length of the fiber. Such microbends are caused by manufacturing defects which are in the form of non-uniformities in the core radius or in the lateral pressure created by the cabling of the fiber. The effect of mode coupling in multimode fibers on pulse broadening can be significant for long fibers. Microbend losses in single-mode fibers can also be excessive if proper care is not taken to minimize them. One way to reduce such losses in single-mode fibers is to choose the V-value near the cut-off value of V_c (e.g., 2.405 for the step-index profile), so that the mode energy is confined mainly to the core.

5.5.4 Joint Losses

In any fiber-optic communication system, there is a need for connecting optical fibers. Such fiber-to-fiber connections may be achieved in two ways. These are (i) permanent joints, referred to as *splices* and (ii) demountable joints, known as *connectors*. The major consideration in making these connections is the optical loss associated with them. The various loss mechanisms associated with fiber-to-fiber connections in multimode fibers will be discussed in Sec. 6.5. Here we will discuss very briefly typical losses associated with single-mode fibers.

When two single-mode fibers are spliced, very often there is a lateral offset of the fiber core axes at the joint. It has been shown that the splice loss due to the lateral offset (Δy) between two compatible (i.e., identical) single-mode fibers may be given by (Ghatak & Thyagarajan 1999)

$$\text{Loss}_{\text{lat}}\ (\text{dB}) \approx 4.34\left(\frac{\Delta y}{w_P}\right)^2 \tag{5.32}$$

where w_P is the Peterman-2 spot size defined by Eq. (5.4). It is important to note here that a larger value of w_P will give less splice loss. However, there is an optimum value of the spot size for a given operating wavelength. The latter is about 5 μm for $\lambda = 1.3$ μm.

Example 5.5 Two identical single-mode fibers are spliced with a lateral offset of 1 μm. If we take the data of Example 5.1, what is the splice loss at $\lambda = 1.30$ and 1.55 μm?

Solution

From Example 5.1, for $\lambda_1 = 1.30$ μm, $w_P = 4.6155$ μm; for $\lambda_2 = 1.55$ μm, $w_P = 5.355$ μm. Therefore,

$$\text{Splice loss at } \lambda_1\ (= 1.30\ \mu\text{m}) \approx 4.34\left(\frac{1}{4.6155}\right)^2 = 0.20\ \text{dB}$$

$$\text{Splice loss at } \lambda_2\ (= 1.55\ \mu\text{m}) \approx 4.34\left(\frac{1}{5.3559}\right)^2 = 0.15\ \text{dB}$$

5.6 DESIGN OF SINGLE-MODE FIBERS

In the design of single-mode fibers, two factors are to be considered. These are (i) dispersion and (ii) attenuation. We know from Figs 5.3 and 5.4 that for silica-based fibers the dispersion minimum is around 1.3 μm, whereas the attenuation minimum is around 1.55 μm. For an ideal design, the dispersion minimum should be at the attenuation minimum. In order to optimize the performance of single-mode fibers, therefore, a number of designs with different refractive index profiles for the core and cladding have been investigated; Fig. 5.6 shows a few of these. The designs may be grouped into four categories: (i) standard or conventional 1.3-μm optimized fibers, (ii) dispersion-shifted fibers (DSFs), (iii) dispersion-flattened fibers (DFFs), and (iv) large effective area fibers (LEAFs).

Commonly used single-mode fibers have more or less step-index profiles. The design is optimized for operation at the minimum dispersion 1.3-μm region. They have either matched cladding (MC) or depressed cladding (DC) as shown in Figs 5.6(a) and 5.6(b). MC fibers have a uniform refractive index in the cladding, which is slightly less than that of the core. The typical MFD is around 10 μm with a Δ of 0.3%. In DC fibers, an extra inner cladding region adjacent to the core is formed, which has lower refractive index than that of the cladding region. A typical MFD of 9 μm with $\Delta_1 = +0.25\%$ and $\Delta_2 = -0.12\%$ has been reported for this type of fiber. In this case, the fundamental mode is more tightly confined by a large index step $(\Delta_1 + \Delta_2)$ and smaller core diameter. If the ratio a_2/a_1 is large enough, the design reduces susceptibility to bending losses.

It is possible to modify the dispersion behaviour of a single-mode fiber by tailoring specific parameters, e.g., the composition of the core material, the core radius, profile parameter, Δ, etc. To do this, the total dispersion of a single-mode fiber has to be recognized as the resultant of material dispersion and waveguide dispersion. Thus, if the design parameters can be so adjusted that the total dispersion is shifted to a longer wavelength, e.g., 1.55 μm, it can provide a both low-dispersion and low-loss fiber. Such fibers are called dispersion-shifted fibers. The triangular profile of Fig. 5.6(c) exhibits a loss of about 0.22 dB km^{-1} at $\lambda = 1.55$ μm. But with this design the cut-off wavelength λ_c is reduced to about 0.85–0.90 μm. Thus the fiber must be operated far from cut-off, which increases its sensitivity to microbend losses. The triangular core profile with a depressed cladding shown in Fig. 5.6(d) in an improved version. A typical MFD of 7 μm at 1.55 μm has been observed.

An alternative approach is to reduce the total dispersion over a broader range from about 1.30 μm to 1.6 μm so that the dispersion curve exhibits two points of zero dispersion. Such fibers are called dispersion-flattened fibers. These are good for the wavelength-division multiplexing (WDM) of optical signals. Figures 5.6(e) and 5.6(f) illustrate two profiles that provide dispersion-flattening. The *W*-profile, shown in Fig. 5.6(e), has the same design as the DC profile but with the diameter of the inner depressed cladding reduced. This design is highly sensitive to bend losses. The light lost through bending in the *W*-profile can be retrapped by further introducing regions of raised index into the structure as shown in Fig. 5.6(f).

In general, the effective area of the core of single-mode fibers is set at 55 μm^2 and the cladding diameter is kept at 125 μm. As we will see in Chapter 12, with such a low effective area, fiber non-linearities impose limits on fiber-optic systems intended

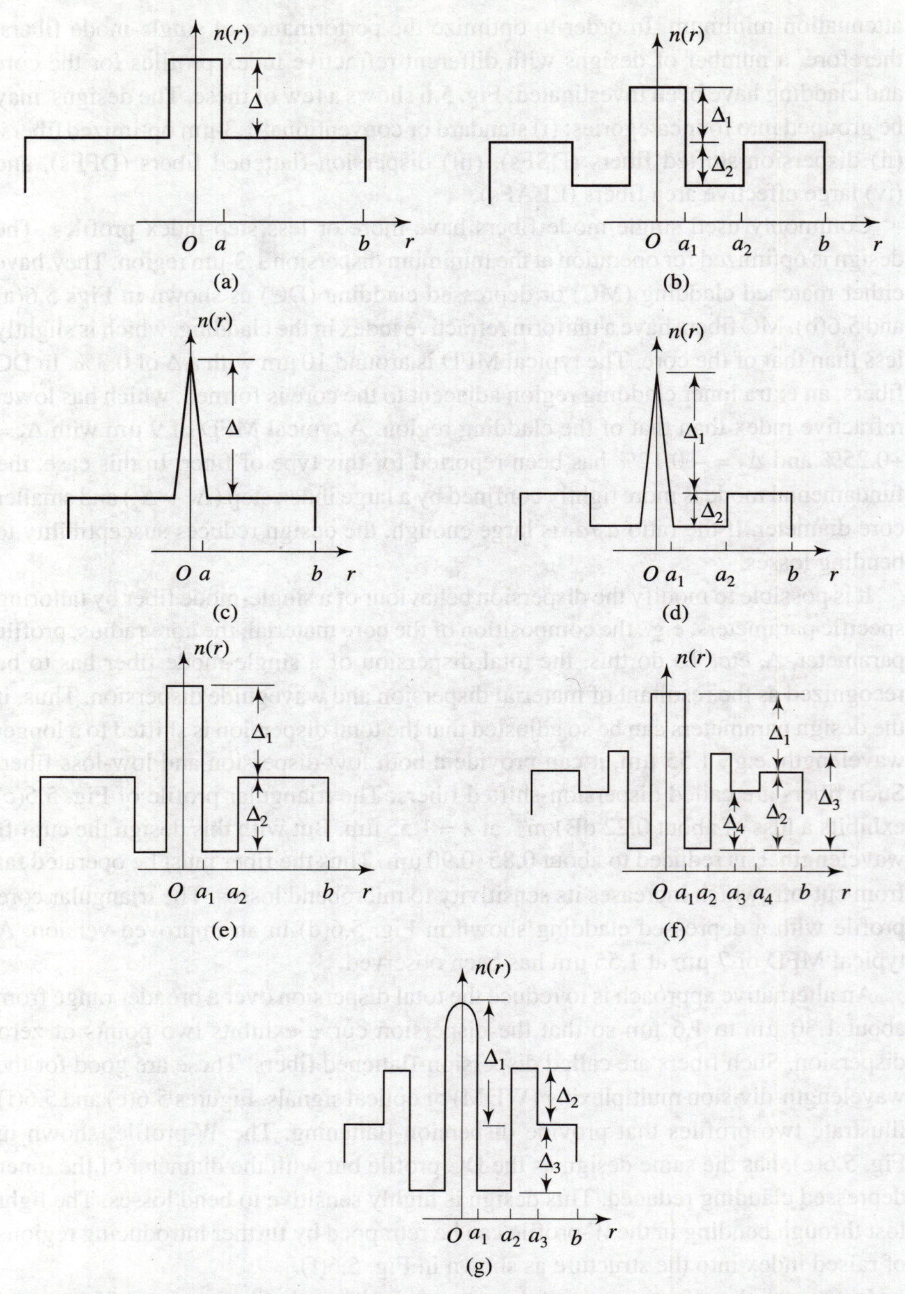

Fig. 5.6 Refractive index profile for single-mode fibers: (a) and (b), standard or conventional 1.3-μm optimized designs; (c) and (d), dispersion-shifted designs; (e) and (f) dispersion-flattened designs; and (g) large effective area design

for operation with dense WDM. An increase in the effective area of the core reduces non-linearities by reducing the light density propagating through the core of the fiber. Figure 5.6(g) shows the refractive index profile of a large effective area fiber. The effective area of the core varies from about 72 to 100 μm^2. Such fibers are most suitable for dense-WDM-based fiber-optic systems.

The fiber-optic communication industry is passing through a phenomenal transformation, which has been driven by demands of very high channel capacity and the need for longer transmission distances. Standard optical fibers are able to satisfy these requirements only partially. Therefore newer designs of optical fibers are being explored and tested. These efforts are coupled with other relevant inventions in the area of optoelectronic devices such an EDFAs, add/drop multiplexers, optoelectronic modulators, optical filters, etc. These will be instrumental in satisfying the present as well as future needs in this area.

SUMMARY

- An optical fiber that supports only the fundamental mode is called a single-mode fiber. In SI fibers, single-mode operation is possible over a normalized frequency range $0 < V < 2.405$. V_c for single-mode operation in GI fibers is given by

$$V_c = 2.405\,(1 + 2/\alpha)^{1/2}$$

  Thus, for single-mode operation at a specific wavelength λ, the V-parameter can be tailored by adjusting the core index, core diameter, Δ, α, etc.
- The mode field diameter (MFD) or the mode spot size of a single-mode fiber is defined by

$$\text{MFD} = 2w = 2\left[\frac{\int_0^\infty r^2\psi^2(r)\,dr}{\int_0^\infty r\psi^2(r)\,dr}\right]^{1/2}$$

  where $\psi(r)$ is the field distribution in the LP_{01} mode. MFD can also be defined (in a more accurate but complex way) as

$$\text{MFD}_P = 2w_P = 2\left[\frac{2\int_0^\infty r\psi^2(r)\,dr}{\int_0^\infty r\left(\dfrac{d\psi}{dr}\right)^2 dr}\right]^{1/2}$$

  where w_P is called the Peterman-2 mode radius.

- In practice it is not possible to keep the core diameter of an SMF uniform throughout its length, hence the two orthogonally polarized components of the fundamental mode propagate with different velocities. This makes the fiber birefringent. Pulse broadening caused by this phenomenon is called polarization mode dispersion (PMD).
- In single-mode fibers, intermodal dispersion is absent. However, the pulse broadening does not vanish altogether, due to group velocity dispersion (GVD). The latter has two components, namely, (i) waveguide dispersion and (ii) material dispersion. The total dispersion parameter D may be written to a good approximation as

$$D = D_w + D_m$$

where D_w and D_m are the waveguide and material dispersion parameters, respectively. These are given by the following expressions:

$$D_w = -\frac{n_2 \Delta}{c\lambda}\left[V \frac{d^2}{dV^2}(bV)\right]$$

and

$$D_m = -\frac{\lambda}{c}\frac{d^2 n}{d\lambda^2}$$

The total dispersion D is zero near 1.32 μm.
- Signal attenuation in optical fibers is caused by material absorption, scattering, and bending. The minimum loss in silica fiber is around 1.55 μm. Splices, if not made properly, may also contribute to losses.
- By varying the refractive index profile in the core and cladding, it is possible to design single-mode fibers with different dispersion properties.

MULTIPLE CHOICE QUESTIONS

5.1 What will be the required cut-off value of the normalized frequency parameter to support a single mode in a graded-index fiber with a parabolic profile?
(a) 0.0 (b) 2.405 (c) 3.401 (d) 3.832

5.2 What is the cut-off wavelength of a step-index single-mode fiber with a core diameter of 8.2 μm and NA = 0.12?
(a) 0.850 μm (b) 1.285 μm (c) 1.320 μm (d) 1.550 μm

5.3 At the Gaussian mode field radius w, the fractional modal power $[P(r)/P_0]$ in the fundamental mode becomes
(a) zero. (b) 0.135. (c) 0.367. (d) 0.50.

5.4 The difference in the propagation constants of the two orthogonally polarized modes in a typical single-mode fiber is 62.83 m^{-1}. What is the beat length?
(a) 62.83 m (b) 1 m (c) 10 cm (d) 1 mm

5.5 A single-mode fiber has an average D_{PMD} of 10 ps/$\sqrt{km}$. If the fiber length is 100 km, what is the PMD-induced pulse broadening?
(a) 1 ps (b) 10 ps (c) 33 ps (d) 100 ps

5.6 The value of $|D_w|$ for the fiber of Question 5.5 is 4 ps nm^{-1} km^{-1} and it is excited by a laser diode emitting at $\lambda = 1.55$ μm and has $\Delta\lambda = 1$ nm. What is the pulse broadening caused by waveguide dispersion?
(a) 100 ps (b) 200 ps (c) 400 ps (d) 500 ps

5.7 A fiber-optic link of length 50 km has a rated 0.2 dB km^{-1} loss. The maximum power required to run a photodetector is 20 nW. What power must be supplied by the source?
(a) 20 nW (b) 0.20 μW (c) 2 μW (d) 1 W

5.8 Two single-mode fibers with mode field diameters of 10 μm and 9 μm are spliced. What is the splice loss in dB?[1]
(a) 0.048 dB (b) 0.24 dB (c) 1 dB (d) 3 dB

5.9 Which of the following refractive index profiles is suitable for achieving the dispersion-flattened design of a single-mode fiber?
(a) Matched cladding (b) Triangular profile
(c) *W*-profile (d) Depressed cladding

5.10 Which of the following fibers are suitable for wavelength-division multiplexing of signals?
(a) Dispersion-optimized (b) Dispersion-shifted
(c) Dispersion-flattened (d) Any fiber

Answers

5.1 (c)	5.2 (b)	5.3 (b)	5.4 (c)	5.5 (d)
5.6 (c)	5.7 (b)	5.8 (a)	5.9 (c)	5.10 (c)

REVIEW QUESTIONS

5.1 (a) What is a single-mode fiber?
(b) Define MFD. How is it related to the *V*-parameter?

5.2 (a) What is meant by the cut-off condition?
(b) A single-mode step-index fiber has a core index of 1.46 and a core diameter of 8 μm. The relative refractive index difference is 0.52%. Calculate the cut-off wavelength for the fiber.

Ans: 1.556 μm

---

[1]Hint: The loss at a joint between two single-mode fibers with mode field radii w_1 and w_2 (assuming they are perfectly aligned) is given by

$$\text{Loss}_{\text{MFD}}\ (\text{dB}) = -20 \log_{10}\left(\frac{2\,w_1 w_2}{w_1^2 + w_2^2}\right)$$

5.3 (a) Define modal birefringence and beat length of a single-mode fiber.
(b) Explain the effect of modal birefringence on pulse propagation in single-mode fibers.

5.4 A typical step-index single-mode fiber has a core diameter of 8.2 μm, and $\Delta = 0.36\%$. Calculate w and w_P for operation at $\lambda = 1.310$ μm and 1.550 μm, given that the effective core indices at these wavelengths are 1.4677 and 1.4682, respectively.
Ans: At $\lambda = 1.31$ μm, $w = 4.452$ μm and $w_P = 4.374$ μm
At $\lambda = 1.55$ μm, $w = 5.0427$ μm and $w_P = 4.9377$ μm

5.5 The modal birefringence of a typical conventional single-mode fiber is in the range 10^{-6}–10^{-5}. Calculate (a) the range of $\delta\beta$ when the fiber is operating at $\lambda = 1.30$ μm and (b) the range of the beat length.
Ans: (a) 4.833–48.33 m^{-1} (b) 13 cm–1.3 m

5.6 (a) What are intermodal and intramodal dispersions?
(b) What are the components of intramodal dispersion in a single-mode fiber?

5.7 A step-index single-mode fiber has a core index of 1.48, relative refractive index difference of 0.27%, and a core radius of 4.4 μm. Estimate the waveguide dispersion for this fiber at $\lambda = 1.32$ μm.
Ans: $D_w = -2.51$ ps nm^{-1} km^{-1}

5.8 Describe the fiber structures utilized to provide (a) dispersion-shifting and (b) dispersion-flattening in single-mode fibers.

5.9 A step-index single-mode fiber has a core index of 1.48 and $\Delta = 1\%$. If the material dispersion at 1.55 μm for this fiber is 7 ps nm^{-1} km^{-1}, what should the radius of the core be so that the total dispersion at this wavelength is zero.
Ans: 2.74 μm. This is a dispersion-shifted fiber.

5.10 (a) What are the causes of attenuation in optical fibers?
(b) Why could bending loss in single-mode fibers be severe? What can be done to minimize this loss?

5.11 Consider the DC fiber shown in Fig. 5.6(b). Assume that the core diameter is 8.2 μm, $n_1 = 1.46$, $\Delta = 0.3\%$, and the inner (depressed) cladding diameter is 25 μm. Further, assume that the mode field distribution is Gaussian, and hence the power at a radius r is given by $P(r) = P_0 \exp(-2r^2/w^2)$, where w is the Gaussian mode field radius. Calculate the fraction of power that may leak at the inner–outer cladding interface for transmitting at $\lambda = 1.3$ μm and $\lambda = 1.55$ μm.
Ans: $[P(r)/P_0]_{\lambda = 1.3\,\mu m} = 8.9387 \times 10^{-7}$ and $[P(r)/P_0]_{\lambda = 1.55\,\mu m} = 3.3869 \times 10^{-5}$

6 Optical Fiber Cables and Connections

After reading this chapter you will be able to understand the following:

- Material requirements for the production of optical fibers
- Production of optical fibers
- Liquid-phase (or melting) methods
- Vapour-phase deposition and oxidation methods
- Design of optical fiber cables
- Jointing of optical fibers and related losses
- Splicing of optical fibers
- Fiber-optic connectors
- Characterization of optical fibers

6.1 INTRODUCTION

What should be the criteria for considering fiber-optic systems as viable replacements of their existing counterparts, e.g., metallic cables in communication systems? Any system that is to be considered as a replacement should give a better performance over the existing one and be economic as far as possible. Thus the criteria in the present case would include, among many other minor factors, the following major factors. It should be possible to

(i) economically produce low-loss optical fibers of long lengths with stable and reproducible transmission characteristics;

(ii) fabricate different types of optical fibers which may vary in size, core and cladding indices, relative refractive index difference, index profiles, operating wavelengths, etc., so that the requirements of different systems can be met;

(iii) fabricate cables, in general and for communication applications in particular, out of optical fibers so that they can be handled easily in the field without damaging the transmission properties of optical fibers; and

(iv) terminate and joint fibers and cables without much difficulty and in a way that restricts, or at least limits, the loss associated with such a process.

This chapter, therefore, is devoted to the discussion of material requirements of optical fibers, fiber fabrication methods, techniques of cabling, splicing, and connecting fibers, and the characterization of fibers.

6.2 FIBER MATERIAL REQUIREMENTS

Light guidance through a step-index optical fiber requires that the refractive indices of the core and cladding be different. Hence two compatible materials that are transparent in the operating wavelength range are required. For graded-index fibers, in order to produce a particular index profile, a variation of the refractive index within the core is also required. This is possible only by varying the dopant concentration. Here again two compatible materials, which are mutually soluble and have similar transmission characteristics, will be required. Further, these materials should be such that long, thin, flexible fibers can be drawn. Considering all these requirements, it appears that the choice of materials for fiber fabrication is limited to either glass or plastic. In plastic fibers, index grading is difficult. They also exhibit high attenuation. Hence plastic fibers are used for short-haul communications. Thus the only material available, at present, for making optical fibers for long-haul applications is glass.

Most low-loss optical fibers are made of oxide glasses, the most widely used material being silica (SiO_2). To produce two compatible transparent materials with different refractive indices, silica is doped with either fluorine or various other oxides such as GeO_2, P_2O_5, B_2O_3, etc. The effect of dopant concentration on the refractive index of a fiber is illustrated in Fig. 6.1. The addition of GeO_2, P_2O_5, Al_2O_3, etc. increases the refractive index of the fiber, whereas the addition of B_2O_3, F, etc. decreases it. The desirable properties of silica-based fibers are that they (i) are resistant to deformation at high temperatures (up to ~1000 °C), (ii) exhibit good chemical durability, (iii) exhibit low attenuation over the operating wavelength region required for fiber-optic communication systems, and (iv) may be fabricated in single-mode or multimode, step-index or graded-index form.

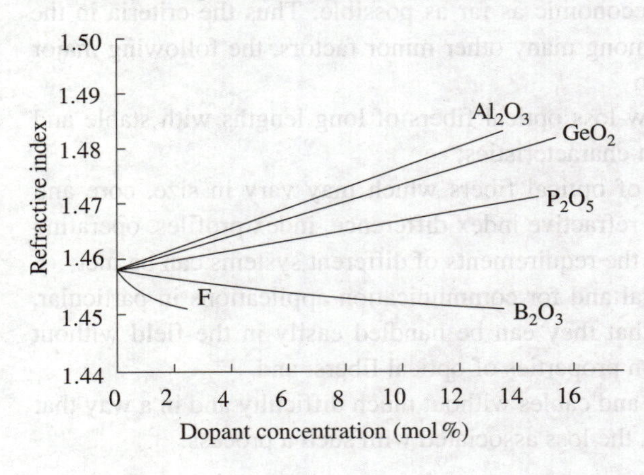

Fig. 6.1 Variation of refractive index of silica glass as a function of concentration of various dopants

A relatively newer class of fibers is being developed using fluoride glasses, which have extremely low transmission losses at mid-infrared wavelengths (0.2–8 μm), with least loss at ~2.55 μm. A typical core glass consists of ZBLAN glass (named after its constituents ZrF_4, BaF_2, LaF_3, AlF_3, and NaF) and ZHBLAN cladding glass (H standing for HaF_4).

Another class of fibers, called active fibers, incorporates some rare-earth elements into the matrix of passive glass. These dopant ions absorb light from the optical source, get excited, and emit fluorescence. Erbium and neodymium have been widely used. Thus it is possible to fabricate fiber amplifiers, using selective doping.

6.3 FIBER FABRICATION METHODS

Fabrication of all-glass fibers is a two-stage process. The first stage consists of producing a pure glass and converting it into a rod or preform. In the second stage, a pulling technique is employed to make fibers of required diameters. Various methods are in use for producing pure glass for optical fibers. These may be grouped into two categories: (i) liquid-phase (or melting) methods and (ii) vapour-phase oxidation methods. These are described, in brief, as follows.

6.3.1 Liquid-phase (or Melting) Methods

These methods employ conventional glass-refining techniques for producing ultra-pure powders of the starting materials, which are oxides such as SiO_2, GeO_2, B_2O_3, Al_2O_3, etc., which decompose into oxides during the melting process. An appropriate mixture of these materials is then melted in silica or platinum crucibles at temperatures varying from 1000 °C to 1300 °C. After the melt has been suitably processed, it is cooled and drawn into rods or tubes (of about 1 m length) of multicomponent glass. The rod of core glass is then inserted into a tube of cladding glass to make a preform. The fiber is drawn from this preform using the apparatus shown in Fig. 6.2. Here, the preform is precision-fed into a cylindrical furnace capable of maintaining high temperature, normally called a drawing furnace. During its passage through the hot zone, its end is softened to the extent that a very thin fiber can be drawn from it. The outer diameter of the fiber is monitored through a feedback mechanism, which controls the feed rate of the preform and also the winding rate of the fiber. The bare fiber is then given a primary protective coating of polymer by passing it through the coating bath. This coating is cured either by UV lamps or thermally. The finished fiber is then wound on a take-up drum.

A fiber 20–30 km long can be drawn from a preform of about 1 m in 2–3 h. Higher pulling rates are limited by the pulling process as well as the subsequent primary coating operation. This method of preparing fibers tends to be a batch process and

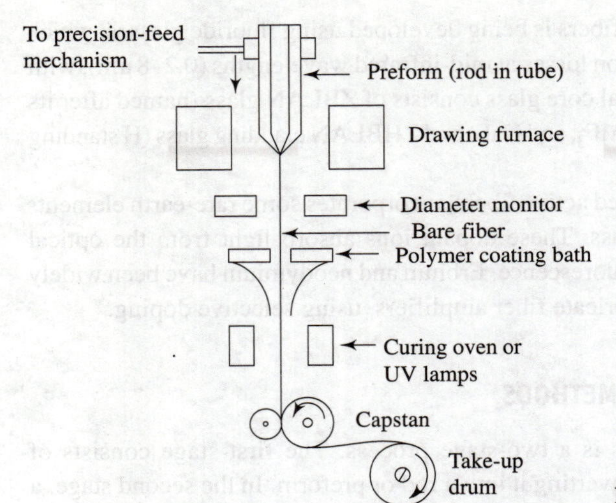

Fig. 6.2 Fiber-drawing apparatus

hence continuous production is not possible. Continuous manufacture is possible using another technique, which is called the *double crucible method*. The apparatus used is shown in Fig. 6.3.

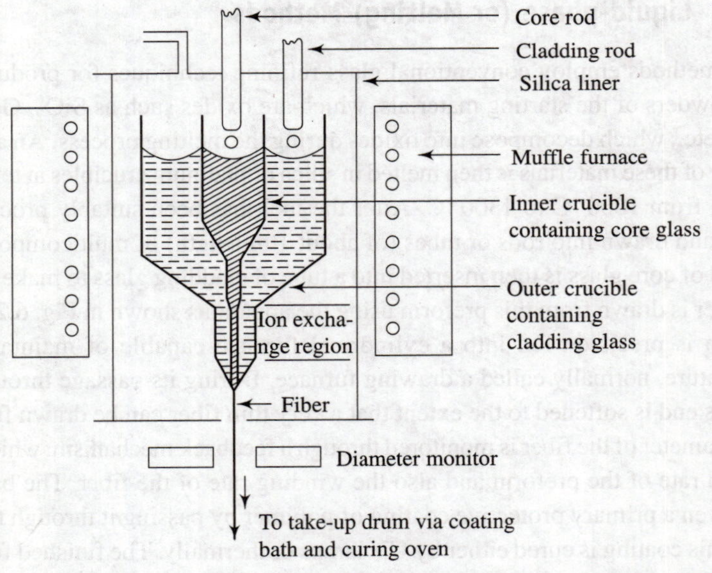

Fig. 6.3 Double crucible method for continuous production of fibers

It consists of two concentric platinum crucibles (also called a double crucible) mounted inside a vertical cylindrical muffle furnace whose temperature may be varied from 800 °C to 1200 °C. The starting material—core and cladding glasses, either directly in the powdered form or in the form of preformed rods—is fed into the two crucibles separately. Both the crucibles have nozzles at their bases from which a clad fiber may be drawn from the melt in a manner similar to that shown in the Fig. 6.2. Index grading may be achieved by diffusion of dopant ions across the core–cladding interface, within the melt. Relatively inexpensive fibers of large core diameters and, therefore, large numerical apertures may be produced continuously by this method. An attenuation level of the order of 3 dB/km for sodium borosilicate glass fiber, which is prepared using this technique, has been reported.

6.3.2 Vapour-phase Deposition Methods

The melting temperatures of silica-rich glasses are too high for liquid-phase melting techniques; therefore, vapour-phase deposition methods are used. Herein, the starting materials are halides of silica (e.g., $SiCl_4$) and of the dopants, e.g., $GeCl_4$, $TiCl_4$, BBr_3, etc., which are purified to reduce the concentration of transition-metal impurities to below 10 ppb. Gaseous mixtures of halides of silica and the dopants are combined in vapour-phase oxidation through either flame hydrolysis or chemical vapour deposition methods. Some typical reactions are given below:

$$\underset{\text{(vapour)}}{SiCl_4} + \underset{\text{(vapour)}}{2H_2O} \xrightarrow{\text{heat}} \underset{\text{(solid)}}{SiO_2} + \underset{\text{(gas)}}{4HCl}$$

$$\underset{\text{(vapour)}}{GeCl_4} + \underset{\text{(vapour)}}{2H_2O} \xrightarrow{\text{heat}} \underset{\text{(solid)}}{GeO_2} + \underset{\text{(gas)}}{4HCl}$$

$$\underset{\text{(vapour)}}{2BBr_3} + \underset{\text{(vapour)}}{3H_2O} \xrightarrow{\text{heat}} \underset{\text{(solid)}}{B_2O_3} + \underset{\text{(gas)}}{4HBr}$$

$$\underset{\text{(vapour)}}{SiCl_4} + \underset{\text{(gas)}}{O_2} \xrightarrow{\text{heat}} \underset{\text{(solid)}}{SiO_2} + \underset{\text{(gas)}}{2Cl_2}$$

$$\underset{\text{(vapour)}}{GeCl_4} + \underset{\text{(gas)}}{O_2} \xrightarrow{\text{heat}} \underset{\text{(solid)}}{GeO_2} + \underset{\text{(gas)}}{2Cl_2}$$

$$\underset{\text{(vapour)}}{TiCl_4} + \underset{\text{(gas)}}{O_2} \xrightarrow{\text{heat}} \underset{\text{(solid)}}{TiO_2} + \underset{\text{(gas)}}{2Cl_2}$$

The oxides resulting from these reactions are normally deposited onto a substrate or within a hollow tube, which is built up as a stack of successive layers. Thus, the concentration of the dopant may be varied gradually to produce the desired index profile. This process results in either a solid rod or a hollow tube of glass, which must be collapsed to produce a solid preform. Fiber may be drawn from this preform using the apparatus shown in Fig. 6.2.

Various techniques have been developed based on the above principle. We will discuss below only four of them.

Outside vapour-phase oxidation (OVPO) method This method uses flame hydrolysis to deposit the required glass composition onto a rotating mandrel (an alumina rod) as shown in Fig. 6.4. The mixture of vapours of the starting materials, e.g., $SiCl_4$, $GeCl_4$, BBr_3, etc., is blown through the oxygen–hydrogen flame. The soot produced by the oxidation of halide vapours is deposited on a cool mandrel. The flame is moved back and forth over the length of the mandrel so that a sufficient number of layers is deposited on it. The concentration of the dopant halides is either varied gradually, if index grading is required, or maintained constant, if step-index fiber is required.

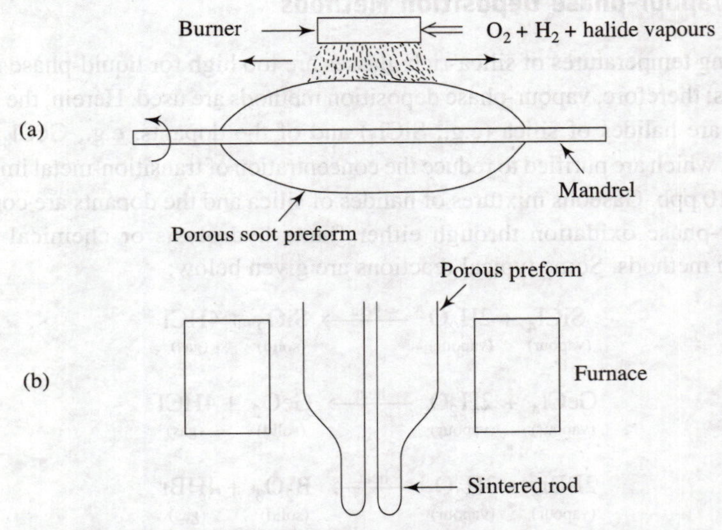

Fig. 6.4 Glass deposition by the OVPO method: (a) soot deposition and (b) preform sintering

After the deposition of the core and cladding layers is complete, the mandrel is removed. The hollow and porous preform left behind is then sintered in a furnace to form a solid transparent glass rod. This is then drawn into a fiber by the apparatus discussed earlier. With a single preform, 30–40 km of fiber can be easily prepared. The index profile can be controlled well using this method, as the flow rate of vapours can be adjusted after the deposition of each layer. An attenuation of less than 1 dB/km at an operating wavelength of 1.3 μm has been reported, but the fibers show an axial dip in the refractive index.

Vapour axial deposition (VAD) method In this method, core and cladding glasses are simultaneously deposited onto the end of a seed rod, which is rotated to maintain azimuthal homogeneity and also pulled up as shown in Fig. 6.5. The porous preform

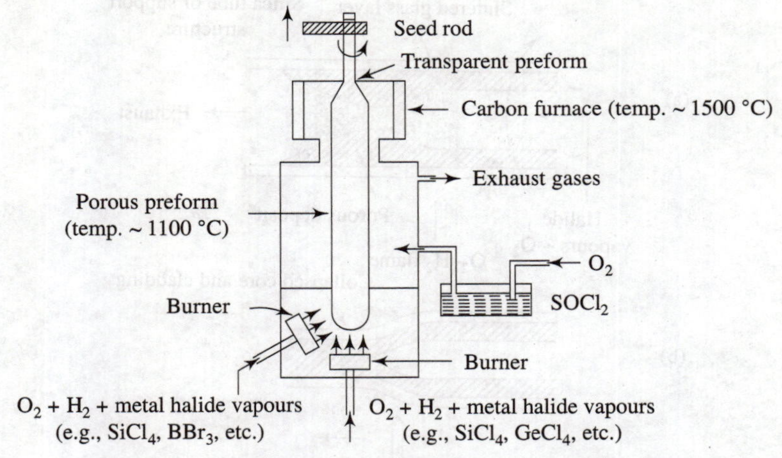

Fig. 6.5 Apparatus for the VAD process

so deposited, while the seed rod is being pulled up, is heated to about 1100 °C in an electric furnace in an atmosphere of O_2 and thionyl chloride. Any water vapour in the preform is removed through the following reaction:

$$SOCl_2 + H_2O \rightarrow SO_2 + 2HCl$$

The porous preform is then heated to about 1500 °C in a carbon furnace, where it is sintered into a transparent solid glass rod.

A good control over the index profile may be achieved with germania-doped cores. Graded-index fiber with an attenuation level less than 0.5 dB/km at 1.3 μm has been produced by this method.

Modified chemical vapour deposition (MCVD) method This is a vapour-phase oxidation process taking place inside a hollow silica tube as shown in Fig. 6.6. The tube has a length of about 1 m and a diameter of about 15 mm. It is horizontally mounted and rotated on a glass-working lathe, with an arrangement (normally an oxygen–hydrogen flame) for heating the outer surface of the tube to about 1500 °C. The reactants in the form of halide vapours and oxygen are passed at a controlled rate through the tube. The halides are oxidized in the hot zone of the tube and the generated soot (glass particles) is deposited on the inner wall. The hot zone (i.e., the flame) is moved back and forth allowing the layer-by-layer deposition of the soot. Index grading may be achieved by varying the concentration of the dopants layer by layer. The tube can form a cladding material or serve as a support structure only for the porous preform. After the deposition is complete, the tube or the porous preform is sintered at a higher temperature (1700–1900 °C) to form a transparent glass rod. Fiber is then drawn from this rod in the usual manner.

At present this is a widely used method for fabricating fibers, as it allows the deposition to occur in a clean environment, with reduced OH impurity. This method

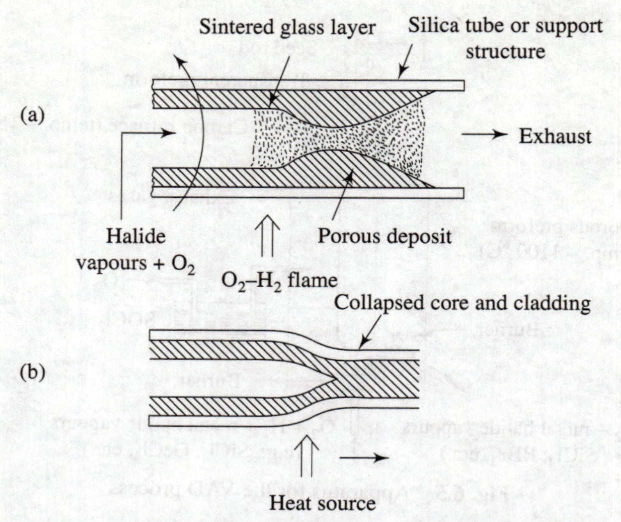

Fig. 6.6 (a) Apparatus for the MCVD process. (b) Preform sintering.

is suitable for preparing a variety of glass compositions for multimode or single-mode step-index (SI) or graded-index fibers. Typically, attenuation of the order of 0.2 dB/km at a wavelength of 1.55 μm has been reported for single-mode germania-doped silica fibers prepared by this method. Further, this technique is also suitable for preparing polarization-maintaining single-mode fibers.

Plasma-activated chemical vapour deposition (PCVD) method The deposition rates of the MCVD process may be increased if microwave-frequency plasma is created in the reaction zone. This process is called plasma-activated chemical vapour deposition and is illustrated in Fig. 6.7. Herein, a microwave cavity (operating at 2.45 GHz) surrounds the substrate tube.

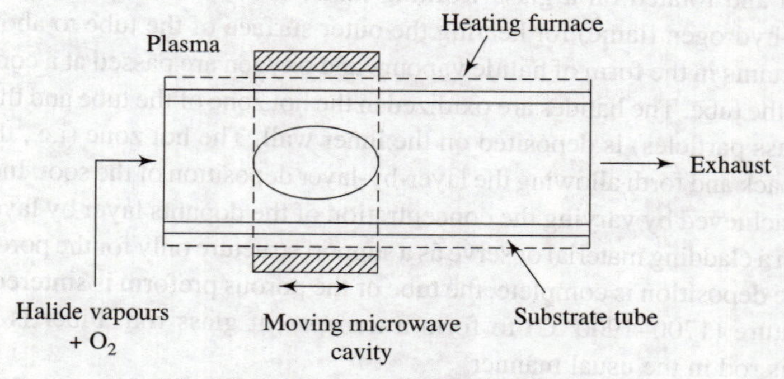

Fig. 6.7 Apparatus for the PCVD process

The halide vapours of the silica-based compound or the dopants along with oxygen are introduced into the tube where they react in the microwave-excited plasma zone. The tube temperature is maintained at about 1900 °C using a stationary furnace. The reaction zone is moved back and forth along the tube enabling circularly symmetrical deposition of glass layers onto the inner wall of the substrate tube. High deposition efficiency and an excellent control of index grading is possible with this method. Attenuation of the order of 0.3 dB/km at $\lambda = 1.55\,\mu m$ has been obtained for dispersion flattened single-mode fibers.

6.4 FIBER-OPTIC CABLES

It is instructive to note that optical glass fibers are brittle and have very small cross-sectional areas (typical outer diameters range from 100 to 250 μm). They are, therefore, highly susceptible to damage during normal handling and use. In order to improve their tensile strength and protect them from external influences, it is necessary to encase them in cables.

What should be the criteria for designing a fiber cable? The exact design of the cable may vary depending on its application; that is, it will depend on whether the cable is required to be used in underground ducts, buried directly, hung from the poles, or laid underwater, etc. Nevertheless, a general criterion to be applied to all the designs may be formulated. Thus the cable design should be such that it (i) protects the optical fiber from damage and breakage, (ii) does not degrade the transmission characteristics of the optical fiber, (iii) prevents the fiber from being subjected to excessive strain and limits the bending radius, (iv) provides a strength member which can improve its mechanical strength, and (v) provides for (in the case of multifiber cables) the identification and jointing of optical fibers within the cable.

Keeping in view the above factors, the primary coated fibers are given a secondary or buffer coating for protection against external influences. It is also possible to place the fibers in an oversized extruded tube normally called a loose buffer jacket. This structure isolates the fiber mechanically from external forces as well as microbending losses. The empty space in the loose tube may be filled with soft, self-healing material that remains stable over a wide range of temperatures. The buffered fibers are then either stranded helically around a central strength member or placed in the slots of a structural member. The structural member may also serve as a strength member if made of load-bearing material.

The common desirable features of strength and structural members are high Young's modulus, high tolerance to strain, flexibility, and low weight per unit length. An additional requirement of the strength member is that it should have high tensile strength. In order to provide cushion to the entire assembly consisting of buffered fiber, and structural and strength members, a coating of extruded plastic is applied or a tape is helically wound. A further thick outer sheath of plastic is necessary to provide the cable with extra protection against mechanical forces such as crushing. Some designs include copper wires in the cable. These wires are used to feed electrical power to the remote online repeaters and also to serve as voice channels during installation and repair. Some cable designs are shown in Figs 6.8–6.11.

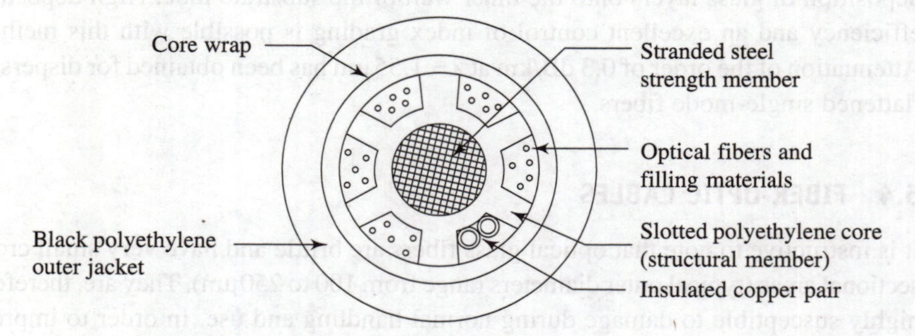

Fig. 6.8 A slotted core cable: In this design, a slotted polyethylene core is extruded over the stranded steel strength member. The buffered fibers are placed in the slots. The design is easy to fabricate and may be adopted for a variety of applications.

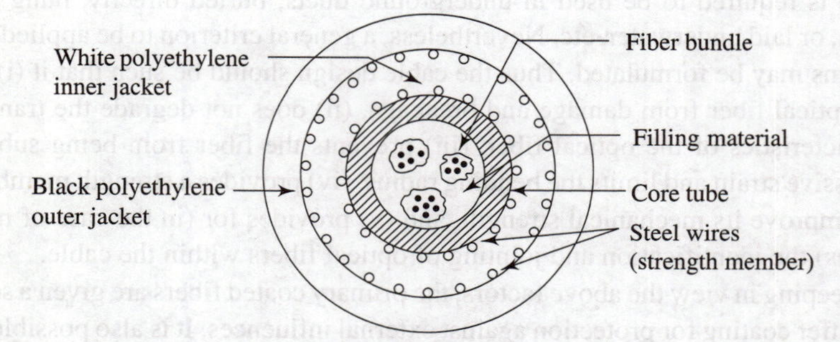

Fig. 6.9 A loose fiber bundle cable: Herein, a tube is extruded over the fiber bundles, each bundle containing several fibers. The steel wires surrounding this tube serve as strength members. This allows a large number of fibers to be accommodated in a compact design.

6.5 OPTICAL FIBER CONNECTIONS AND RELATED LOSSES

The continuous length of an optical fiber along a communication link is determined by three factors, namely, (i) the continuous length of fiber that can be produced by prevalent manufacturing methods, (ii) the length of the cable that can be produced and installed as a continuous section along the link, and (iii) the cable length between the repeaters. This uninterrupted length of optical fiber along a link, therefore, is not more than 10 km. Thus, for establishing long-haul transmission links, optical fibers

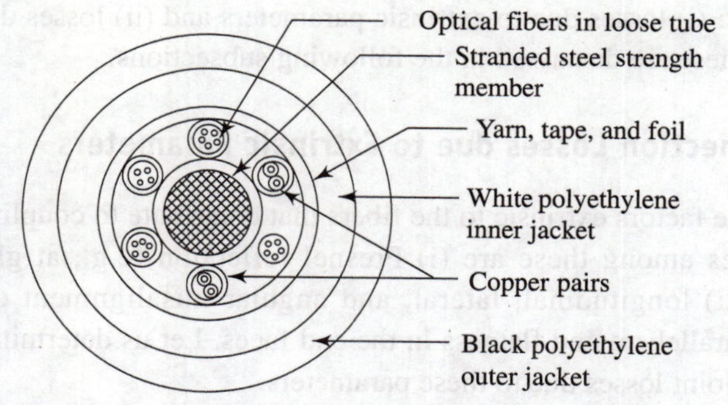

Fig. 6.10 A multifiber cable: In this design, buffered fibers are placed in loose tubes. Out of the six tubes shown, four contain optical fibers and two contain insulated copper pairs. A central steel strength member has been provided. This type of cable is suitable for underground ducts.

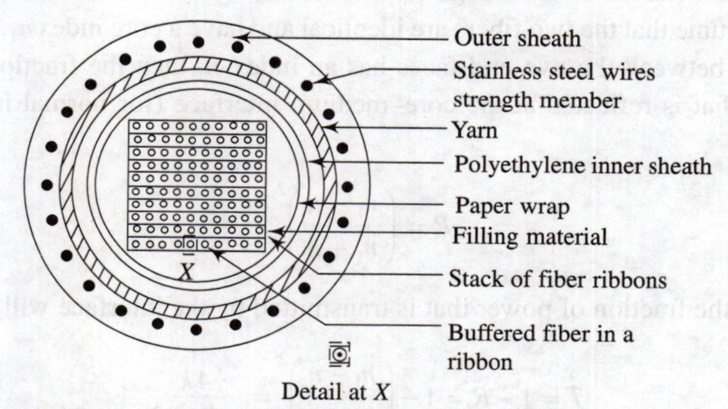

Fig. 6.11 A multifiber ribbon cable: This design permits a large number of fibers to be placed in a single cable. AT&T has developed a cable design which can accommodate 144 fibers in the form of a stack of 12 ribbons, each ribbon containing 12 optical fibers. Ribbon cables are being developed, which can accommodate several hundred fibers.

are required to be connected. The fiber-to-fiber connection may be achieved in two ways: using (i) splices, which are permanent joints between two fibers (splicing is analogous to the electrical soldering of two metallic wires), and (ii) connectors, which are demountable joints (analogous to a plug-in-socket arrangement). The major consideration in making these connections is the optical loss associated with them. Thus before we discuss the techniques used for connecting optical fibers through splices or connectors, let us briefly review the loss mechanisms associated with fiber-to-fiber connections. Connection losses may be grouped into two categories: (i) losses due to extrinsic parameters and (ii) losses due to intrinsic parameters. These are discussed in the following subsections.

6.5.1 Connection Losses due to Extrinsic Parameters

There are some factors extrinsic to the fibers that contribute to coupling losses. The important ones among these are (i) Fresnel reflection (e.g., at glass–air–glass interfaces), (ii) longitudinal, lateral, and angular misalignment of fibers, and (iii) lack of parallelism and flatness in the end faces. Let us determine the order of magnitude of joint losses due to these parameters.

Loss due to Fresnel reflection When light passes from one medium to another, a part of it is reflected back into the first medium. This phenomenon is called Fresnel reflection. Therefore, even if the end faces of the fibers (to be connected) are perfectly flat and the axes of the fibers are perfectly aligned; there will be some loss at the joint due to Fresnel reflection. The magnitude of this loss may be determined as follows.

If we assume that the two fibers are identical and have a core index n_1, and that the medium in between the two end faces has an index n, then the fraction of optical power, R, that is reflected at the core–medium interface (for normal incidence) is given by

$$R = \left(\frac{n_1 - n}{n_1 + n}\right)^2 \tag{6.1}$$

Therefore, the fraction of power that is transmitted by the interface will be given by

$$T = 1 - R = 1 - \left(\frac{n_1 - n}{n_1 + n}\right)^2 = \frac{4k}{(k+1)^2} \tag{6.2}$$

where $k = n_1/n$. As there are two interfaces (glass–medium–glass) at a joint, the coupling efficiency η_F in the presence of Fresnel reflection for two compatible fibers will be given by

$$\eta_F = \frac{4k}{(k+1)^2}\frac{4k}{(k+1)^2} = \frac{16k^2}{(k+1)^4} \tag{6.3}$$

However, if the cores of the two fibers have the same size (i.e., same core diameter) but different refractive indices, say, n_1 and n_1', then the coupling efficiency η_F' in this case will be given by

$$\eta_F' = \frac{4k}{(k+1)^2}\frac{4k'}{(k'+1)^2} \tag{6.4}$$

where $k = n_1/n$ and $k' = n_1'/n$.

Thus the loss, in decibels (dB), at a joint due to Fresnel reflection will be given by

$$L_F = -10\log_{10}(\eta_F) \tag{6.5a}$$

or

$$L'_F = -10\log_{10}(\eta'_F) \tag{6.5b}$$

Typically, if $n_1 = 1.5$ and $n = 1$ (for air), $L_F = 0.36$ dB. Normally, in order to minimize such losses, index-matching fluid is used in the gap between fiber ends.

Example 6.1 Two compatible multimode SI fibers are jointed with a small air gap. The fiber axes and end faces are perfectly aligned. Determine the refractive index of the fiber core if the joint is showing a loss of 0.47 dB.

Solution

Using Eq. (6.5a), we get

$$L_F = -10\log_{10}(\eta_F) = 0.47 \text{ dB}$$

This gives

$$\eta_F = 0.897 = \frac{16k^2}{(k+1)^4} \quad \text{[from Eq. (6.3)]}$$

For air, $n = 1$; therefore,

$$k = \frac{n_1}{n} = n_1$$

Thus

$$n_1^2 - 2.22n_1 + 1 = 0$$

Taking the positive root, $n_1 = 1.59$. The negative root gives n_1 less than 1, which is not possible.

Fiber-to-fiber misalignment losses In an optical fiber connection, the alignment of the two fibers to be connected is very important. The three types of misalignment that may occur are shown in Fig. 6.12. It is possible that there is (i) separation between the fiber ends along the common axis (end separation or longitudinal misalignment), (ii) a lateral offset between the axes of the two fibers (lateral misalignment), and (iii) an angle between the axes of the two fibers (angular misalignment). In each of these cases, the loss at a joint is determined by the optical coupling efficiency between the two fibers. It has been shown (Tsuchiya et al. 1977) that for the three types of misalignment shown in parts (a), (b), and (c) of Fig. 6.12, the coupling efficiencies for two compatible multimode step-index fibers are given by Eqs (6.6), (6.7), and (6.8), respectively. The coupling efficiency η_{long} for a longitudinal misalignment Δx between the two fibers is given by

$$\eta_{\text{long}} = \left[1 - \frac{\Delta x \,\text{NA}}{4an}\right] \tag{6.6}$$

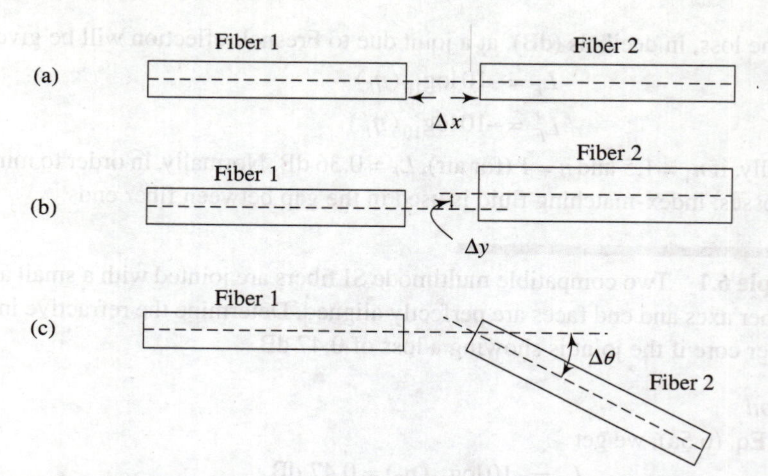

Fig. 6.12 (a) Longitudinal, (b) lateral, and (c) angular misalignment of fiber 2 with respect to fiber 1

where a is the core radius and NA is the numerical aperture of both fibers. The coupling efficiency η_{lat} for a lateral offset Δy between the axes of the two fibers is given by

$$\eta_{lat} \approx \frac{2}{\pi}\left[\cos^{-1}\left(\frac{\Delta y}{2a}\right) - \left(\frac{\Delta y}{2a}\right)\left\{1 - \left(\frac{\Delta y}{2a}\right)^2\right\}^{1/2}\right] \tag{6.7}$$

Here $\cos^{-1}(\Delta y/2a)$ is expressed in radians. Finally, the coupling efficiency η_{ang} for an angular misalignment $\Delta\theta$ between the axes of the two fibers is given by

$$\eta_{ang} \approx \left[1 - \frac{n\Delta\theta}{\pi \text{NA}}\right] \tag{6.8}$$

The main assumptions in deriving these formulae may be summarized as follows: (i) All the modes are uniformly excited. (ii) In Eq. (6.6), the optical wave propagating through the fiber is assumed to be expressed by a meridional ray and the gap (Δx) between the two fibers is assumed to be less than $a/\tan\phi_0$, where $\phi_0 = \text{NA}/n$. (iii) In Eq. (6.7), it has been assumed that the overlapped area between both the cores gives the coupling efficiency η_{lat} approximately, and the change in the optical ray angular component is small. This relation is valid for $0 \le \Delta y \le 2a$. (iv) In Eq. (6.8) the propagation angle (in radians) in the second fiber is restricted to $\Psi \le (2\Delta)^{1/2}$, where Δ is the relative refractive index difference of the fiber.

The loss, in decibels, due to the three types of misalignment described above may be determined using the following relations. The loss due to longitudinal misalignment is given by

$$L_{long} = -10\log_{10}\eta_{long} \tag{6.9}$$

the loss due to lateral misalignment is given by

$$L_{lat} = -10\log_{10}\eta_{lat} \tag{6.10}$$

and the loss due to angular misalignment is given by

$$L_{ang} = -10\log_{10}\eta_{ang} \tag{6.11}$$

Experimental results of these three types of misalignment losses for typical fibers are shown in Figs 6.13(a)–6.13(c).

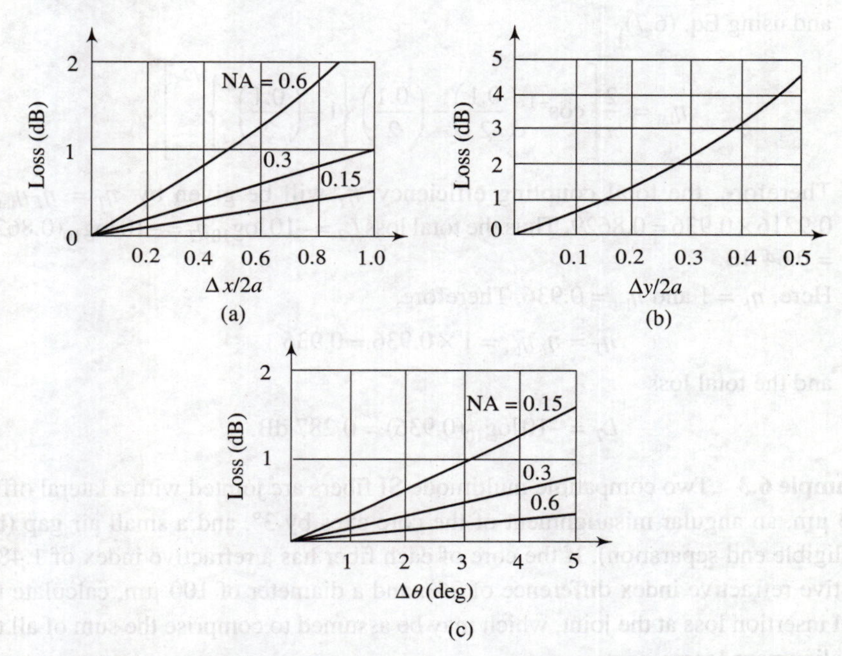

Fig. 6.13 Measured losses for (a) longitudinal, (b) lateral, and (c) angular misalignment

Losses due to other factors The other extrinsic factors that may cause losses at a joint are related to the state of the end faces of the two fibers. For example, the end faces of the fibers may not be orthogonal with respect to the fiber axes. Further, they may not be flat. The mathematical expressions for such losses are rather difficult to arrive at. Nevertheless, such factors are taken care of during the cleaving and polishing of the fiber end faces before making a connection.

Example 6.2 Two compatible multimode SI fibers are jointed with a lateral offset of 10% of the core radius. The refractive index of the core of each fiber is 1.50. Estimate the insertion loss at the joint when (a) there is small air gap and (b) an index-matching fluid is inserted between the fiber ends.

Solution

It is given that $\Delta y/a = 10\% = 0.1$ and $n_1 = 1.50$. With the air gap ($n = 1$), $k = n_1/n = 1.5$, and with index-matching fluid ($n = n_1$), $k = 1$.

(a) Using Eq. (6.3)

$$\eta_F = \frac{16(1.5)^2}{(1.5+1)^4} = 0.9216$$

and using Eq. (6.7),

$$\eta_{\text{lat}} = \frac{2}{\pi}\left[\cos^{-1}\left(\frac{0.1}{2}\right) - \left(\frac{0.1}{2}\right)\left\{1 - \left(\frac{0.1}{2}\right)^2\right\}^{1/2}\right]$$

Therefore, the total coupling efficiency η_T will be given by $\eta_T = \eta_F \eta_{\text{lat}} = 0.9216 \times 0.936 = 0.8629$. Thus the total loss $L_T = -10\log_{10}\eta_T = -10\log_{10}(0.8629) = 0.64$ dB.

(b) Here, $\eta_F = 1$ and $\eta_{\text{lat}} = 0.936$. Therefore,

$$\eta_T = \eta_F \eta_{\text{lat}} = 1 \times 0.936 = 0.936$$

and the total loss

$$L_T = -10\log_{10}(0.936) = 0.287 \text{ dB}.$$

Example 6.3 Two compatible multimode SI fibers are jointed with a lateral offset of 3 μm, an angular misalignment of the core axes by 3°, and a small air gap (but negligible end separation). If the core of each fiber has a refractive index of 1.48, a relative refractive index difference of 2%, and a diameter of 100 μm, calculate the total insertion loss at the joint, which may be assumed to comprise the sum of all the misalignment losses.

Solution

It is given that $\Delta x \approx 0$ (negligible), $\Delta y = 3$ μm, $\Delta\theta = 3° = 0.052$ rad, $n_1 = 1.48$, $\Delta = 2\% = 0.02$, $2a = 100$ μm, $n = 1$ (for air), and therefore $k = n_1/n = n_1 = 1.48$.

Using Eq. (6.3), we get

$$\eta_F = \frac{16 \times (1.48)^2}{(1.48+1)^4} = 0.9264$$

Using Eq. (6.7), we get

$$\eta_{\text{lat}} = \frac{2}{\pi}\left[\cos^{-1}\left(\frac{3}{100}\right) - \frac{3}{100}\left\{1 - \left(\frac{3}{100}\right)^2\right\}^{1/2}\right] = 0.962$$

and using Eq. (6.8), we get

$$\eta_{\text{ang}} = \left[1 - \frac{1 \times 0.052}{\pi \times 0.296}\right]$$

as $$\text{NA} = n_1\sqrt{2\Delta} = 1.48\sqrt{2 \times 0.02} = 0.296$$

$$\eta_{\text{ang}} = 0.944$$

Therefore the total coupling efficiency η_T will be given by

$$\eta_T = \eta_F \eta_{\text{lat}} \eta_{\text{ang}} = 0.9264 \times 0.962 \times 0.944 = 0.8412$$

Thus the total loss

$$L_T = -10\log_{10}\eta_T = 0.75 \text{ dB}$$

6.5.2 Connection Losses due to Intrinsic Parameters

If the fibers to be connected are not compatible, that is, they have different geometrical and/or optical parameters, then power may be lost at the joint. In this context, the following are important:

(i) the core diameter,
(ii) the numerical aperture or the relative refractive index difference, and
(iii) the refractive index profile.

The mismatch of these parameters is illustrated in Fig. 6.14. In a multimode step-index or graded-index fiber, if we assume that all the modes are uniformly excited and that all the parameters of the two fibers are same except their diameters, then the coupling efficiency η_{cd} may be estimated by the ratio of the core areas. Thus

$$\eta_{\text{cd}} = \begin{cases} \dfrac{\pi d_2^2/4}{\pi d_1^2/4} = \left(\dfrac{d_2}{d_1}\right)^2 \text{ for } d_2 < d_1 & \text{(6.12a)} \\ 1 \text{ for } d_2 \geq d_1 & \text{(6.12b)} \end{cases}$$

where d_1 and d_2 are the core diameters of the transmitting and receiving fibers, respectively. The corresponding loss (in dB) is given by

$$L_{\text{cd}} = -10\log_{10}\eta_{\text{cd}} \tag{6.13}$$

Diameter discrepancies of the order of 5% can result in a loss of the order of 0.42 dB.

When there is a mismatch in the numerical aperture of the two fibers, the light cone transmitted by one fiber either overfills or underfills the receiving fiber. If we assume that NA_1 and NA_2 are, respectively, the numerical apertures of the transmitting and receiving fibers and all their other parameters are same, then the coupling efficiency [see Fig. 6.14(b)] is given by

$$\eta_{\text{NA}} = \begin{cases} \left(\dfrac{\text{NA}_2}{\text{NA}_1}\right)^2 \text{ for } \text{NA}_1 > \text{NA}_2 & \text{(6.14a)} \\ 1 \text{ for } \text{NA}_1 \leq \text{NA}_2 & \text{(6.14b)} \end{cases}$$

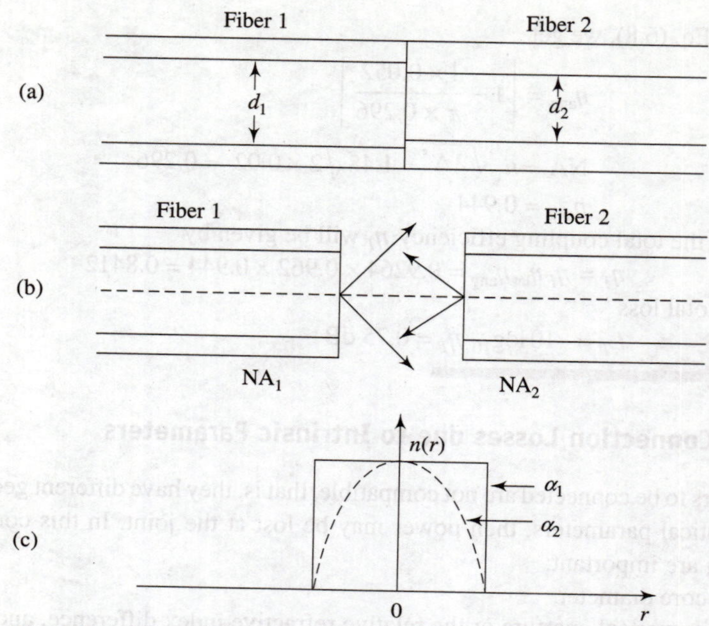

Fig. 6.14 Mismatch of intrinsic parameters: (a) core diameter, (b) numerical aperture, and (c) refractive index profile

The corresponding loss (in dB) is given by

$$L_{\mathrm{NA}} = -10 \log_{10} \eta_{\mathrm{NA}} \tag{6.15}$$

Further, the coupling efficiency due to a mismatch of the refractive index profiles [see Fig. 6.14(c)] is given by

$$\eta_\alpha = \begin{cases} \left(\dfrac{1 + 2/\alpha_1}{1 + 2/\alpha_2}\right) & \text{for } \alpha_1 > \alpha_2 \quad (6.16\text{a}) \\ 1 \quad \text{for } \alpha_1 \le \alpha_2 & \quad (6.16\text{b}) \end{cases}$$

and the corresponding loss is given by

$$L_\alpha = -10 \log_{10} \eta_\alpha \tag{6.17}$$

where α_1 and α_2 are the profile parameters for the transmitting and receiving fibers, respectively. Thus if the transmitting fiber has a step-index profile ($\alpha = \infty$) and the receiving fiber has a parabolic profile, and if both fibers have the same core diameter and axial NA, then, for an index-matched joint ($k = 1$), there is a loss of 3 dB. However, if the direction of light propagation is reversed, there will be no loss at this joint.

Example 6.4 A 80/125 μm graded-index (GI) fiber with a NA of 0.25 and α of 2.0 is jointed with a 60/125 μm GI fiber with an NA of 0.21 and α of 1.9. The fiber axes

are perfectly aligned and there is no air gap. Calculate the insertion loss at a joint for the signal transmission in the forward and backward directions.

Solution

Fiber 1:

Core diameter: $d_1 = 80$ μm, $NA_1 = 0.25$, $\alpha_1 = 2.0$.

Fiber 2:

Core diameter: $d_2 = 60$ μm, $NA_2 = 0.21$, $\alpha_2 = 1.9$.

In the *forward direction*, that is, when the signal is propagating from fiber 1 to fiber 2,

$$\eta_{cd} = \left(\frac{d_2}{d_1}\right)^2 = \left(\frac{60}{80}\right)^2 = 0.5625$$

$$\eta_{NA} = \left(\frac{NA_2}{NA_1}\right)^2 = \left(\frac{0.21}{0.25}\right)^2 = 0.7056$$

$$\eta_\alpha = \left(\frac{1 + 2/\alpha_1}{1 + 2/\alpha_2}\right) = \left(\frac{1 + 2.0/2.0}{1 + 2.0/1.9}\right) = 0.9743$$

Therefore, total coupling efficiency at the joint, $\eta_T = \eta_{cd}\eta_{NA}\eta_\alpha$. Substituting the values of η_{cd}, η_{NA}, and η_α, we get

$$\eta_T = 0.5625 \times 0.7056 \times 0.9743 = 0.3867$$

Therefore, the total loss at the joint will be

$$L_T = -10\log_{10}\eta_T = -10\log_{10}(0.3867) = 4.1259 \text{ dB}$$

In the *backward direction*, η_{cd}, η_{NA}, and η_α are all unity and hence there will be no loss.

Example 6.5 A 60/120 μm graded-index fiber with a numerical aperture of 0.25 and a profile parameter of 1.9 is jointed with a 50/120 μm graded-index fiber with a numerical aperture of 0.20 and a profile parameter of 2.1. If the fiber axes are perfectly aligned and there is no air gap, calculate the insertion loss at the joint in the forward and backward directions.

Solution

In the forward direction,

$$\eta_{cd} = \left(\frac{50}{60}\right)^2 = 0.6944$$

$$\eta_{NA} = \left(\frac{0.20}{0.25}\right)^2 = 0.64$$

and $\eta_\alpha = 1$

Therefore, $\eta_T = 0.6944 \times 0.64 \times 1 = 0.444$

and $L_T = -10\log_{10}\eta_T = 3.52$ dB

In the backward direction, $\eta_{cd} = 1$ and $\eta_{NA} = 1$.

But $$\eta_\alpha = \left(\frac{1+2/2.1}{1+2/1.9}\right) = 0.95$$

Therefore, $\eta_T = 1 \times 1 \times 0.95 = 0.95$

and $L_T = -10\log_{10}(0.95) = 0.218$ dB

6.6 FIBER SPLICES

A fiber splice is a permanent joint formed between two optical fibers. Splicing is required (i) when the length of the system span is more than the manufactured cable length and (ii) when the cable is broken and needs to be repaired. The primary objective of splicing is to establish transmission continuity in the fiber-optic link. This can be done in two ways, namely, through (i) fusion splices or (ii) mechanical splices.

In order to achieve a low-loss splice, it is essential for the fiber ends (to be joined) to be smooth, flat, and perpendicular to the core axes. This is normally achieved using a cleaving tool (a blade of hard metal or diamond). The technique is called 'scribe and break' or 'score and break'. It involves scoring the fiber under tension with a cleaving tool, as shown in Fig. 6.15. This generates a crack in the fiber surface that propagates in the transverse direction and a flat fiber end is produced.

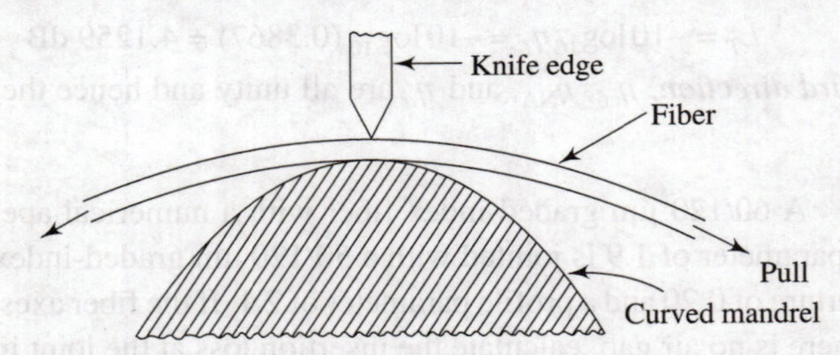

Fig. 6.15 'Score and break' technique of cleaving optical fibers

6.6.1 Fusion Splices

A good quality permanent joint may be obtained by fusion or welding the prepared fiber ends. A widely used heating source for fusion is the electric arc. The set-up for arc fusion is shown in Fig. 6.16. Herein, the prepared fiber ends are placed in a precision alignment jig. The alignment is done with the help of an inspection microscope (not shown). After the initial setting, a short arc discharge is applied to 'fire polish' the fiber ends. This removes any defects due to imperfect cleaving. In the final step, the

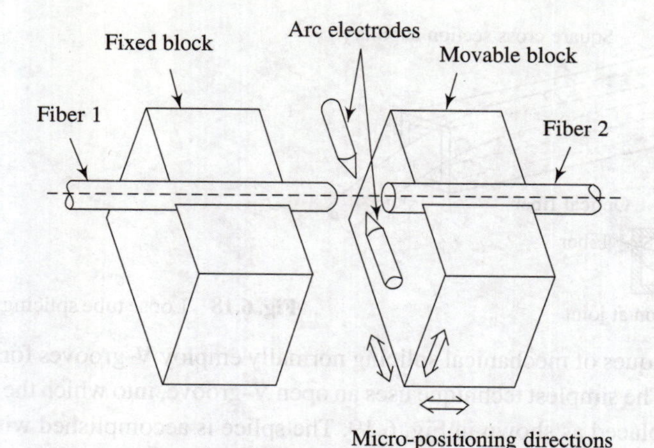

Fig. 6.16 Fusion splicing apparatus

two ends are pressed together and fused with a stronger arc, thus producing a fusion splice. A possible drawback of such a splicing mechanism is that the heat produced by the welding arc may weaken the fiber in the vicinity of the splice.

6.6.2 Mechanical Splices

There are several mechanical techniques of splicing fibers. These normally use appropriate fixtures for aligning the fibers and holding them together. A popular technique, known as the snug tube splice, uses a glass or ceramic capillary with an inner diameter just large enough to accommodate the optical fibers, as shown in Fig. 6.17. The prepared fiber ends are gently inserted into the capillary and a transparent adhesive (e.g., epoxy resin) is injected through a transverse hole. The adhesive ensures both mechanical bonding and index-matching. A stable low-loss splice may be obtained in this way but it poses stringent limits on the capillary diameters.

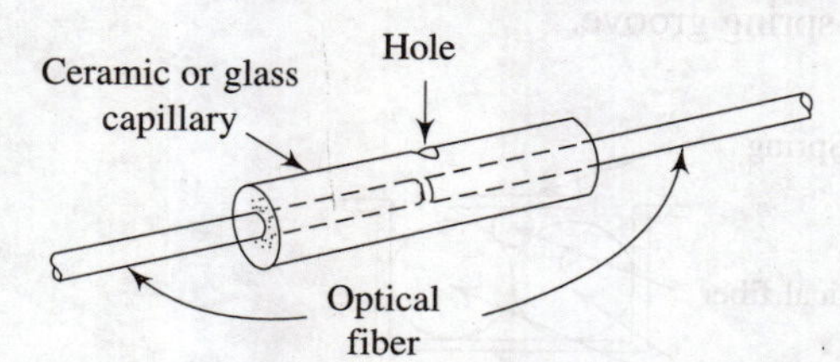

Fig. 6.17 Capillary splicing technique

A slightly different technique uses an oversized metallic capillary of square cross section, as shown in Fig. 6.18. The capillary is first filled with the transparent adhesive, after which the prepared fiber ends are inserted into it. The two fiber ends are forced against one of the four inner corners of the capillary.

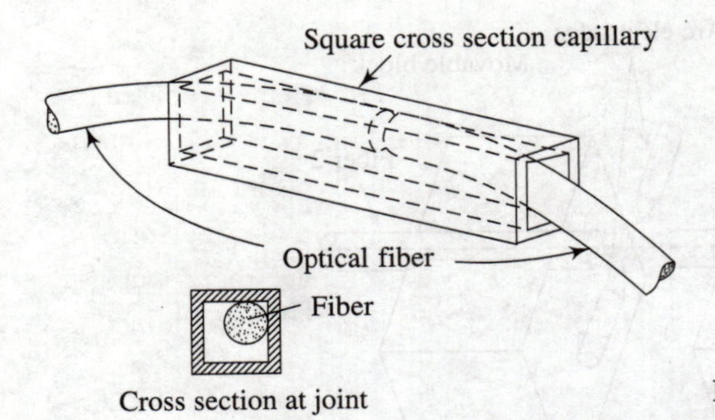

Fig. 6.18 Loose tube splicing technique

Other techniques of mechanical splicing normally employ V-grooves for securing optical fibers. The simplest technique uses an open V-groove, into which the prepared fiber ends are placed as shown in Fig. 6.19. The splice is accomplished with the aid of epoxy resin.

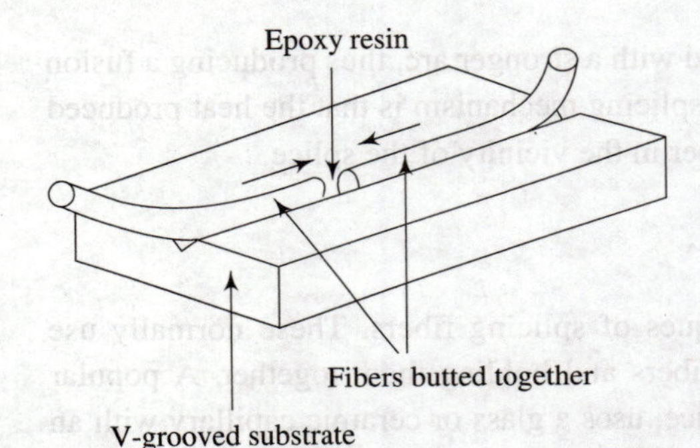

Fig. 6.19 V-groove splicing technique

It is also possible to obtain a suitable groove by placing two precision pins (of appropriate diameter) close to each other. The fibers may then be placed in the cusp as shown in Fig. 6.20. A transparent adhesive ensures bonding as well as index-matching, and a flat spring on the top applies pressure ensuring that fibers remain in their positions. Such a groove is called a spring groove.

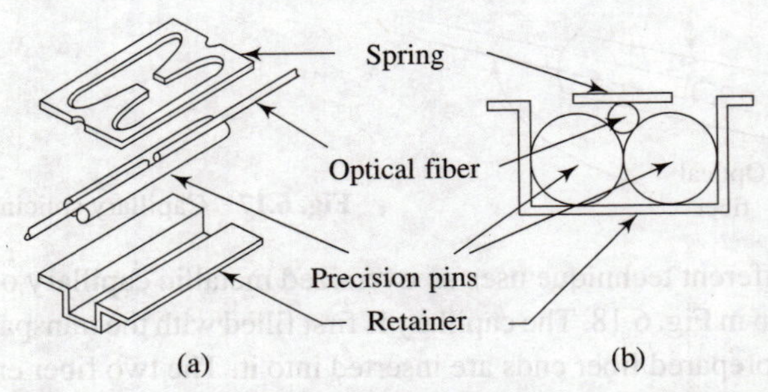

Fig. 6.20 Spring-groove splicing technique: (a) exploded view illustrating the spring, fibers on pins, and retainer; (b) cross-sectional view

There is yet another technique that utilizes the V-groove principle to realize what is known as an elastomeric splice, shown in Fig. 6.21. In this method, the prepared fiber ends are sandwiched between two elastomeric internal parts, one of which contains a V-groove. An outer sleeve holds these two parts compressed so that the fibers are held tightly in alignment. Index-matching gel is employed to improve its performance. Originally, the technique was developed for coupling multimode fibers, but it can also be used for single-mode fibers as well as fibers with different core diameters.

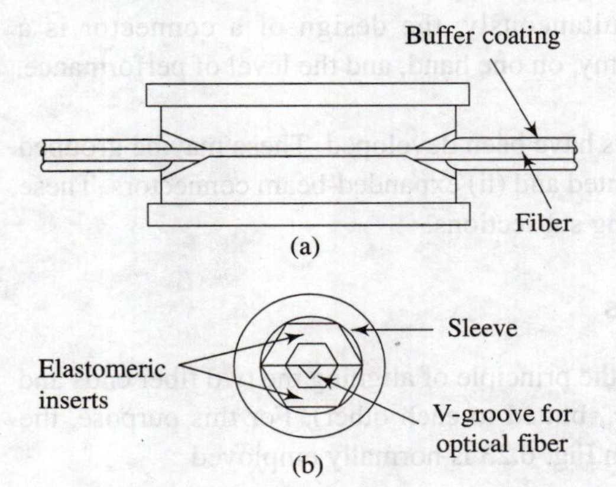

Fig. 6.21 An elastomeric splice: (a) longitudinal section, (b) cross section

Splicing with most of these techniques, if properly carried out, results in splice loss of about 0.1 dB for multimode fibers. Some of these can also be used for splicing single-mode fibers.

6.6.3 Multiple Splices

For ribbon cables containing linear arrays of fibers, the following technique has been used. In this method, shown in Fig. 6.22, the fiber ends are individually prepared, and then placed in a grooved substrate. Adhesive is then used for bonding and index-matching. A cover plate retains the fibers in their position and also maintains mechanical stability.

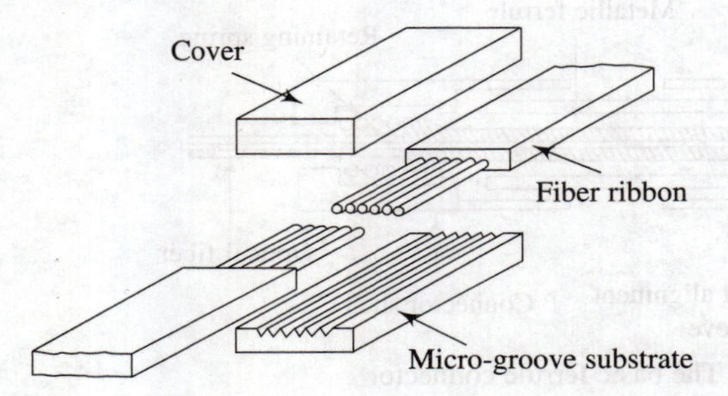

Fig. 6.22 Multiple splicing technique

6.7 FIBER-OPTIC CONNECTORS

A fiber-optic connector is a device which is used to efficiently couple and decouple two, or two groups of, fibers. The criteria for designing a connector are that it must (i) allow for repeated connection and disconnection without problems of fiber alignment and/or damage to fiber ends, (ii) be insensitive to environmental factors such as moisture and dust, and capable of bearing load on the cable, and (iii) have low insertion losses (which should be repeatable) and low cost. Since it is difficult to optimize all three parameters simultaneously, the design of a connector is a compromise between ease and economy, on one hand, and the level of performance, on the other.

A number of fiber-optic connectors have been developed. These may be grouped in two categories, namely, (i) butt-jointed and (ii) expanded-beam connectors. These are discussed, in brief, in the following subsections.

6.7.1 Butt-jointed Connectors

Butt-jointed connectors are based on the principle of aligning the two fiber ends and keeping them in close proximity (i.e., butted to each other). For this purpose, the plug-in-socket configuration shown in Fig. 6.23 is normally employed.

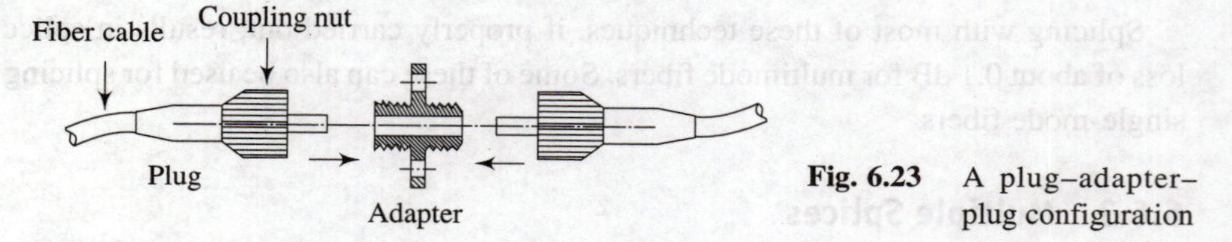

Fig. 6.23 A plug–adapter–plug configuration

The mechanical connection between the plug and the adapter on both the ends is made with the help of either threaded nuts or bayonet locks. Some connectors employ standard BNC or SMA configurations. The design of connectors differs mainly in the technique of aligning fiber ends. The simplest connector design is shown in Fig. 6.24.

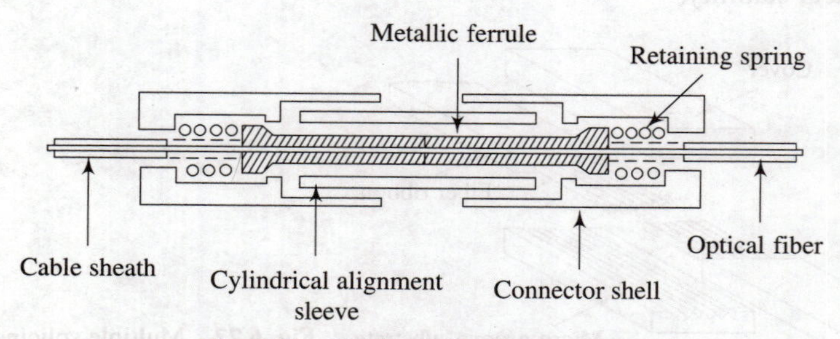

Fig. 6.24 The basic ferrule connector

It consists of metal plugs (normally called ferrules), which are precision-drilled along the central axis. The prepared fiber ends (to be connected) are placed in these holes. They are then permanently bonded to the ferrules by an epoxy resin. A spring retains the ferrule in its position. The two opposite ferrules are aligned by a coaxial cylindrical alignment sleeve.

Another plug–adapter–plug design is shown in Fig. 6.25. Instead of metal ferrules, it employs ceramic capillary ferrules. Ceramic has better thermal, mechanical, and chemical resistance than metallic or plastic.

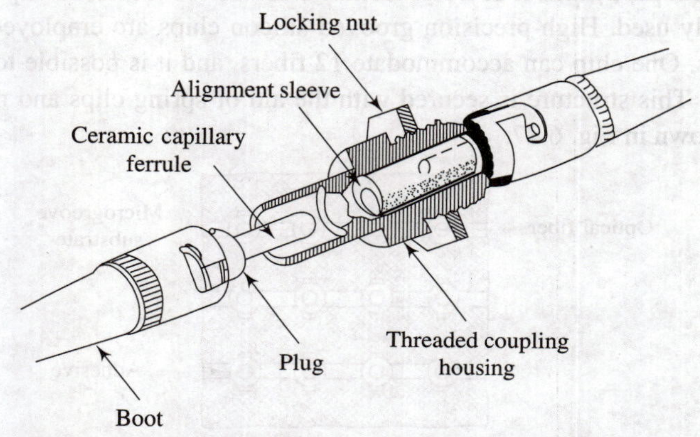

Fig. 6.25 Typical connector design employing ceramic ferrules

6.7.2 Expanded-beam Connectors

An alternative design of connectors is based on expanded-beam coupling, illustrated in Fig. 6.26.

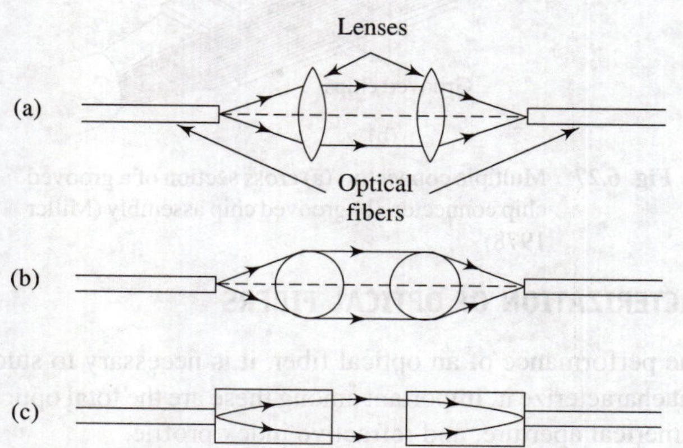

Fig. 6.26 Expanded-beam coupling using (a) a convex microlens, (b) a spherical microlens, and (c) GRIN rod lenses

This technique uses two microlenses for collimating and refocusing light from one fiber end to another. As the beam diameter is expanded, the requirement of lateral alignment of the two plugs in an adapter becomes less critical as compared to butt-jointed connectors. Fresnel reflection losses may increase in this case but are normally reduced with the help of antireflection coating on the lenses.

6.7.3 Multifiber Connectors

In order to couple a number of fibers from two multifiber cables, multiple connectors are normally used. High-precision grooved silicon chips are employed to position fiber arrays. One chip can accommodate 12 fibers, and it is possible to stack many such chips. This structure is secured with the aid of spring clips and metal-backed chips as shown in Fig. 6.27.

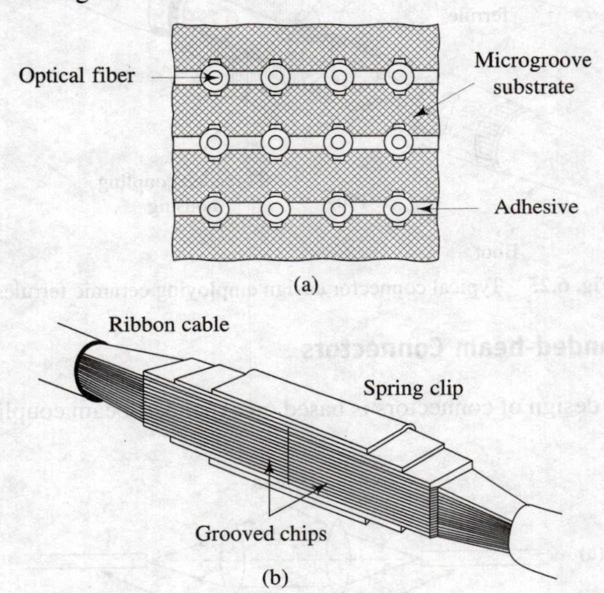

Fig. 6.27 Multiple connector: (a) cross section of a grooved chip connector, (b) grooved chip assembly (Miller 1978)

6.8 CHARACTERIZATION OF OPTICAL FIBERS

To evaluate the performance of an optical fiber, it is necessary to study the various parameters that characterize it. Important among these are the total optical attenuation, dispersion, numerical aperture, and refractive index profile.

A number of methods have been developed for measuring each of these parameters. It is not possible to review all of them here. We will discuss, in a generalized manner, only a few of them.

6.8.1 Measurement of Optical Attenuation

Three types of measurement techniques have been developed for measuring attenuation in optical fibers. They measure (i) total attenuation, (ii) absorption loss, and (iii) scattering loss. The overall or total attenuation is of interest to the system designer, whereas the contribution to this total by the absorption loss and scattering loss mechanisms is important in the development of low-loss optical fibers.

A commonly used method for measuring total fiber attenuation is called the cutback or differential technique and is based on the following principle. Power P_0 is launched at one end of a long length L_1 of the test fiber; the power P_1 received at the other end is measured. The fiber is then cut back to a smaller length L_2 and the power P_2 received at the other end is again measured. Assuming identical launch conditions at a particular wavelength λ, the optical attenuation per unit length α (in, say, dB/km) may be given by the following relation:

$$\alpha = \frac{10}{L_1 - L_2} \log_{10} \frac{P_2}{P_1} \tag{6.18}$$

What should be the criterion for designing the equipment for studying this parameter by the cutback method?

First, a polychromatic continuous source of radiation (containing sufficient power at all the wavelengths of interest) is required. As the attenuation is to be studied for all wavelengths, a wavelength-isolation device (e.g., a monochromator) is required to follow the source. Then suitable optics has to be designed for launching the optical power at one end of the test fiber. At the other end of it, again suitable optics is required so that most of the power transmitted by the fiber is received by the detector. The detector signal should be processed and then output read on a meter or recorded on a chart recorder. Accordingly, the modules may be arranged as shown in Fig. 6.28.

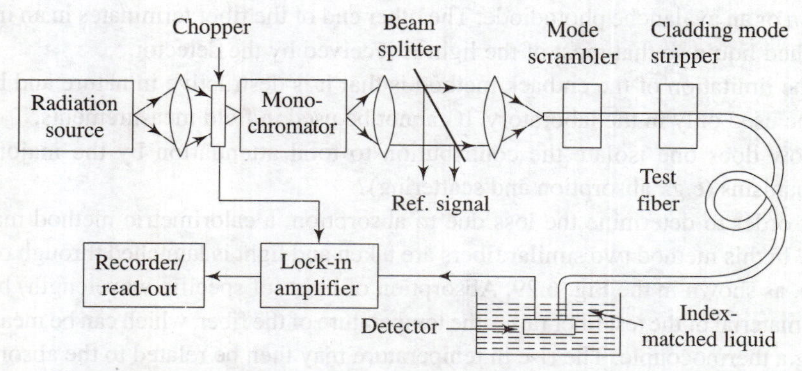

Fig. 6.28 Experimental set-up for the measurement of total attenuation

The importance of each of these modules may be understood if one investigates the dependence of the signal S developed by the detector on pertinent variables:

$$S = P(\lambda)M(\theta,\lambda)\ (n_a \sin\alpha_m)T(\alpha_{ab},\alpha_{sc},L)D(\lambda) \tag{6.19}$$

The different terms may be identified as follows. $P(\lambda)$ is the power furnished by the source at a specific wavelength λ; $M(\theta,\lambda)$ is a function governing the solid angle θ seen by the monochromator and its transmittance with wavelength λ; $n_a \sin \alpha_m$ gives the numerical aperture of the fiber, n_a is the refractive index of the medium surrounding the launching end of the fiber; $T(\alpha_{ab},\alpha_{sc},L)$ is a function determining the transmittance of the fiber. This is dependent on the absorption loss per unit length α_{ab}, the scattering loss per unit length α_{sc}, and the total length L of the fiber. Finally, $D(\lambda)$ is the responsivity of the detector as a function of wavelength λ.

Thus, the source should have high radiance and be continuous. A black body radiator, e.g., a tungsten halogen lamp or a high-pressure discharge lamp, e.g., a xenon arc lamp may be used. A monochromator should collect as much light as possible. The components in the monochromator should have high transmittance in the regions of investigation. If a grating is used as a monochromator, overlapping orders may cause problems and hence an order sorting filter may also be required at the exit slit of the monochromator. In order to improve the *S/N* ratio, the signal from the source is generally chopped at a low frequency and at the receiver end a lock-in amplifier is used to perform phase-sensitive detection.

A beam splitter may be placed as shown in Fig. 6.28 for obtaining a reference signal as well as for viewing the optics. If viewing the optics is not required, a rotating sector mirror may be used in place of a beam splitter and the chopper in between the source and the monochromator may be omitted. This will provide a greater energy throughput to the optical fiber as well as a reference signal for comparison. A mode scrambler has been used to obtain equilibrium mode distribution. The fiber is also put through a cladding mode stripper, which is a device for removing the light launched into the fiber cladding. At the receiver end, the optical power is detected using either a *p-i-n* or an avalanche photodiode. The other end of the fiber terminates in an index-matched liquid so that most of the light is received by the detector.

The limitation of the cutback method is that it is destructive in nature and hence can be used only in the laboratory. It cannot be used in field measurements.

How does one isolate the contribution to total attenuation by the major loss mechanisms (e.g., absorption and scattering)?

In order to determine the loss due to absorption, a calorimetric method may be used. In this method two similar fibers are taken and light is launched through one of them, as shown in the Fig. 6.29. Absorption of light (of specific wavelength) by the bulk material of the test fiber raises the temperature of the fiber, which can be measured using a thermocouple. The rise in temperature may then be related to the absorption loss.

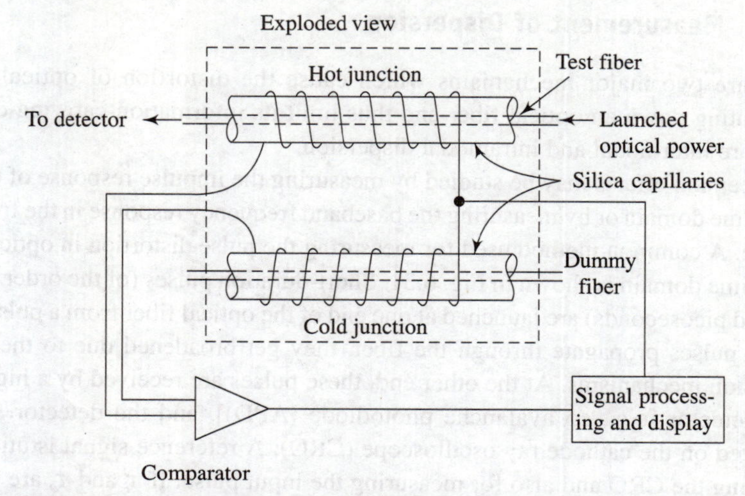

Fig. 6.29 Measurement of the temperature of an optical fiber using a thermocouple

The power loss due to scattering alone may be measured employing a scattering cell as shown in Fig. 6.30. Light from a powerful source is launched into the optical fiber through appropriate launch optics. A certain length (say, L) of the fiber is enclosed inside the scattering cell. All the six inner surfaces of the cell are fitted with six photovoltaic detectors. These detectors measure the optical power (P_{sc}) scattered by the enclosed length L of the fiber. The scattering loss α_{sc} (dB/km) may be expressed by the following relation:

$$\alpha_{sc} = \frac{10}{L\,(\text{km})} \log_{10}\left(\frac{P_0}{P_0 - P_{sc}}\right) \text{dB/km} \tag{6.20}$$

where P_0 is the power launched.

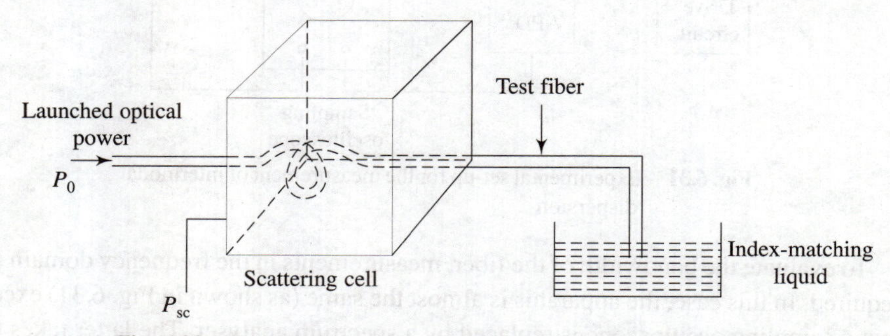

Fig. 6.30 Experimental set-up for measurement of scattering loss

6.8.2 Measurement of Dispersion

There are two major mechanisms which cause the distortion of optical signals propagating down an optical fiber and thus limit its information-carrying capacity. These are intermodal and intramodal dispersion.

Dispersion effects may be studied by measuring the impulse response of the fiber in the time domain or by measuring the baseband frequency response in the frequency domain. A common method used for measuring the pulse distortion in optical fibers in the time domain is shown in Fig. 6.31. Short-duration pulses (of the order of a few hundred picoseconds) are launched at one end of the optical fiber from a pulsed laser. As the pulses propagate through the fiber, they get broadened due to the various dispersion mechanisms. At the other end, these pulses are received by a high-speed photodetector [e.g., an avalanche photodiode (APD)], and the detector signal is displayed on the cathode ray oscilloscope (CRO). A reference signal is utilized for triggering the CRO and also for measuring the input pulse. If τ_i and τ_o are the half-widths of the input and output pulses and if the shape of the pulses is assumed to be Gaussian, then the impulse response of the fiber is given by

$$\tau = \frac{(\tau_o^2 - \tau_i^2)^{1/2}}{L} \quad \text{(in, say, ns/km)} \tag{6.21}$$

where L is the length of the fiber.

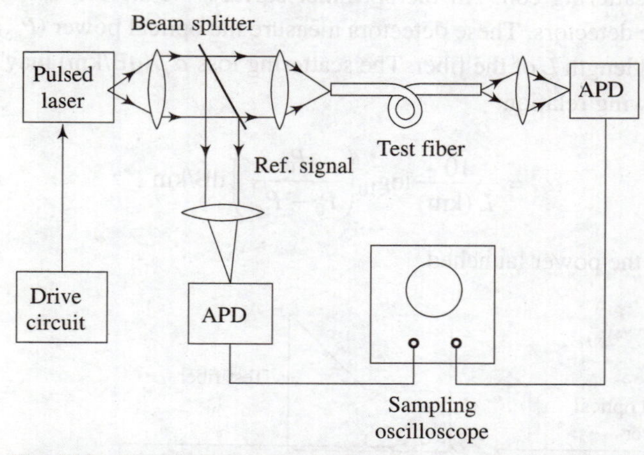

Fig. 6.31 Experimental set-up for the measurement of intermodal dispersion

To evaluate the bandwidth of the fiber, measurements in the frequency domain are required. In this case, the apparatus is almost the same (as shown in Fig. 6.31) except that a sampling oscilloscope is replaced by a spectrum analyser. The latter takes the Fourier transform of the output pulse in the time domain and displays its constituent frequency components. To measure the intramodal or chromatic dispersion, a polychromatic source is required in place of a laser.

6.8.3 Measurement of Numerical Aperture

The numerical aperture (NA) is an important characterizing parameter, as it is directly related to the light-gathering capacity of the fiber. This also decides the number of modes propagating through the multimode fibers. For a step-index fiber, NA is given by the relation

$$\text{NA} = n_a \sin\alpha_m = (n_1^2 - n_2^2)^{1/2} \tag{6.22}$$

where α_m is the angle of acceptance, n_a is the refractive index of the medium in which the fiber is placed, and n_1 and n_2 are the refractive indices of the core and cladding, respectively. For a graded-index fiber, NA is not constant but varies with the distance r from the core axis. The local NA at a radial distance r is given by

$$\text{NA}(r) = n_a \sin\alpha_m(r) = [n_1^2(r) - n_2^2]^{1/2} \tag{6.23}$$

From Eqs (6.22) and (6.23), it becomes clear that the NA can be calculated if the refractive index profile of the fiber is known. However, this method is seldom used.

A commonly used method, shown in Fig. 6.32 involves the measurement of the far-field pattern of the fiber. Light from a powerful source such as a laser is launched at one end of the fiber. The other end is held in the chuck of the fiber holder on a rotating stage. As the tip of the fiber is rotated, the intensity of light reaching the detector falls off on either side and an approximately Gaussian curve results. The angle at which the intensity falls to 5% of its maximum gives the value of α_m.

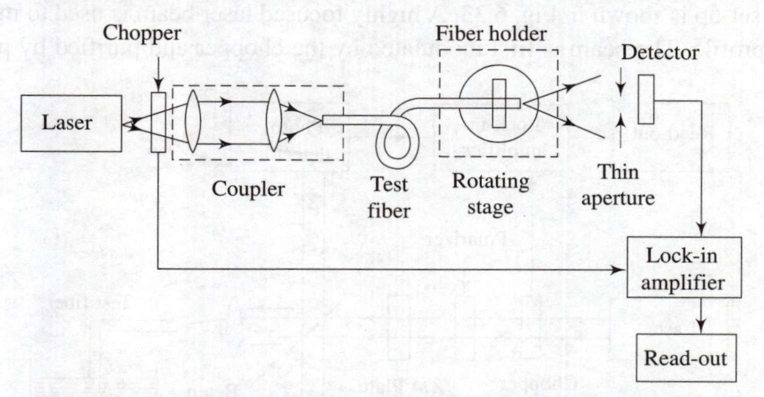

Fig. 6.32 Experimental set-up for evaluating the NA of an optical fiber

Alternatively, the light from the laser source may be made to fall at different angles on one end of the fiber and the output at the other end may be measured with the help of a detector. Again, an approximately Gaussian curve results when the output power is plotted as a function of the angle of rotation. Again, the angle for which the power falls to 5% of its maximum value gives the value of α_m.

6.8.4 Measurement of Refractive Index Profile

The refractive index (RI) profile of an optical fiber plays an important role in its characterization. The knowledge of this profile helps in determining the NA of the fiber and the number of guided modes propagating within the fiber. It also enables one to predict the impulse response and hence the information-carrying capacity of the fiber.

There are several methods for measuring the RI profile. We discuss below the end-reflection method, which is based on the following principle. When a focused beam of light is incident normally on the flat end face of a fiber, a part of the light is reflected back. The fraction R of the light reflected at the fiber–medium interface is given by the Fresnel reflection coefficient

$$R = P_r/P_i = \left(\frac{n_1 - n}{n_1 + n}\right)^2 \tag{6.24}$$

where P_r and P_i are the reflected and incident powers, n_1 is the RI at the striking point of the fiber, and n is the RI of the medium surrounding the fiber. For a small variation in the value of n_1,

$$\Delta R = 4n\left\{\frac{n_1 - n}{(n_1 + n)^3}\right\}\Delta n_1 \tag{6.25}$$

Thus, the variation in the reflected light intensity can be used to calculate the RI.

The set-up is shown in Fig. 6.33. A highly focused laser beam is used to measure the RI profile. The beam is first modulated by the chopper and purified by passing

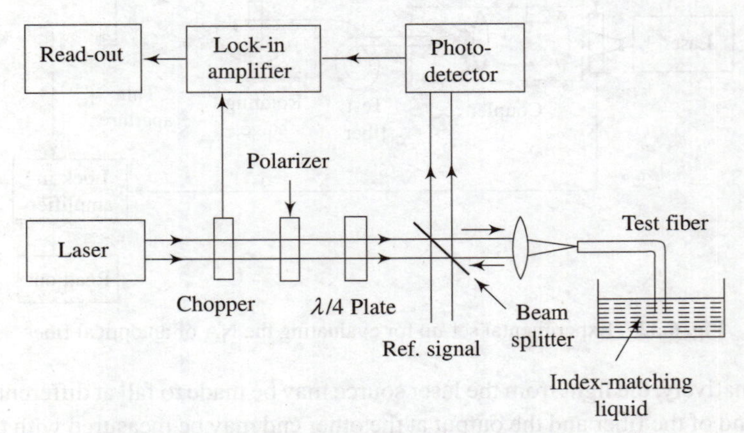

Fig. 6.33 Experimental set-up for studying the RI profile

through a polarizer and quarter-wave plate combination. This combination also decouples the incident light from the reflected light. This light is then focused on the polished flat end of the test fiber. The other end of the fiber is dipped into an index-matching liquid so that light is not reflected back from this end. The light reflected from the flat end of the fiber is directed onto the detector via the beam splitter. The modulated output of the detector is amplified and recorded on the recorder.

6.8.5 Field Measurements: OTDR

The methods that have been discussed so far are primarily suited to the laboratory environment. However, a technique that can measure attenuation, connector and splicing losses, and can also locate faults in optical fiber links in the field is required. A method that finds wide applications in the field is called optical time domain reflectometry (OTDR) or the backscatter technique.

A schematic diagram of the OTDR apparatus is shown in Fig. 6.34. Herein, a powerful beam of light is launched through a bidirectional coupler into one end of the fiber and the backscattered light is detected using an APD receiver. The received signal is integrated and amplified, and the averaged signals for successive points

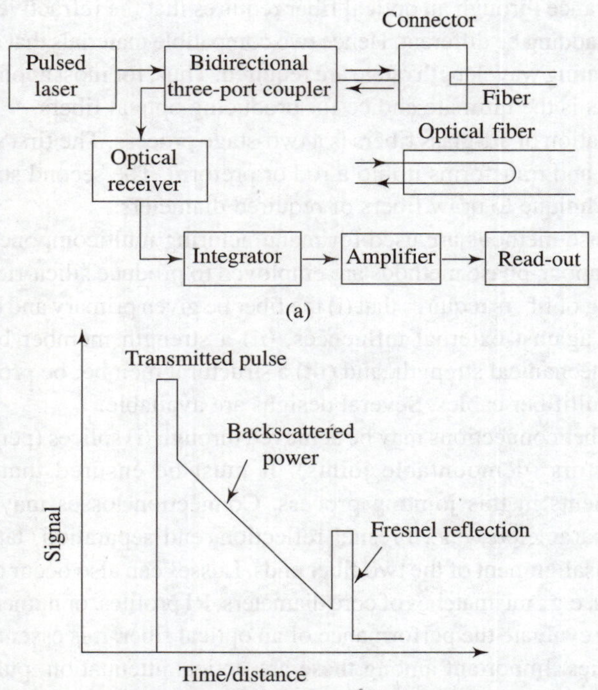

Fig. 6.34 (a) OTDR apparatus. (b) Backscatter plot for an ideal fiber.

within the fiber are presented on the recorder or the cathode ray tube (CRT). The information displayed on the chart of the recorder or the screen of the CRT is the signal strength along the y-axis and time along the x-axis. The time is usually multiplied by the velocity of propagation to give an indication of the distance. The display expected for an ideal optical fiber of finite length is shown in Fig. 6.34(b). Any deviation from this is due to some kind of fault.

OTDR provides information about the location dependence of the attenuation. The slope of the plot shown in Fig. 6.34(b) simply gives the attenuation per unit length for the fiber. In this way, it is superior to other methods of measuring attenuation, which provide the average loss over the whole length. Further, it gives information about the splice or connector losses and the location of any faults on the link. Finally, the overall link length can be calculated from the time difference between the Fresnel reflections from the two ends of the fiber. Furthermore, it requires access to only one end of the fiber for performing measurements.

SUMMARY

- Light guidance through an optical fiber requires that the refractive indices of the core and cladding be different. Hence two compatible materials that are transparent in the operating wavelength range are required. Thus, for most applications, silica-based glass is the ultimate choice for producing optical fibers.
- The fabrication of all-glass fibers is a two-stage process. The first stage produces pure glass and transforms it into a rod or preform. The second stage employs a pulling technique to draw fibers of required diameters.
- Liquid-phase methods are used for manufacturing multicomponent glass fibers whereas vapour-phase methods are employed to produce silica-rich glass fibers.
- The cabling of fibers requires that (i) the fiber be given primary and buffer coatings to protect against external influences, (ii) a strength member be provided to improve mechanical strength, and (iii) a structural member be provided to place them in multifiber cables. Several designs are available.
- Fiber-to-fiber connections may be achieved through (i) splices (permanent joints) or connectors (demountable joints). It must be ensured that there are no misalignments in this jointing process. Connection losses may occur due to extrinsic parameters, e.g., Fresnel reflection, end separation, lateral offset, or angular misalignment of the two fiber ends. Losses can also occur due to intrinsic parameters, e.g., mismatches of core diameters, RI profiles, or numerical apertures.
- In order to evaluate the performance of an optical fiber, it is essential to measure its properties. Important among these are optical attenuation, pulse dispersion, numerical aperture, RI profile, etc. In the field, however, optical time domain reflectometry (OTDR) is an essential tool.

MULTIPLE CHOICE QUESTIONS

6.1 Increase in the concentration of GeO_2 in SiO_2 will
(a) decrease the RI. (b) increase the RI.
(c) change RI randomly. (d) not change RI at all.

6.2 What type of optical fibers can be drawn from a solid preform (formed by collapsing a solid rod or hollow tube deposited by the vapour-phase oxidation method)?
(a) Multimode SI fibers (b) Multimode GI fiber
(c) Single-mode fibers (d) All of these

6.3 In a multifiber cable, the strength member
(a) must be placed along the central axis of the cable.
(b) must be placed in a coaxial cylindrical configuration.
(c) can be placed anywhere within the cable.
(d) is not required at all.

6.4 An air gap is introduced while splicing two compatible fibers with core indices of 1.46. What is the loss due to Fresnel reflection at the joint?
(a) Zero (b) 0.154 dB (c) 0.309 dB (d) 0.36 dB

6.5 Two optical fibers with numerical apertures 0.17 and 0.20 are to be spliced. What will be the loss at the joint in the forward direction?
(a) Zero (b) 1.41 dB (c) 1.82 dB (d) 2.50 dB

6.6 For the optical fibers of Question 6.5, what will be the joint loss in the backward direction?
(a) Zero (b) 1.41 dB (c) 1.82 dB (d) 2.50 dB

6.7 A 62.5/125 μm SI fiber is to be spliced to a 50/125 μm SI fiber. Both the fibers have a core index of 1.50. What will be the joint loss in the forward direction?
(a) Zero (b) 0.97 dB (c) 1.94 dB (d) 2.45 dB

6.8 A multimode SI fiber with a core RI of 1.50 is spliced with an identical fiber. What is the NA of the fiber if the splice loss is 0.7 dB, which is mainly due to a 5° angular misalignment of the fiber core axes?
(a) 0.17 (b) 0.21 (c) 0.28 (d) 0.30

6.9 If two optical fibers with different diameters are to be spliced, which of the following mechanical splices will be most suitable?
(a) Snug tube splice (b) Loose tube splice
(c) Spring-groove splice (d) V-groove splice

6.10 With an OTDR, it is possible to know
(a) the location dependence of attenuation.
(b) the overall link length.
(c) splice and connector losses.
(d) all of the above.

Answers

6.1 (b)	6.2 (d)	6.3 (c)	6.4 (c)	6.5 (a)
6.6 (b)	6.7 (c)	6.8 (c)	6.9 (d)	6.10 (d)

REVIEW QUESTIONS

6.1 (a) Describe the double crucible method for producing optical fibers. What are the limitations of this method?

(b) Distinguish between the outside vapour-phase oxidation method and the inside vapour-phase oxidation method for manufacturing optical fibers. Compare the salient features of both the methods.

6.2 Discuss the design of multifiber cables that employ (a) a central strength member, (b) a structural member that also acts as a strength member, and (c) fiber ribbons.

6.3 (a) Distinguish between a splice and a connector.

(b) How can one avoid or reduce loss due to Fresnel reflection at a joint?

(c) Distinguish between fusion and mechanical splicing of optical fibers. Discuss the advantages and drawbacks of these techniques.

6.4 A fusion splice is made for a broken multimode step-index fiber. The splice shows a loss of 0.36 dB, which appears to be mainly due to an air gap. Calculate the refractive index of the fiber core.

Ans: 1.5

6.5 Discuss, with the aid of suitable diagrams, the three types of fiber-to-fiber misalignment which may contribute to insertion loss at a joint.

6.6 A multimode step-index fiber with a core refractive index of 1.46 is fusion spliced. However, the joint exhibits an insertion loss of 0.6 dB, which has been found to be entirely due to a 4° angular misalignment of the core axes. Find the numerical aperture of the fiber.

Ans: 0.25

6.7 Discuss, with the aid of suitable diagrams, the following techniques of mechanical splicing: (a) snug tube splice, (b) spring-groove splice, and (c) elastomeric splice.

6.8 A mechanical splice in a multimode step-index fiber has a lateral offset of 12% of the fiber core diameter. The refractive index of the core is 1.5 and an index-matching fluid with a refractive index of 1.47 is inserted in the splice between the two fiber ends. Determine the insertion loss of the splice. Assume that there are no other types of misalignment.

Ans: 0.7144 dB

6.9 Discuss with the aid of suitable diagrams, the design of the following connectors: (a) ferrule connector and (b) expanded-beam connector

6.10 The fraction of light reflected (R) at an air–fiber interface is given by Eq. (6.24). Show that for a small variation in the value of core index n_1 the change in R is given by

$$\Delta R = 4n\left\{\frac{n_1 - n}{(n_1 + n)^3}\right\}\Delta n_1$$

6.11 What is meant by OTDR? Discuss, with the aid of a diagram, how this method may be used in field measurements? In addition, mention the merits of this technique.

Part II: Optoelectronics

7 Optoelectronic Sources

After reading this chapter you will be able to understand the following:

- Fundamental aspects of semiconductor physics
- The *p-n* junction
- Injection efficiency
- Injection luminescence and the light-emitting diode (LED)
- Internal and external quantum efficiencies
- LED designs
- Modulation response of LEDs
- Basics of lasers
- Laser action in semiconductors
- Modulation response of injection laser diodes (ILDs)
- ILD designs
- Source-fiber coupling

7.1 INTRODUCTION

In fiber-optic systems, electrical signals (current or voltage) at the transmitter end have to be converted into optical signals as efficiently as possible. This function is performed by an optoelectronic source. What should be the criterion for selecting such a source? Ideally, the size and shape of the source should be compatible with the size of an optical fiber so that it can couple maximum power into the fiber. The response of the source should be linear; i.e., the optical power generated by the source should be directly proportional to the electrical signal supplied to it. Further, it should provide sufficient optical power so that it overcomes the transmission losses down the link. It must emit monochromatic radiation at the wavelength at which the optical fiber exhibits low loss and/or low dispersion. Finally, it must be stable, reliable, and cheap as far as possible. There are two types of sources which, to a large extent, fulfil these requirements. These are (i) incoherent optoelectronic sources (e.g., light-emitting diodes, LEDs) and (ii) coherent optoelectronic sources (e.g., injection laser diodes, ILDs).

In order to understand the principle of operation, efficiency, and design of these devices, it is essential to be familiar with the properties of semiconductors, *p-n* homojunctions and heterojunctions, the light emission process, etc. This chapter, therefore, begins with the discussion of such fundamental aspects of semiconductor physics, followed by the efficiency and design aspects of light-emitting diodes. The basic principles of laser action and injection laser diodes are taken up next. We conclude with source-fiber coupling.

7.2 FUNDAMENTAL ASPECTS OF SEMICONDUCTOR PHYSICS

According to the band theory of solids, materials may be classified into three categories from the point of view of electrical conduction. These are (i) conductors, (ii) insulators, and (iii) semiconductors. This distinction may be understood with the aid of an energy band diagram. Within a material, the permitted electron energy levels fall into bands of allowed energy as shown in Fig. 7.1. Herein, the vacuum level E_o represents the energy of an electron at rest just outside the surface of the solid. The highest band of allowed energy levels inside the material, which extends from the vacuum level E_o down to energy E_c, is called the conduction band (CB). The energy width of this band, $\chi = E_o - E_c$, is called the electron affinity of the material. The next highest allowed band is known as the valence band (VB). The energy corresponding to the top of the VB is depicted as E_v. These two bands are separated by an energy gap (called the forbidden gap), or a band gap in which no energy levels exist. The energy gap $E_g = E_c - E_v$.

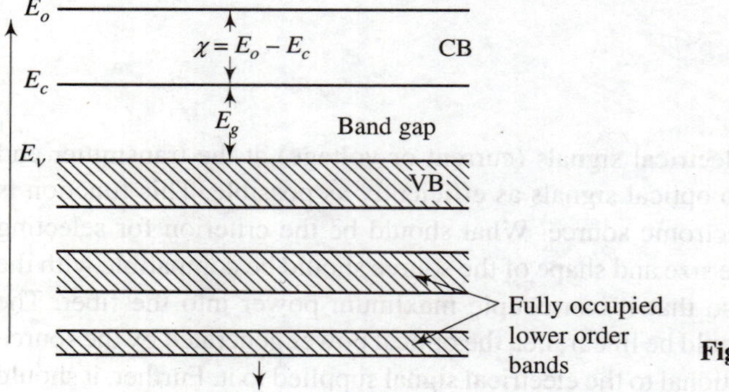

Fig. 7.1 Energy band diagram of a solid

A material with no energy gap between the conduction and valence bands or with overlapping bands is a good conductor. A material with a completely empty CB separated from a completely filled VB by a large band gap is an insulator. If this band gap is small, then the material is a semiconductor. The resistivity of these three classes of materials lies in the following range of values:

Conductor: 10^{-6}–10^{-4} Ω cm
Insulator: 10^{10}–10^{20} Ω cm
Semiconductor: 10^{-2}–10^{8} Ω cm

7.2.1 Intrinsic and Extrinsic Semiconductors

In a pure semiconducting crystal, at absolute zero, the VB is completely filled and the CB is devoid of electrons. However, as the temperature is increased, some electrons from the top of the VB are thermally excited to the lower levels of the CB, thus giving rise to a concentration of n free electrons per unit volume in the CB. This process of electron excitation leaves behind an equal concentration per unit volume, p, of vacancies of electrons in the VB as shown schematically in Fig. 7.2. This vacancy of an electron is called a hole, and it carries a positive charge of magnitude equal to that of an electronic charge. Both the charge carriers, i.e., free electrons and holes are mobile within the material, so both contribute to electrical conductivity. Such semiconductors in which there are equal number of electrons and holes are called intrinsic semiconductors.

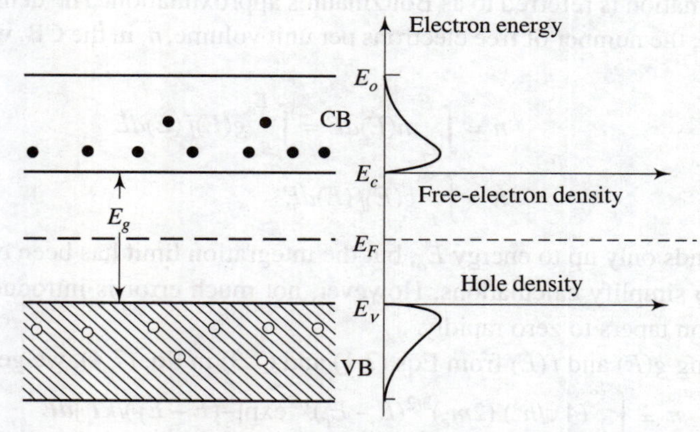

Fig. 7.2 Thermal excitation of electrons from the valence to the conduction band giving rise to concentration distributions of electrons and holes in the CB and VB, respectively. Solid circles represent electrons and hollow circles represent holes.

Let us calculate the density of charge carriers, namely, electrons and holes in an intrinsic semiconductor. For such a calculation, obviously, we need two parameters: (i) the density of states function $g(E)$ which may be defined as the number of energy states per unit energy, per unit volume and (ii) the probability function $f(E)$ that each of these energy states is occupied by an electron. The density of states function $g(E)$ is given by

$$g(E) = (4\pi/h^3)(2m_e)^{3/2}(E - E_c)^{1/2} \qquad (7.1)$$

where h is Planck's constant, m_e is the effective mass of an electron, and E is the energy at which this density is sought.

The probability that a particular energy level at energy E is occupied at a temperature T (K) is given by the Fermi–Dirac distribution function $f(E)$ as follows:

$$f(E) = 1/[\exp\{(E - E_F)/kT\} + 1] \qquad (7.2)$$

where E_F is called the Fermi energy and k is Boltzmann's constant. In Fig. 7.2, a reference energy level (dashed line) in the middle of the band gap has been shown. This is known as the Fermi level and the corresponding energy is E_F. An electron state at E_F, should one exist there, would have a 50% probability of being occupied. The difference of energy between a vacuum level and a Fermi level is called the work function ϕ. Thus $\phi = E_o - E_F$.

Coming to Eq. (7.2), if the lower edge of the CB is about $4kT$ above the Fermi level, i.e., if $E - E_F > 4kT$, we can neglect the unity term in the denominator. Thus Eq. (7.2) may be written as

$$f(E) = \exp[-(E - E_F)/kT] \qquad (7.3)$$

This approximation is referred to as Boltzmann's approximation. The density of free electrons, i.e., the number of free electrons per unit volume, n, in the CB, will then be given by

$$n = \int_{E_c}^{E_o} n(E)dE = \int_{E_c}^{E_o} g(E)f(E)dE$$

or

$$n \approx \int_{E_c}^{\infty} g(E)f(E)dE \qquad (7.4)$$

The CB extends only up to energy E_o, but the integration limit has been extended to ∞ in order to simplify calculations. However, not much error is introduced, as the Fermi function tapers to zero rapidly.

Substituting $g(E)$ and $f(E)$ from Eqs (7.1) and (7.3) in Eq. (7.4), we get

$$n \simeq \int_{E_c}^{\infty} (4\pi/h^3)(2m_e)^{3/2}(E - E_c)^{1/2}\exp[-(E - E_F)/kT]dE$$

Solving this, we get

$$n = 2(2\pi m_e kT/h^2)^{3/2}\exp[(E_F - E_c)/kT] \qquad (7.5a)$$

$$= N_c \exp[(E_F - E_c)/kT] \qquad (7.5b)$$

where $N_c = 2(2\pi m_e kT/h^2)^{3/2}$ is known as the effective density of states in the CB.

Similarly, the density of holes, p, in the VB may be calculated using the integral

$$p = \int p(E)dE \approx \int_{-\infty}^{E_y} g(E)[1 - f(E)]\, dE \qquad (7.6)$$

Here $[1 - f(E)]$ represents the probability of electron states being unoccupied in the VB. In other words, it is the probability of occupation of the states by holes. Now,

$$1 - f(E) = 1 - 1/[\exp\{(E - E_F)/kT\} + 1]$$

$$= \exp[(E - E_F)/kT]/[\exp\{(E - E_F)/kT\} + 1]$$

In the VB, E is lower than E_F and hence the term $\exp[(E - E_F)/kT]$ is much smaller compared to 1 in the denominator. Therefore

$$1 - f(E) \approx \exp[(E - E_F)/kT] \tag{7.7}$$

The density of states function for holes in the VB is given by

$$g(E) = (4\pi/h^3)\,(2m_h)^{3/2}\,(E_v - E)^{1/2} \tag{7.8}$$

where m_h is the effective mass of a hole. Substituting the values of $g(E)$ and $[1 - f(E)]$ from Eqs (7.8) and (7.7), respectively, in Eq. (7.6), we get

$$p = (4\pi/h^3)(2m_h)^{3/2}\int_{-\infty}^{E_v}(E_v - E)^{1/2}\exp[(E - E_F)/kT]dE$$

$$= 2(2\pi m_h kT/h^2)^{3/2}\exp[(E_v - E_F)/kT] \tag{7.9a}$$

$$= N_v \exp[(E_v - E_F)/kT] \tag{7.9b}$$

where $N_v = 2(2\pi m_h kT/h^2)^{3/2}$ is known as the effective density of states in the VB.

On the assumption that, in an intrinsic semiconductor, all the electrons in the CB are obtained from the thermal excitation of the electrons from the VB, we can equate the electron and hole densities:

$$n = p = n_i \tag{7.10}$$

where n_i is called the intrinsic carrier density. Taking the product of n and p by substituting their values from Eqs (7.5) and (7.9), we get

$$n_i^2 = np = N_c N_v \exp[(E_F - E_c - E_F + E_v)/kT]$$

$$= N_c N_v \exp(-E_g/kT) \quad (\text{as } E_c - E_v = E_g) \tag{7.11}$$

Therefore,

$$n_i = n = p = (N_c N_v)^{1/2}\exp(-E_g/2kT)$$

$$= 2(2\pi kT/h^2)^{3/2}\,(m_e m_h)^{3/4}\exp(-E_g/2kT) \tag{7.12}$$

The conduction property of an intrinsic semiconductor may be modified by adding minute quantities of appropriate impurities. Let us take the case of silicon (Si) as an intrinsic semiconductor. Its band gap is 1.1 eV and it is tetravalent. If it is doped with a pentavalent impurity such as phosphorus, P (i.e, P substituting for Si in the crystal structure), then four electrons of P are used for covalent bonding with Si and the fifth loosely bound electron is available for conduction. This generates an occupied level just below the bottom of the CB called the *donor level*. Such dopant impurities which can donate electrons to the CB are called *donors*. This process of doping (Si with P) gives rise to an increase in the free-electron concentration in the CB as shown in Fig. 7.3. Now the majority of charge carriers in the semiconductor are (negative) electrons, and hence it is called an *n-type semiconductor*.

Taking the case of Si again, it is also possible to dope it with a trivalent impurity such as boron (B). In this case, the three electrons of B (substituting for Si) make covalent bonds, and a vacancy of one electron, i.e., a hole, is created. This produces an unoccupied *acceptor level* just above the top of the VB. This level is so called because it accepts electrons from the VB, thereby increasing the hole concentration

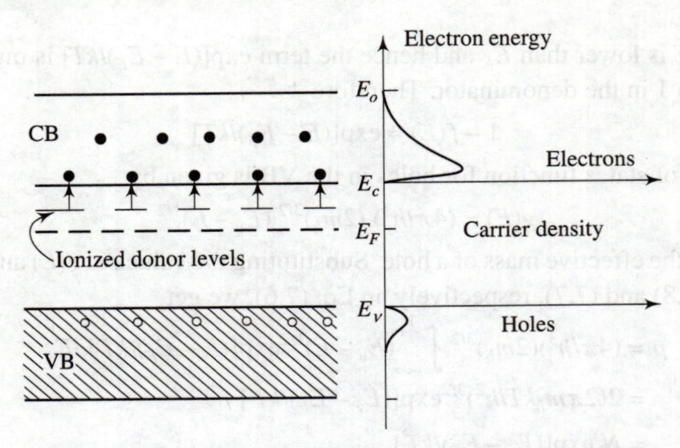

Fig. 7.3 Energy band diagram of an *n*-type semiconductor

in the VB, as shown in Fig. 7.4. The majority of charge carriers are now (positive) holes, and hence it is called a *p-type semiconductor*.

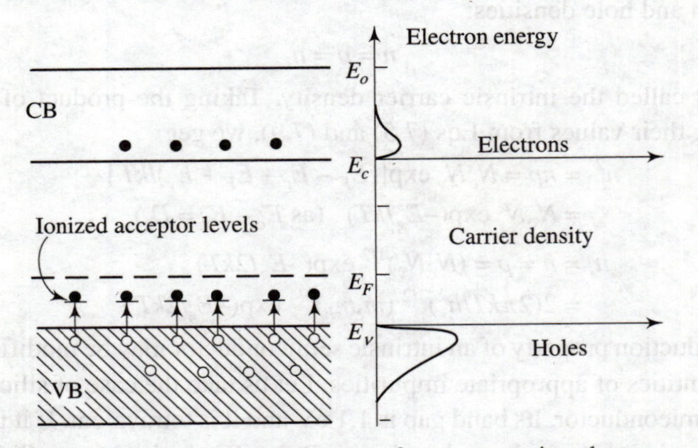

Fig. 7.4 Energy band diagram of a *p*-type semiconductor

The materials which become *n*- or *p*-type after doping are called *extrinsic semiconductors* because in this case, the doping concentration, rather than the temperature, is the main factor determining the number of free charge carriers available for conduction purposes.

As can be seen in Fig. 7.3, the increase in free-electron concentration in the *n*-type material causes the position of the Fermi level to be raised within the band gap. Conversely, the position of the Fermi level is lowered in the *p*-type material (see Fig. 7.4).

If the doping concentrations are not very high, the product of electron and hole densities remains almost independent of the doping concentration. That is,

$$np = n_i^2 = N_c N_v \exp(-E_g/kT) \qquad (7.13)$$

This simply means that in an extrinsic semiconductor, there are *majority carriers* (either electrons in the n-type semiconductor or holes in the p-type material) and *minority carriers* (either holes in the n-type or electrons in the p-type material).

Example 7.1 Calculate the intrinsic carrier concentration in a semiconductor GaAs at room temperature (RT = 300 K) from the following data: $m_e = 0.07m$, $m_h = 0.56m$, $E_g = 1.43$ eV, where m is the mass of an electron in free space.

Solution

$$n_i = 2\left(\frac{2\pi kT}{h^2}\right)^{3/2} (m_e m_h)^{3/4} \exp(-E_g/2kT).$$

$m = 9.11 \times 10^{-31}$ kg, $k = 1.38 \times 10^{-23}$ J K^{-1}, $h = 6.626 \times 10^{-34}$ J s, 1 eV $= 1.6 \times 10^{-19}$ J

Therefore,

$$n_i = 2\left[\frac{2\pi \times 1.38 \times 10^{-23} \times 300}{(6.626 \times 10^{-34})^2}\right]^{3/2} [0.07 \times 0.56 \times (9.11 \times 10^{-31})^2]^{3/4}$$

$$\times \exp\left[\frac{1.43 \times 1.6 \times 10^{-19}}{2 \times 1.38 \times 10^{-23} \times 300}\right]$$

$$= 2.2 \times 10^{12} \text{ m}^{-3}$$

7.3 THE p-n JUNCTION

7.3.1 The p-n Junction at Equilibrium

It is possible to fabricate an abrupt junction between a p-type region and an n-type region in the same single crystal of a semiconductor. Such a junction is called a *p-n junction*. We may assume (though it is not a practice) that this junction has been formed by cementing two isolated pieces of p-type and n-type materials. So when this contact is made, holes from the p-region will diffuse into the n-region, as their concentration is higher in the p-region as compared to the n-region. Similarly, the electrons from the n-region will diffuse into the p-region. The diffusion of holes from the p-region leaves behind ionized acceptors, thereby creating a negative space charge near the junction as shown in Fig. 7.5(a). The diffusion of electrons from the n-region creates a positive space charge near the junction.

This double space charge sets up an internal electric field (directed from the n- to the p-side) in a narrow region on either side of the junction. At equilibrium (that is, with no applied voltages or thermal gradients), it has the effect of obstructing the further diffusion of majority carriers. This induced field establishes a contact or diffusion potential V_D between the two sides and, as a consequence, the energy bands

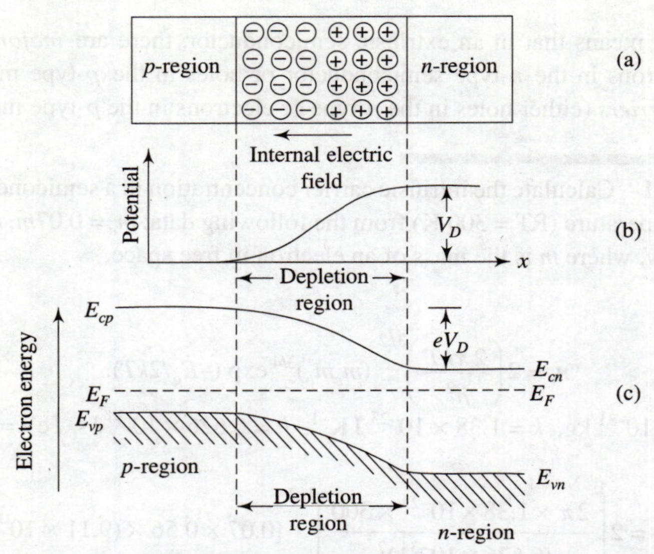

Fig. 7.5 Schematic illustration of (a) the formation of a *p-n* junction, (b) the potential gradient across the depletion region, and (c) the energy band diagram of a *p-n* junction

of the *p*-side are displaced relative to those of the *n*-side as shown in Fig. 7.5(c). The effect of the varying potential is that the region around the junction is almost depleted of its majority carriers. In fact, this region is normally referred to as the *depletion region*. The carrier densities on the two sides of a *p-n* junction, in equilibrium, are shown in Fig. 7.6. The following notation has been used: the equilibrium concentration of majority holes in the *p*-region = p_{p0}, minority electrons in the *p*-region = n_{p0}, majority electrons in the *n*-region = n_{n0}, minority holes in the *n*-region = p_{n0}.

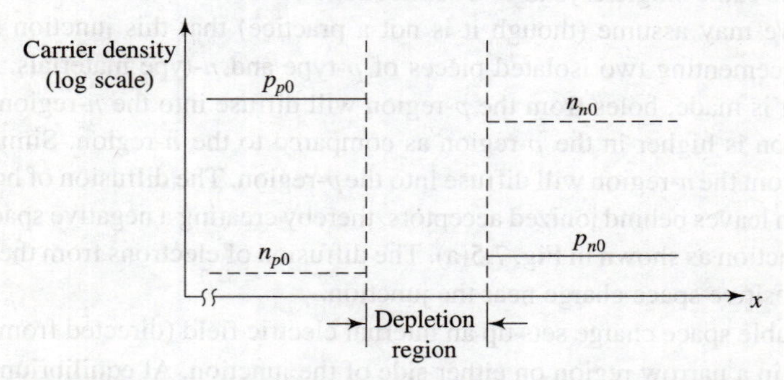

Fig. 7.6 Carrier densities on the two sides of a *p-n* junction, in equilibrium

Employing Eq. (7.5b), a relation between the diffusion potential and the doping concentration may be obtained. Thus, adopting the above notation, and that of Fig. 7.5, we may write the electron concentration in the CB of the *p*-region as

$$n_{p0} = N_c \exp[-(E_{cp} - E_{Fp})/kT] \tag{7.14}$$

Similarly, the electron concentration in the *n*-region will be given by

$$n_{n0} = N_c \exp[-(E_{cn} - E_{Fn})/kT] \tag{7.15}$$

Since the Fermi level is constant in both the regions, in equilibrium, we have $E_{Fp} = E_{Fn} = E_F$ (say).

Therefore, the elimination of N_c gives us

$$n_{p0}/n_{n0} = \exp[(E_{cn} - E_{cp})/kT] \tag{7.16}$$

or

$$E_{cp} - E_{cn} = kT \ln\left(\frac{n_{n0}}{n_{p0}}\right) = eV_D$$

or

$$V_D = (kT/e)\ln(n_{n0}/n_{p0}) \tag{7.17}$$

At normal operating temperature, the majority carrier concentrations are almost equal to the dopant concentrations. Thus, if the acceptor and donor concentrations per unit volume are N_a and N_d, respectively, then $p_{p0} = N_a$ and $n_{n0} = N_d$. From Eq. (7.13), we know that $np = n_i^2$, i.e.,

$$n_{n0}p_{n0} = n_{p0}p_{p0} = n_i^2 \tag{7.18}$$

Therefore we may write

$$n_{p0} = n_i^2/p_{p0} = n_i^2/N_a \tag{7.19}$$

and

$$p_{n0} = n_i^2/n_{n0} = n_i^2/N_d \tag{7.20}$$

Using Eqs (7.18)–(7.20), we may write Eq. (7.17) as

$$V_D = (kT/e)\ln(N_a N_d/n_i^2) \tag{7.21}$$

Equation (7.17) can also be used to express the relationship between the electron concentration on either side of the junction. Thus,

$$n_{p0} = n_{n0}\exp(-eV_D/kT) \tag{7.22}$$

Similarly, one may arrive at the following expression for the hole concentrations on the two sides of the *p-n* junction.

$$p_{n0} = p_{p0}\exp(-eV_D/kT) \tag{7.23}$$

Example 7.2 Consider a GaAs *p-n* junction in equilibrium at room temperature (RT = 300 K). Assume that the acceptor and donor impurity concentrations are 5×10^{23} m^{-3} and 5×10^{21} m^{-3}, respectively. Calculate the diffusion potential V_D.

Solution

It is given that $N_a = 5 \times 10^{23}$ m^{-3} and $N_d = 5 \times 10^{21}$ m^{-3}

$$\frac{kT}{e} = \frac{1.38 \times 10^{-23} \times 300}{1.6 \times 10^{-19}} = 0.025875 \text{ V}$$

From Eq. (7.21), we have

$$V_D = \left(\frac{KT}{e}\right) ln \left(\frac{N_a N_d}{n_i^2}\right)$$

We can take the value of n_i for GaAs from Example 7.1 to be 2.2×10^{12} m^{-3}. Thus,

$$V_D = (0.025875)\lambda n \left[\frac{5 \times 10^{23} \times 5 \times 10^{21}}{(2.2 \times 10^{12})^2}\right]$$
$$= 1.234 \text{ V}$$

7.3.2 The Forward-biased *p-n* Junction

When an external voltage source is connected across a *p-n* junction such that the *p*-side is connected to the positive terminal and the *n*-side is connected to the negative terminal of the voltage source as shown in Fig. 7.7(a), the junction is said to be forward-biased. As the depletion region is very resistive as compared to the bulk region on the two sides, almost all of the applied voltage V appears across this region. This lowers the height of the potential barrier to $V_D - V$ as shown in Fig. 7.7(b). Consequently, the majority carriers are injected into the bulk regions on the opposite sides of the depletion region to become minority carriers there. Thus the minority carrier densities adjacent to the depletion layer rise to new values n_p and p_n, and a concentration gradient of excess minority carriers is established as shown in Fig. 7.8.

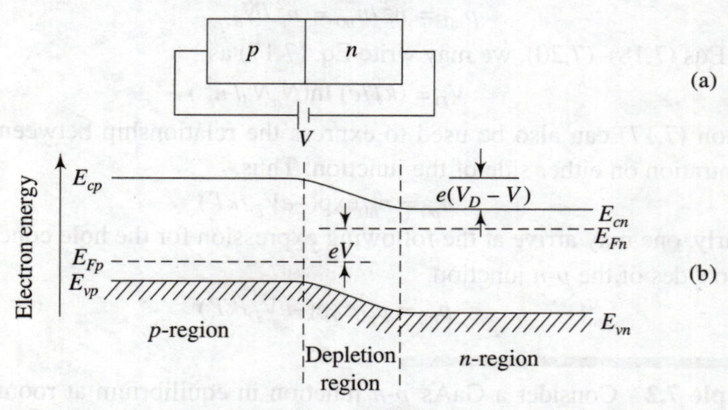

Fig. 7.7 (a) Forward-biased *p-n* junction and (b) energy- level diagram under forward bias (note the splitting of the Fermi level)

The appropriate expressions for the new densities of minority carriers (with forward bias) are given as follows:

$$n_p = n_{n0}\exp[-e(V_D - V)/kT] \tag{7.24}$$

and

$$p_n = p_{p0}\exp[-e(V_D - V)/kT] \tag{7.25}$$

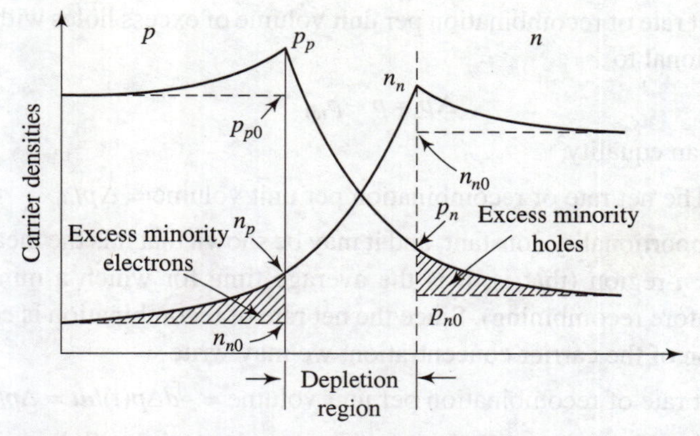

Fig. 7.8 Carrier densities in the p and n bulk regions of a forward-biased p-n junction

These equations may be modified with the aid of Eqs (7.22) and (7.23) to

$$n_p = n_{p0}\exp(eV/kT) \tag{7.26}$$

and

$$p_n = p_{n0}\exp(eV/kT) \tag{7.27}$$

In this non-equilibrium situation, let us denote the local instantaneous values of the densities of free electrons and holes by n and p, respectively (irrespective of the side). Thus, excess hole concentration on the n-side outside the depletion region may be written as

$$\Delta p = p - p_{n0} \tag{7.28}$$

As the bulk regions are supposed to be free of space charge, there will be equal excess concentration of majority electrons on the n-side. That is,

$$\Delta p = p - p_{n0} = n - n_{n0} \tag{7.29}$$

Similarly, the excess concentration of minority electrons on the p-side may be given by

$$\Delta n = n - n_{p0} = p \; p_{p0} \tag{7.30}$$

What happens to the excess minority carriers that are injected by the forward bias? Let us consider the n-region. Here, injected holes diffuse away from the depletion layer and in the process recombine with the excess electrons. The electrons lost in this way are replaced by the external voltage source, so that a current flows in the external circuit. A similar process takes place in the p-region.

7.3.3 Minority Carrier Lifetime

In the bulk region of a forward-biased p-n junction, the net rate of recombination of carriers will be proportional to the local excess carrier concentration. Thus, in the *n*-region, the net rate of recombination per unit volume of excess holes with electrons will be proportional to

$$\Delta p = p - p_{n0}$$

In the terms of an equality,

$$\text{The net rate of recombination per unit volume} = \Delta p / \tau_h$$

where τ_h is a proportionality constant, and it may be shown that it is the mean lifetime of holes in the *n*-region (that is, it is the average time for which a minority hole remains free before recombining). Since the net rate of recombination is equal to the rate of reduction of the carrier concentration, we may write

$$\text{The net rate of recombination per unit volume} = -d\Delta p(t)/dt = \Delta p/\tau_h \quad (7.31)$$

where $\Delta p(t)$ is the excess concentration of minority holes at time t. Solving Eq. (7.31), we get

$$\Delta p(t) = \Delta p(0) \exp(-t/\tau_h) \quad (7.32)$$

where $\Delta p(0)$ is the excess carrier concentration at $t = 0$. Then the mean lifetime of the excess minority holes will be given by

$$\langle t \rangle = \left(\int_0^\infty t \Delta p(0) \exp(-t/\tau_h)\, dt \right) \Bigg/ \left(\int_0^\infty \Delta p(0) \exp(-t/\tau_h)\, dt \right) = \tau_h \quad (7.33)$$

Similarly, in the *p*-region, we can write

$$\text{The net rate of recombination per unit volume} = \Delta n/\tau_e \quad (7.34)$$

where τ_e is the mean lifetime of minority electrons in the *p*-region.

7.3.4 Diffusion Length of Minority Carriers

Consider again a forward-biased *p-n* junction. The net rate of flow of minority holes per unit area due to diffusion in the *n*-region has been found to be proportional to the concentration gradient of holes, that is,

$$\text{The flux of minority holes} = -D_h d(\Delta p)/dx$$

Similarly,

$$\text{The flux of minority electrons in the } p\text{-region} = -D_e d(\Delta n)/dx$$

Here D_h and D_e are the hole and electron diffusion coefficients. These are related to the hole and electron mobilities μ_h and μ_e, respectively, by Einstein's relations:

$$D_e = \mu_e kT/e \quad (7.35)$$

$$D_h = \mu_h kT/e \quad (7.36)$$

Now let us concentrate on the flow of holes that are injected into the *n*-region. Consider an element of thickness Δx and cross-sectional area A at a distance x from the depletion layer edge (as shown in Fig. 7.9 by dashed lines). Then, the net rate at which holes accumulate in the elemental volume ΔxA is given by

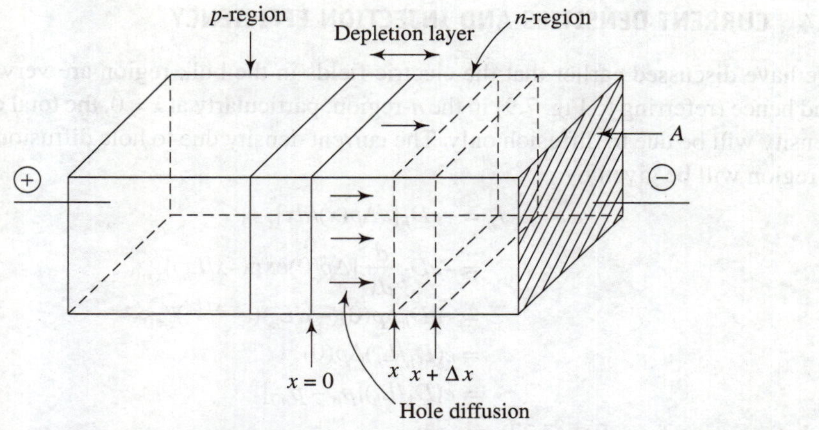

Fig. 7.9 Hole diffusion in the *n*-region of a forward-biased *p-n* junction

$$D_h(d\Delta p/dx)_x A - D_h(d\Delta p/dx)_{x+\Delta x} A = -D_h(d^2\Delta p/dx^2)\Delta x A$$

In the steady state, this rate will be equal to the rate of recombination of excess holes, within this volume. From Eq. (7.31), the rate of recombination in the elemental volume may be written as

$$(-\Delta p/\tau_h)\Delta x A$$

Equating the two rates, we obtain

$$-D_h(d^2\,\Delta p/dx^2)\Delta x A = (-\Delta p/\tau_h)\Delta x A$$

$$d^2(\Delta p)/dx^2 - \Delta p/\tau_h D_h = 0 \tag{7.37}$$

Subject to the boundary conditions $\Delta p = \Delta p(0)$ at $x = 0$ and $\Delta p \to 0$ as $x \to \infty$, Eq. (7.37) may be integrated to give

$$\Delta p(x) = \Delta p(0)\exp(-x/\sqrt{D_h\tau_h}) \tag{7.38a}$$

or

$$\Delta p(x) = \Delta p(0)\exp(-x/L_h) \tag{7.38b}$$

where

$$L_h = \sqrt{D_h\tau_h} \tag{7.39}$$

is known as the *diffusion length* of minority holes in the *n*-region. If we put $x = L_h$ in Eq. (7.38b), we see that $\Delta p(L_h) = \Delta p(0)e^{-1}$. Thus, the hole diffusion length L_h may be defined as that distance inside the *n*-region at which the concentration of minority holes reduces to $1/e$ of its value at the depletion layer edge (i.e., $x = 0$).

Similarly, we can arrive at an expression for the diffusion of electrons in the *p*-region:

$$\Delta n(x') = \Delta n(0)\exp(-x'/L_e) \tag{7.40}$$

where

$$L_e = \sqrt{D_e\tau_e} \tag{7.41}$$

is the diffusion length of minority electrons in the *p*-region.

7.4 CURRENT DENSITIES AND INJECTION EFFICIENCY

We have discussed earlier that the electric fields in the bulk region are very small, and hence (referring to Fig. 7.9) in the n-region, particularly at $x = 0$, the total current density will be due to diffusion only. The current density due to hole diffusion in the n-region will be given by

$$\begin{aligned} J_h &= -eD_h[d\Delta p(x)/dx]_{x=0} \\ &= -eD_h \frac{d}{dx}[\Delta p(0)\exp(-x/L_h)]_{x=0} \\ &= -eD_h\Delta p(0)(-1/L_h)(e^{-x/L_h})_{x=0} \\ &= e(D_h/L_h)\Delta p(0) \\ &= e(D_h/L_h)[p_n - p_{n0}] \end{aligned}$$

Substituting p_n from Eq. (7.27), we get

$$\begin{aligned} J_h &= e(D_h/L_h)[p_{n0}\exp(eV/kT) - p_{n0}] \\ &= e(D_h/L_h)p_{n0}[\exp(eV/kT) - 1] \end{aligned} \tag{7.42}$$

Similarly, we can obtain an expression for the current density J_e due to electron diffusion in the p-region.

$$J_e = e(D_e/L_e)\, n_{p0}[\exp(eV/kT) - 1] \tag{7.43}$$

The total current density crossing the junction would, therefore, be given by

$$\begin{aligned} J &= J_e + J_h \\ &= e(D_e n_{p0}/L_e + D_h p_{n0}/L_h)\,[\exp(eV/kT) - 1] \\ &= J_s[\exp(eV/kT) - 1] \end{aligned} \tag{7.44}$$

where

$$J_s = e(D_e n_{p0}/L_e + D_h p_{n0}/L_h) \tag{7.45}$$

is called the saturation current density. The total diffusion current I flowing across an ideal junction would then be given by

$$\begin{aligned} I &= JA = J_sA[\exp(eV/kT) - 1] \\ &= I_s[\exp(eV/kT) - 1] \end{aligned} \tag{7.46}$$

where $I_s = J_sA$ is the saturation current.

An important case arises when one side is doped more heavily than the other side. This case is represented in Fig. 7.8. Here, the p-side is shown to possess a higher doping level. In such a case, the forward-biased current is mainly carried by the holes injected into the lightly doped n-region. These holes recombine with electrons to emit what is known as *recombination radiation* from the n-side. This device works as an optoelectronic source. Here, we can define the injection efficiency η_{inj} as the ratio of current density due to holes to the total current density. Thus,

$$\eta_{\text{inj}} = \frac{J_h}{(J_e + J_h)} = \frac{1}{1 + J_e/J_h} \tag{7.47}$$

Substituting the values of J_h and J_e from Eqs (7.42) and (7.43), respectively, we may write

$$\eta_{\text{inj}} = 1/[1 + (D_e/D_h)(L_h/L_e)(n_{p0}/n_{n0})]$$
$$= 1/[1 + (D_e/D_h)(L_h/L_e)(N_d/N_a)] \quad (7.48)$$

It is clear from the above equation that, if η_{inj} is to approach unity, the ratio N_d/N_a has to be very small, that is, the acceptor concentration N_a should be much larger than N_d.

A similar expression for injection efficiency may be arrived at if the n-side is doped more heavily than the p-side. In this case,

$$\eta_{\text{inj}} = \frac{J_e}{(J_e + J_h)} = \frac{1}{1 + J_h/J_e}$$
$$= 1/[1 + (D_h/D_e)(L_e/L_h)(N_a/N_d)] \quad (7.49)$$

Example 7.3 Calculate the injection efficiency of a GaAs diode in which $N_a = 10^{23}\ \text{m}^{-3}$ and $N_d = 10^{21}\ \text{m}^{-3}$. Assume that at RT = 300 K, $\mu_e = 0.85\ \text{m}^2\,\text{V}^{-1}\,\text{s}^{-1}$, $\mu_h = 0.04\ \text{m}^2\,\text{V}^{-1}\,\text{s}^{-1}$, and $L_e \approx L_h$.

Solution

It is given that $N_a = 10^{23}\ \text{m}^{-3}$ and $N_d = 10^{21}\ \text{m}^{-3}$. This means that the p-side is doped more heavily as compared to the n-side. Hence, in this case,

$$\eta_{\text{inj}} = \frac{1}{\left[1 + \dfrac{D_e}{D_h}\dfrac{L_h}{L_e}\dfrac{N_d}{N_a}\right]}$$

Now, $$D_e = \mu_e \frac{kT}{e} = \frac{(0.85\ \text{m}^2\,\text{V}^{-1}\text{s}^{-1}) \times (1.38 \times 10^{-23}\,\text{J}\,\text{K}^{-1}) \times (300\ \text{K})}{1.6 \times 10^{-19}\ \text{C}}$$
$$= 0.02199\ \text{m}^2\,\text{J}^{-1}$$

and $$D_h = \mu_h \frac{\text{kT}}{e} = \frac{(0.04\ \text{m}^2\ \text{V}^{-1}\text{s}^{-1}) \times (1.38 \times 10^{-23}\,\text{J}\,\text{K}^{-1}) \times (300\ \text{K})}{1.6 \times 10^{-19}\ \text{C}}$$
$$= 1.035 \times 10^{-3}\ \text{m}^2\,\text{J}^{-1}$$

Therefore, $$\eta_{\text{inj}} = \frac{1}{1 + \dfrac{0.02199}{1.035 \times 10^{-3}} \times \dfrac{1}{1} \times \dfrac{10^{21}}{10^{23}}} = 0.8247$$

7.5 INJECTION LUMINESCENCE AND THE LIGHT-EMITTING DIODE

In the previous section, we have discussed that if a p-n junction diode is forward-biased, the majority carriers from both sides cross the junction and enter the opposite sides. This results in an increase in the minority carrier concentration on the two sides. The excess minority carrier concentration, of course, depends on the impurity levels on the two sides. This process is known as *minority carrier injection*. The injected carriers diffuse away from the junction, recombining with majority carriers as they do so. This recombination process of electrons with holes may be either non-radiative, in which the energy difference of the two carriers is released into the lattice as thermal energy, or radiative, in which a photon of energy equal to or less than the energy difference of the carriers is radiated. The phenomenon of emission of radiation by the recombination of injected carriers is called *injection luminescence*. A *p-n* junction diode exhibiting this phenomenon is referred to as a *light-emitting diode*.

Some probable radiative recombination processes are illustrated in Fig. 7.10. Radiation may be emitted via (i) the recombination of an electron in the CB with a hole in the VB (normally referred to as direct band-to-band transition), shown in Fig. 7.10(a), (ii) the downward transition of an electron in the CB to an empty acceptor level, shown in Fig. 7.10(b), and (iii) the transition of an electron from a filled donor level to a hole in the VB, shown in Fig. 7.10(c).

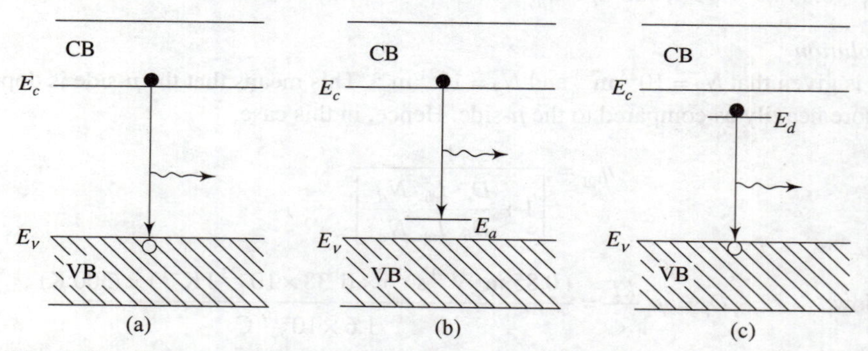

Fig. 7.10 Some probable radiative recombination mechanisms on either the *n*- or the *p*-side as the case may be

7.5.1 Spectrum of Injection Luminescence

What is the spectral distribution of the emitted radiation? In order to simplify things, we assume that radiation is primarily emitted via direct band-to-band transitions. If the transition takes place from the electron level at the bottom of the CB to the hole level at the top of the VB, the emitted photon will have energy

$$E_{ph} = hc/\lambda = E_c - E_v = E_g \tag{7.50}$$

where h is Planck's constant, c is the speed of light, and λ is the wavelength of emitted radiation. However, there is a distribution of electron energy levels in the CB and that of holes in the VB. Thus, depending on the energy levels involved, there will be a range of photon energies that are emitted by the LED.

A simplified calculation (see Review Question 7.4) shows that the spectral distribution of the radiated power P as a function of E_{ph} is given by the following expression:

$$P = \alpha(E_{ph} - E_g)\exp[-(E_{ph} - E_g)/kT] \quad (7.51)$$

where α is a constant. The theoretical plot of relative power versus E_{ph} is shown in Fig. 7.11(a). From this relation it is obvious that the peak power would be emitted at a photon energy $E_{ph} = E_g + kT$ and the full width at half maximum power would be $2.4kT$. However, the observed spectrum of real LEDs is much more symmetrical as shown in Fig. 7.11(b). The wavelength λ of the emitted radiation is given by

$$\lambda = hc/E_{ph} \quad (7.52)$$

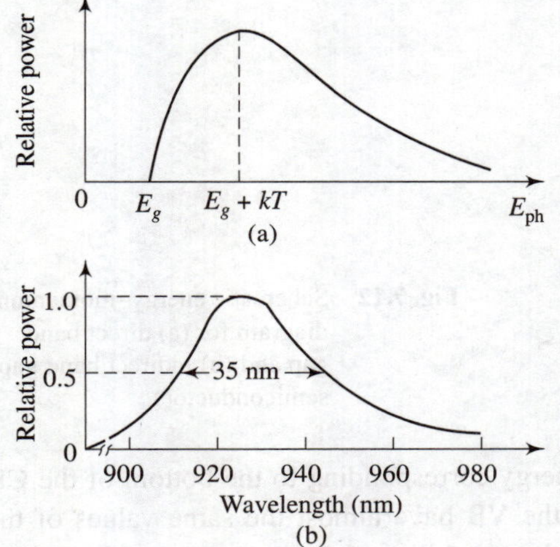

Fig. 7.11 (a) Theoretical spectral power distribution as a function of photon energy. (b) Actual power distribution for a typical GaAs LED.

The spread in wavelength $\Delta\lambda$ may be written as

$$\Delta\lambda = -(hc/E_{ph}^2)\Delta E_{ph} \quad (7.53)$$

and the relative spectral width of the source may be written as

$$\lambda = |\Delta\lambda/\lambda| = \Delta E_{ph}/E_{ph} = 2.4kT/E_{ph} \quad (7.54)$$

This expression leads us to roughly predict the values of γ and $\Delta\lambda$ for LEDs emitting at different wavelengths at room temperature. These are given in Table 7.1.

Table 7.1 Calculated spectral width values of LEDs

λ_{max}(μm)	E_{ph}(eV)	γ	Approx. $\Delta\lambda$ (nm)
0.85	1.455	0.0426	36
1.30	0.952	0.0652	85
1.55	0.798	0.0778	120

7.5.2 Selection of Materials for LEDs

In order to encourage the radiative recombination giving rise to injection luminescence, it is essential to select a proper semiconductor for making an LED. There are two types of semiconducting materials, namely, (i) direct band gap semiconductors and (ii) indirect band gap semiconductors. The energy–momentum diagrams for these two types of materials are shown in Fig. 7.12.

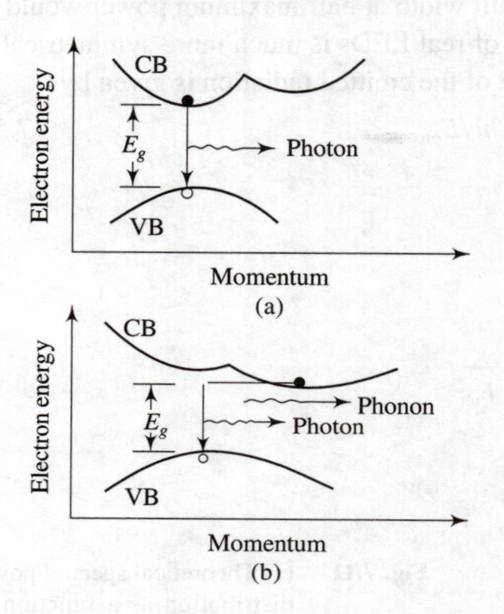

Fig. 7.12 Schematic energy–momentum diagram for (a) direct band gap and (b) indirect band gap semiconductors

In direct band gap materials, the energy corresponding to the bottom of the CB and that corresponding to the top of the VB have almost the same values of the crystal momentum. Thus there is a high probability of the direct recombination of electrons with holes, giving rise to the emission of photons. The materials in this category include GaAs, GaSb, InAs, etc. In indirect band gap materials, the energy corresponding to the bottom of the CB has excess crystal momentum as compared to that corresponding to the top of the VB. Here, the electron–hole recombination requires the simultaneous emission of a photon and a phonon (crystal lattice vibration) in order to conserve the momentum. The probability of such a transition is, therefore, low. The materials in this category include Si, Ge, GaP, etc. These materials are, therefore, not preferred for making LEDs. Among the direct band gap materials, GaAs is the most preferred semiconductor for fabricating LEDs. Its band gap $E_g = 1.43$ eV and it can be doped with *n*- as well as *p*-type impurities. It is also possible to make a heterojunction (to be discussed is the next section) of GaAs with AlAs to prepare a ternary alloy GaAlAs. The band gap of GaAlAs may be varied by varying the percentage of AlAs.

7.5.3 Internal Quantum Efficiency

The internal quantum efficiency η_{int} of an LED may be defined as the ratio of the rate of photons generated within the semiconductor to the rate of carriers crossing the junction. η_{int} will depend, among other things, on the relative probability of the radiative and non-radiative recombination processes. Thus, considering the *n*-side of a forward-biased *p-n* junction, the total rate of recombination of excess carriers per unit volume is given by Eq. (7.31); that is,

$$-dp/dt = -(dp/dt)_{rr} - (dp/dt)_{nr} = \Delta p/\tau_h \tag{7.55}$$

where

$$-(dp/dt)_{rr} = \Delta p/\tau_{rr} \tag{7.56}$$

represents the rate of radiative recombination per unit volume and

$$(-dp/dt)_{nr} = \Delta p/\tau_{nr} \tag{7.57}$$

represents the rate of non-radiative recombination per unit volume. τ_{rr} and τ_{nr} in the above relations are the minority carrier lifetimes for radiative and non-radiative recombinations, respectively. Employing Eqs (7.55)–(7.57), we get

$$1/\tau_h = 1/\tau_{rr} + 1/\tau_{nr} \tag{7.58}$$

Thus the internal quantum efficiency in the bulk *n*-region is given by

$$\eta_{int} = -(dp/dt)_{rr}/[-(dp/dt)_{rr} - (dp/dt)_{nr}] = (1/\tau_{rr})/[1/\tau_{rr} + 1/\tau_{nr}]$$

$$= \frac{1}{1 + \tau_{rr}/\tau_{nr}} \tag{7.59}$$

Therefore, in order to increase η_{int}, the ratio τ_{rr}/τ_{nr} should be as low as possible. Similar arguments hold for the *p*-region. Typically, the ratio τ_{rr}/τ_{nr} for an indirect band gap material, e.g., Si, is of the order of 10^5, whereas that for a direct band gap material, e.g., GaAs, is of the order of unity. Thus, η_{int} for the two cases is, respectively, of the order of 10^{-5} and 0.5.

7.5.4 External Quantum Efficiency

The external quantum efficiency η_{ext} of an LED may be defined as the ratio of the rate of photons emitted from the surface of the semiconductor to the rate of carriers crossing the junction.

In order to determine the order of magnitude of η_{ext} let us look at the configuration of an LED based on a p-n^+ homojunction. This is shown schematically in Fig. 7.13. Here, n^+ denotes that the *n*-region is more heavily doped as compared to the *p*-region so that the current is mainly carried by the electrons, and the injection efficiency is given by Eq. (7.49). On forward-biasing, the electrons cross the junction and reach the *p*-region, where within one or two diffusion lengths (L_e), they recombine with the holes to produce photons. The photons so generated in a thin layer, represented by

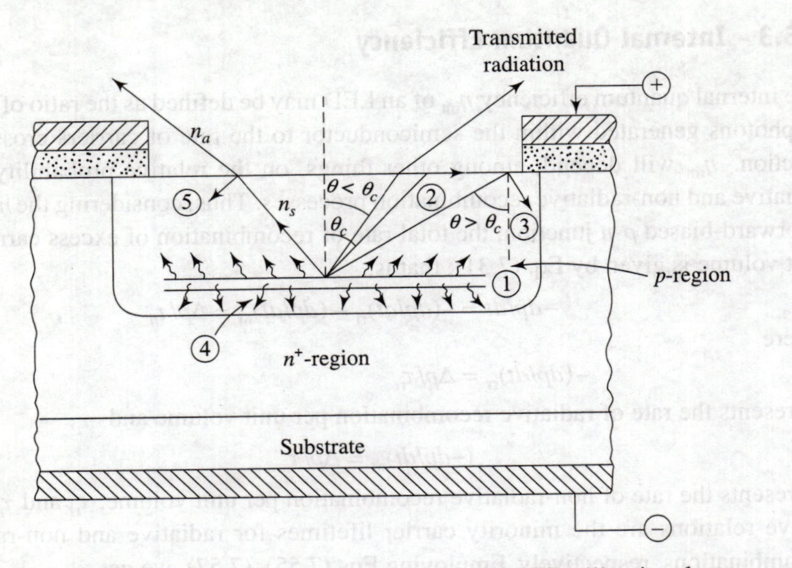

Fig. 7.13 Exploded view of a surface-emitting LED: (1) prime layer generating optical radiation, (2) critical ray, (3) total internal reflection, (4) backside emission, (5) Fresnel reflection

(⇝) in Fig. 7.13, are radiated in all directions. Therefore it behaves like a double-sided Lambertian emitter. (The radiation pattern of a Lambertian source is explained in Appendix A7.1.) Let us assume that the optical power radiated per unit solid angle from the entire emissive area along the normal to the emitting surface is P_0. Then the total radiant power or the flux, ϕ_s, emitted within the semiconductor from both sides of this layer will be given by

$$\phi_s = 2\int_{\theta=0}^{\pi/2} P_0 \cos\theta\,(2\pi)(\sin\theta)d\theta$$

$$\phi_s = 2\pi P_0 \tag{7.60}$$

This entire flux cannot be collected at the surface of the LED. The prime reason for this is that the rays striking the semiconductor–air interface at an angle greater than the critical angle θ_c (for this interface) will be total internally reflected. This is depicted by (2) and (3) in Fig. 7.13. Hence only those rays reaching the emitting surface at an angle of incidence $\theta < \theta_c$ will be transmitted. Further, the radiation emitted towards the backside, depicted by (4) in Fig. 7.13, cannot be collected. Therefore, the fraction F of the total optical power that can be collected at the semiconductor–air surface will be given by

$$F = (1/2\pi P_0)\int_0^{\theta_c} P_0\cos\theta(2\pi)(\sin\theta)d\theta = \sin^2\theta_c/2$$

If n_s and n_a are the refractive indices of the semiconductor and the surrounding medium, respectively,

$$\sin\theta_c = n_a / n_s$$

Hence, $$F = n_a^2 / 2 n_s^2 \quad (7.61)$$

There are two more factors which will further reduce this fraction. First, a small fraction of the light is reflected at the semiconductor–air interface. This is known as Fresnel reflection and is represented by (5) in Fig. 7.13. For normal incidence, the fraction that is reflected is given by the Fresnel reflection coefficient

$$R = [(n_s - n_a)/(n_s + n_a)]^2 \quad (7.62)$$

Therefore, the transmission factor t will be given by

$$t = 1 - R = 4 n_a n_s / (n_s + n_a)^2 \quad (7.63)$$

This factor varies with the angle of incidence. However, this variation is little. The second factor causing loss is the self-absorption of radiation within the semiconductor. This depends on the absorption coefficient of the semiconductor for the wavelength of emission, and the length of traversal inside the semiconductor. However, the effect of this loss mechanism is reduced by keeping the distance of the emitting layer from the surface as short as possible. If we assume that a_s is the fraction of light absorbed within the semiconductor while traversing from the generation layer to the emitting surface, the fraction that is transmitted may be given by

$$T = 1 - a_s \quad (7.64)$$

Thus, combining the factors F, t, and T, we get the external quantum efficiency of the LED:

$$\begin{aligned} \eta_{\text{ext}} &= \eta_{\text{int}} F t T \\ &= \eta_{\text{int}} (1 - a_s)\, 2 n_a^3 / n_s (n_s + n_a)^2 \end{aligned} \quad (7.65)$$

Typically, if we take the case of a GaAs LED emitting in air, then $n_a = 1$ and $n_s = 3.7$, and assuming that $\eta_{\text{ext}} = 0.5$ and $a_s = 0.1$,

$$\eta_{\text{ext}} = 0.5(0.9)2/3.7(4.7)^2 = 0.011$$

This tells us about the low efficiencies that are observed from normal LEDs.

7.6 THE HETEROJUNCTION

In the previous section, our discussion has been centred on LED configuration, which is essentially based on a *p-n* homojunction (i.e., a junction formed by doping the same semiconductor, e.g., GaAs, with *p*- and *n*-type impurity atoms). The efficiency of such a configuration, from the point of view of its application in fiber-optic communication systems, is too low. LEDs with higher efficiencies may be fabricated using what are known as heterojunctions. Such junctions may be formed between two semiconductors which have the same lattice parameters (so that they may be grown together as a single crystal) but different band gaps. For example, a heterojunction may be formed between GaAs and its ternary alloy $Ga_{1-x}Al_x As$. The mole fraction x of AlAs ($E_g = 2.16$ eV) with respect to $GaAs_{1-x}$ ($E_g = 1.43$ eV) determines the band gap of the alloy and the corresponding wavelength of peak emission.

The heterojunction may be employed to sandwich a layer of narrow band gap material, e.g., *n*- or *p*-type GaAs, between layers of wider band gap materials, e.g., *P*- and *N*-type GaAlAs, to form a double-hetero structure (DH). This is shown schematically in Fig. 7.14. When forward-biased, the holes from *P*-GaAlAs are injected into *n*-GaAs, but are prevented from going into *N*-GaAlAs by a potential barrier at J_2. Similarly, the electrons from *N*-GaAlAs are injected into *n*-GaAs but are prevented from going further by the potential barrier at J_1. Thus, a large number of carriers are confined in the central layer of *n*-GaAs, where they recombine to produce optical radiation of wavelength corresponding to the band gap of *n*-GaAs.

As most of the activity takes place in the central layer, it is called an *active layer*.

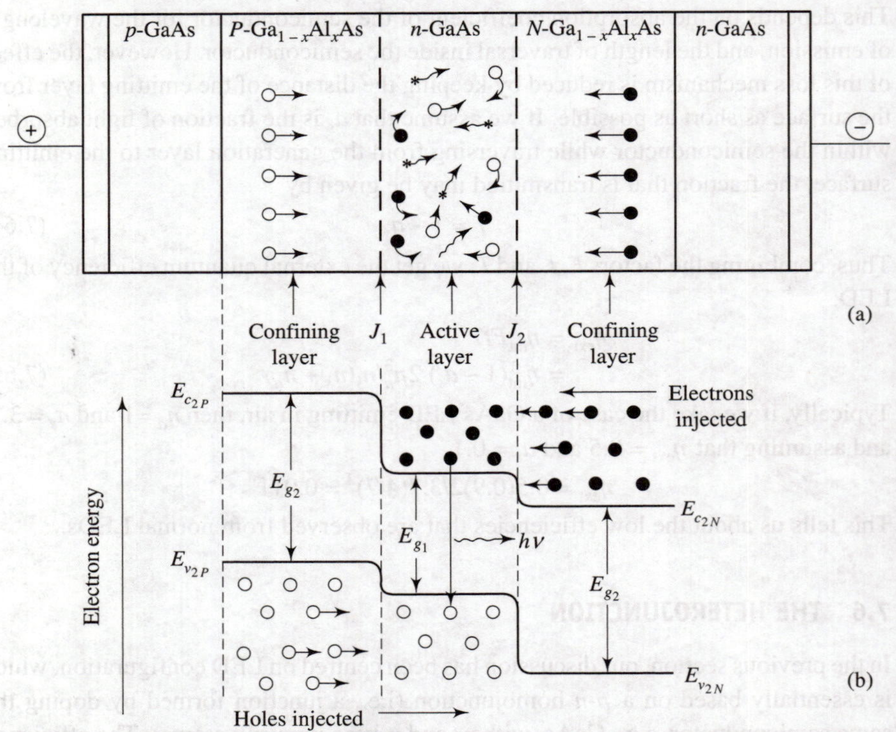

Fig. 7.14 (a) A schematic diagram of a DH LED under forward bias and (b) the corresponding energy-level diagram. Here, *P*- and *N*- denote acceptor and donor impurity doped wider band gap materials. J_1 and J_2 are heterojunctions between *P*- and *n*-type materials and *n*- and *N*-type materials, respectively. Solid circles (●) and hollow circles (○) represent electrons and holes, respectively, and the asterisks (*) denote radiative recombination. E_{g1} and E_{g2} are the band gaps of GaAs and $Ga_{1-x}Al_x$ As, respectively.

The radiation may be collected either through the edge or through one of the surfaces. The corresponding design will be discussed in the next section. An important point to be mentioned here is that such a structure gives rise to a higher rate of radiative recombination and, hence, a brighter LED. Further, the radiation generated by band-to-band transitions in the active layer cannot excite the carriers in the adjoining layers because E_{g1} is lower than E_{g2}. Thus, the confining layers of wider band gap material are transparent to this radiation. This effect may be used in designing surface-emitting LEDs. The limitation of GaAs/$Ga_{1-x}Al_xAs$ based LEDs is that the range of wavelengths (0.80–0.90 μm) emitted by them is outside the wavelength limits of lowest attenuation and zero total dispersion of optical fibers. Therefore, such emitters cannot be used in long-haul communication systems. However, quaternary alloy indium-gallium-arsenide-phosphide/indium-phosphide ($In_xGa_{1-x}As_yP_{1-y}$/InP) based systems have emerged as better candidates for fiber-optic systems, in the sense that a wavelength range of 0.93–1.65 μm can easily be achieved from them. Thus, highly efficient DH LEDs emitting longer wavelengths may be fabricated employing such materials for the active region and InP or quaternary alloys of larger band gaps for the confining layers.

7.7 LED DESIGNS

Two basic structures of LED are in use. These are (i) surface-emitting LED (SLED) and (ii) edge-emitting LED (ELED). Configurations based on GaAs/GaAlAs have been used in short-haul applications, whereas those based on InGaAsP/InP have been employed in medium-range fiber links. Relatively recently, a third device known as a superluminescent diode (SLD) has also been increasingly used in communications. The description of these three types of LEDs, in brief, follows.

7.7.1 Surface-emitting LEDs

When optical radiation emitted in the active layer of a DH shown in Fig. 7.14(a) is taken out from one of the surfaces, the configuration becomes a SLED. A common configuration suitable for fiber-optic communications is shown in Fig. 7.15. It utilizes a *P-n* (or *p*) *-N* planar DH junction. A well is etched into the GaAs substrate layer to avoid reabsorption of light emitted from the substrate side and to accommodate the fiber. It is also called a Burrus-type structure after the scientist who pioneered this design for fiber-optic communications. To increase the carrier density and, hence, recombination rate inside the active region, the light-emitting area is restricted to a small region (typically, a circle of diameter 20–50 μm). This is achieved by confining the injection current to this region through the electrical isolation of the rest of the area by a dielectric (e.g., SiO_2) layer or some other means. The heat generated by the

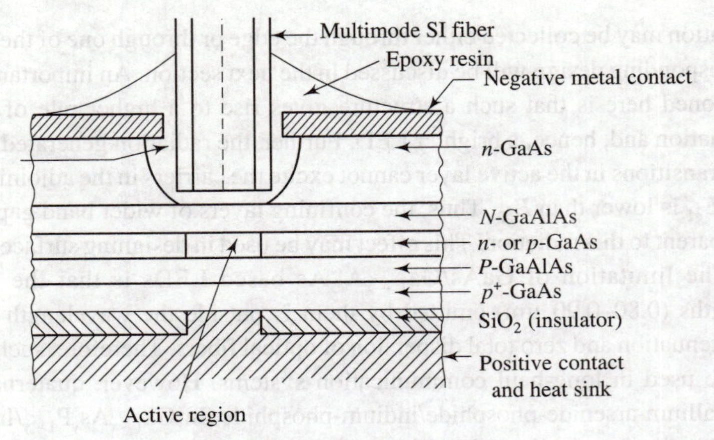

Fig. 7.15 A Burrus-type SLED

device is conducted away by mounting a heat sink near the hot region. A configuration based on GaAs/GaAlAs is shown in Fig. 7.15. A small fraction of AlAs is introduced into the GaAs active layer to tune the wavelength emitted by the device in the range 0.80–0.90 μm. A similar structure based on InP/InGaAsP may be fabricated to emit wavelengths in the range 0.93–1.65 μm. Epoxy resin is used to couple the optical fiber to the emitting surface of the LED. This also reduces the loss due to index mismatch at the semiconductor–air interface.

7.7.2 Edge-emitting LEDs

The DH ELED, shown in Fig. 7.16, is another basic structure providing high radiance for fiber-optic communication. The structure consists of five epitaxial layers of GaAs/GaAlAs. The active layer consists of smaller band gap $Ga_{1-x}Al_xAs$ (here, x is small, typically around 0.1 mole fraction). The positive contact is in the form of a stripe (the rest of the contact being isolated by the SiO_2 layer). The recombination radiation generated in the active region is guided by internal reflection at the heterojunctions and is brought out at the front-end facet of the diode. The rear-end facet is made reflecting while the front-end facet is coated with an antireflection coating, so that the laser action due to optical feedback is suppressed. The self-absorption of radiation in the active layer is reduced because its thickness is made very small. Much of the guided radiation propagates through the confining layers, which have a wider band gap. Therefore, they do not absorb this radiation. An important effect of the optical guidance of emitted radiation is that the output beam has low divergence (typically ~30°) in the vertical direction. This increases the efficiency of coupling the LED with the optical fiber. Stripe geometry ELEDs based on InP/In GaAsP materials and with improved designs for coupling to single-mode fibers have also been made.

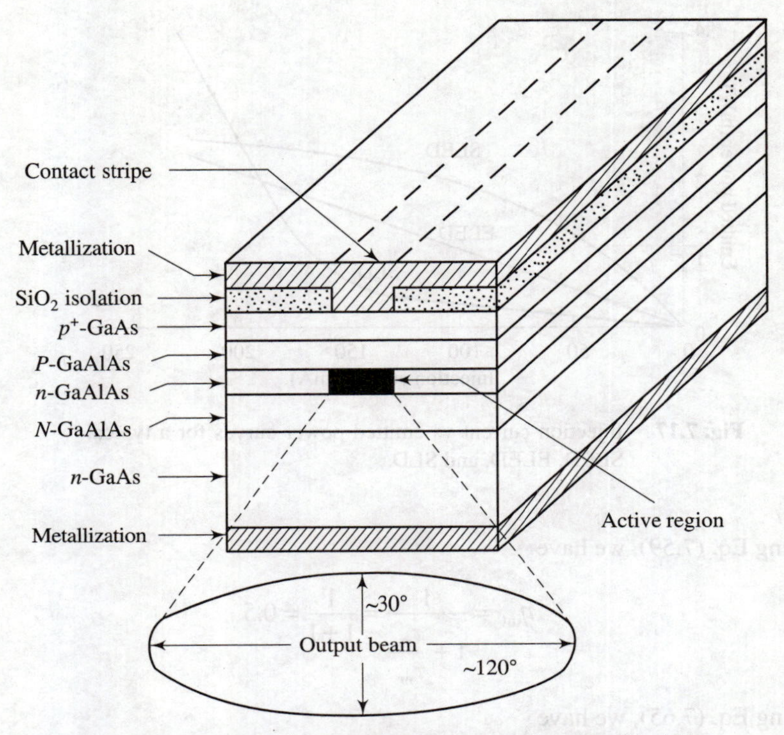

Fig. 7.16 Stripe geometry DH ELED based on GaAs/Ga AlAs

7.7.3 Superluminescent Diodes

The structural features of SLDs are similar to that of DH ELEDs. At low injection current, therefore, an SLD behaves as an ELED, but at high operating current, population inversion similar to that of an injection laser diode (ILD, to be discussed in the next section) is created. Hence this device is able to amplify light, but as it does not have positive feedback, it radiates spontaneous emission. Therefore, an SLD radiates a more powerful and narrower beam than a regular LED. However, its radiation is not coherent. The injection current versus output optical power characteristics for the three types of diodes are shown in Fig. 7.17.

Example 7.4 A Burrus-type *p-n* GaAs LED is coupled to a step-index fiber of core diameter larger than the emitting area of the LED, using transparent bonding cement The refractive indices of the bonding cement and GaAs are, respectively, 1.5 and 3.7. (a) If the mean lifetimes corresponding to radiative and non-radiative recombinations are taken to be the same for GaAs and equal to 100 ns, calculate the internal quantum efficiency of the LED. (b) Calculate the external quantum efficiency, assuming negligible self-absorption within the semiconductor.

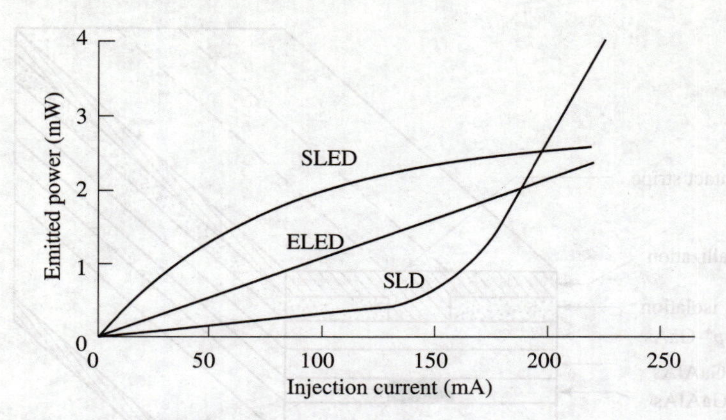

Fig. 7.17 Injection current vs emitted power curves for a typical SLED, ELED, and SLD

Solution

(a) Using Eq. (7.59), we have

$$\eta_{\text{int}} = \frac{1}{1 + \dfrac{\tau_{\text{rr}}}{\tau_{\text{nr}}}} = \frac{1}{1+1} = 0.5$$

(b) Using Eq. (7.65), we have

$$\eta_{\text{ext}} = \eta_{\text{int}}\,(1 - a_s)\,\frac{2\,n_a^3}{n_s\,(n_s + n_a)^2}$$

$$= 0.50(1)\,\frac{2 \times (1.5)^3}{3.7\,(3.7 + 1.5)^2}$$

$$= 0.0337$$

7.8 MODULATION RESPONSE OF AN LED

The modulation response of an LED is governed by the carrier lifetime τ, which represents the total recombination time of charge carriers. It can be defined by the relation

$$\tau = \frac{n}{R_{\text{rr}} + R_{\text{nr}}} \tag{7.66}$$

where n is the charge carrier density, R_{rr} is the rate of radiative recombination, and R_{nr} is the rate of non-radiative recombination. In general, $R_{\text{rr}} = R_{\text{sp}} + R_{\text{st}}$, where R_{sp} is the rate of spontaneous recombination and R_{st} is the rate of stimulated recombination. In the case of an LED, there is no stimulated recombination, and hence $R_{rr} = R_{sp}$.

When the LED is forward-biased, electrons and holes are injected in pairs and they also recombine in pairs. Therefore, in order to study carrier dynamics, the rate equation for one type of charge carriers is enough. We take the case of electrons. The rate equation may be written as follows:

$$\frac{dn}{dt} = \frac{I}{eV} - \frac{n}{\tau} \tag{7.67}$$

where I is the total injected current and V is the volume of the active region.

Let us consider sinusoidal modulation of the LED; that is, the injected current $I(t)$ at time t is given by

$$I(t) = I_0 + I_m \exp(j\omega_m t) \tag{7.68}$$

where the first term, I_0, is the bias current and the second term is the modulation current with I_m as the amplitude and ω_m as the frequency of modulation. Equation (7.67) is a linear differential equation, and hence its solution can be written as

$$n(t) = n_0 + n_m \exp(j\omega_m t) \tag{7.69}$$

where $n_0 = I_0 \tau / eV$ and n_m is given by

$$n_m(\omega_m) = \frac{I_m \tau / eV}{1 + j\omega_m t} \tag{7.70}$$

The corresponding power radiated by the source may be given by

$$P(t) = P_0 + P_m \exp(j\omega_m t) \tag{7.71}$$

The modulated power P_m is linearly related to $|n_m|$. The transfer function $H(\omega_m)$ of an LED may be defined as

$$H(\omega_m) = \frac{n_m(\omega_m)}{n_m(0)} = \frac{1}{1 + j\omega_m \tau} \tag{7.72}$$

The 3-dB (optical) modulation bandwidth of an LED is the modulation frequency at which $|H(\omega_m)|$ is reduced by a factor of 2; that is,

$$(\nu_{3\text{-dB}})_{\text{opt}} = \sqrt{3}\,\frac{1}{2\pi\tau} \tag{7.73}$$

Typical values of $\nu_{3\text{-dB}}$ are in the range of 50–140 MHz. The corresponding electrical bandwidth is given by

$$(\nu_{3\text{-dB}})_{\text{el}} = \frac{1}{2\pi\tau} \tag{7.74}$$

7.9 INJECTION LASER DIODES

Laser is an acronym for light amplification by stimulated emission of radiation. In order to understand the configuration of devices based on laser action, it is essential to know the basic processes governing the absorption and spontaneous and stimulated emission of radiation, first in the simplest atomic system and then in the very complex semiconducting materials.

To begin with, let us consider a hypothetical system consisting of atoms of only two energy levels with energies E_1 and E_2. When a photon of energy $E_2 - E_1 = h\nu$, where h is Planck's constant and ν is the frequency, interacts with an atom of such a system, there exist two possibilities: (i) If the atom is in the ground state, with energy E_1, the photon may be absorbed so that it is excited to the upper level of energy E_2. Subsequent de-excitation may give rise to the emission of radiation in a random manner. This is called *spontaneous emission* and is shown in Fig. 7.18(a). (ii) If the atom is already in the excited state, then the incident photon may stimulate a downward transition with the emission of radiation. Photons emitted in such a manner have been found to be coherent with the stimulating photon; that is, both the stimulating and the stimulated photons have the same energy, same momentum, and same state of polarization. This phenomenon is called *stimulated emission* and is depicted in Fig. 7.18(b).

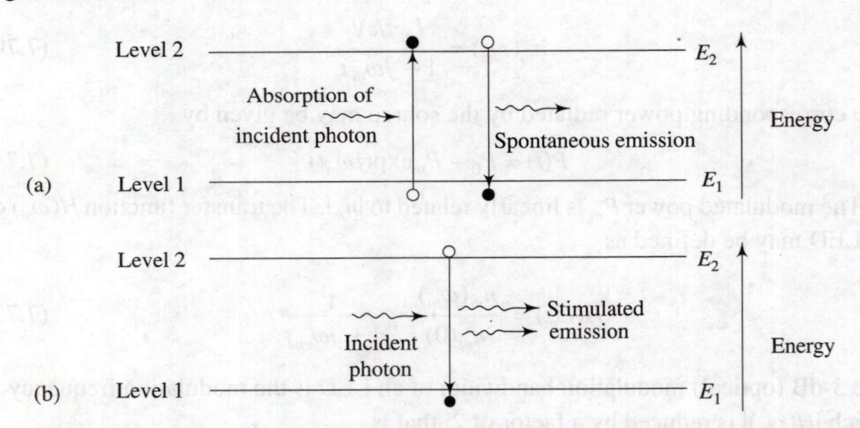

Fig. 7.18 (a) Absorption and spontaneous emission and (b) stimulated emission in a two-level atomic system. Hollow circles (○) and filled circles (●) depict the initial and final states of transitions.

Let us assume that the collection of atoms in a system is in thermodynamic equilibrium; then it must give rise to radiation identical to black-body radiation. The radiation density per unit range of spectral frequency about the frequency ν is then given by

$$\rho_\nu = (8\pi h\nu^3/c^3)\,[1/\{\exp(h\nu/kT) - 1\}] \tag{7.75}$$

where c is the speed of light, k is Boltzmann's constant, and T is the temperature (in kelvin).

Now if the population densities of atoms whose electrons at any instant are at energy levels E_1 and E_2 are N_1 and N_2, respectively, then employing Boltzmann statistics one can show that

$$N_1/N_2 = (g_1/g_2)\exp[(E_2 - E_1)/kT] = (g_1/g_2)\exp(h\nu/kT) \quad (7.76)$$

where g_1 and g_2 are the degeneracies of levels 1 and 2, respectively, that is, g_1 and g_2 are the number of sublevels within the energy levels E_1 and E_2, respectively. Since the atom can absorb a photon only when it is in level *1*, the rate of absorption of a photon by such a system will be proportional to the population density N_1 at energy level E_1 and also to the radiation density ρ_ν. Then the rate of absorption (or the rate of upward transition) may be given by $B_{12} N_1 \rho_\nu$, where B_{12} is the proportionality constant. The downward transition may occur either via spontaneous emission or via stimulated emission. Spontaneous emission is a random process and hence its rate is proportional to the population density N_2 at energy level E_2, whereas stimulated emission requires the presence of an external photon, and hence the rate of stimulated emission is proportional to N_2 as well as ρ_ν. The total rate of downward transition is the sum of the rates of spontaneous and stimulated emissions, which is given by $A_{21}N_2 + B_{21}N_2\rho_\nu$. The constants A_{21}, B_{21}, and B_{12} are called *Einstein's coefficients*. They denote, respectively, the probabilities of spontaneous emission, stimulated emission, and absorption. The relation among these constants may be established as follows.

Under thermodynamic equilibrium, the rate of upward transitions (from level 1 to 2) must equal the rate of downward transitions (from level 2 to 1) (Einstein 1917). Thus, we must have

$$B_{12}N_1\rho_\nu = A_{21}N_2 + B_{21}N_2\rho_\nu \quad (7.77)$$

or $\rho_\nu[B_{12}N_1 - B_{21}N_2] = A_{21}N_2$

or $\rho_\nu[(B_{12}/B_{21})(N_1/N_2) - 1] = A_{21}/B_{21}$

or $\rho_\nu = (A_{21}/B_{21})/[(B_{12}/B_{21})(N_1/N_2) - 1]$

Substituting the value of N_1/N_2 from Eq. (7.76) in the above expression, we get

$$\rho_\nu = (A_{21}/B_{21})/[(g_1B_{12}/g_2B_{21})\exp(h\nu/kT) - 1] \quad (7.78)$$

Comparing Eqs (7.75) and (7.78), we see that for the two equations to be valid we must have

$$B_{12} = (g_2/g_1)B_{21} \quad (7.79)$$

and

$$A_{21}/B_{21} = 8\pi h\nu^3/c^3 \quad (7.80)$$

Equations (7.79) and (7.80) are called *Einstein's relations*.

If the degeneracies of the two levels are equal, i.e., $g_1 = g_2$, then $B_{12} = B_{21}$, which means that the probabilities of absorption and stimulated emission are equal.

Now let us find the ratio of the rate of stimulated emission to that of spontaneous emission for our simplified two-level system. Equation (7.80) enables us to show that

$$\text{Rate of stimulated emission/rate of spontaneous emission} = B_{21}\rho\nu/A_{21}$$
$$= 1/[\exp(h\nu/kT) - 1] \quad (7.81)$$

An incandescent lamp operating at a temperature of 2000 K would emit a peak wavelength $\lambda = 1.449$ μm and a corresponding frequency $\nu = c/\lambda = 3 \times 10^8/1.449 \times 10^{-6} = 2.07 \times 10^{14}$ Hz. The ratio of the rate of stimulated emission to that of spontaneous emission for this frequency would be equal to

$$1/[\exp(6.626 \times 10^{-34} \times 2.07 \times 10^{14}/1.381 \times 10^{-23} \times 2000) - 1] = 7.02 \times 10^{-3}$$

This result simply shows that in an atomic system under thermodynamic equilibrium, spontaneous emission is a dominant mechanism.

In order to produce stimulated emission, it is essential to create a non-equilibrium situation in which the population of atoms in the upper energy level is greater than that in the lower energy level, that is, $N_2 > N_1$. This non-equilibrium condition is called *population inversion.* Now, to achieve this condition we need to excite atoms from the lower to the upper level by some external means. This process of excitation is known as *pumping*. Normally an external source of intense radiation is employed for pumping. However, in semiconductor lasers, electrical excitation is used.

7.9.1 Condition for Laser Action

In a two-level atomic system that is pumped externally, stimulated emission cannot become a dominant process because it has to compete with stimulated absorption. Therefore, either three-level or four-level atomic systems are used for achieving laser action. These are shown in Figs 7.19(a) and 7.19(b).

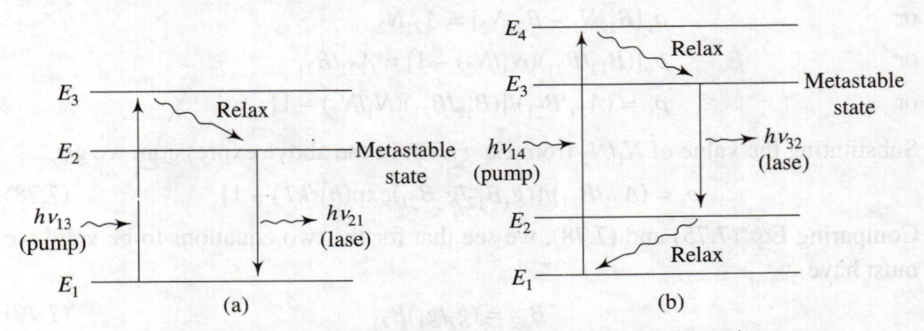

Fig. 7.19 Laser action in a (a) three-level and (b) four-level atomic system

Let us first consider the three-level system shown in Fig. 7.19(a). Assume that this system is pumped with light of photon energy $h\nu_{13}$, so that a large number of atoms in the ground state absorb this radiation and are raised to the excited state E_3. Let us choose this atomic system such that the transition from E_3 to E_2 is faster and preferably non-radiative and that from E_2 to E_1 is much slower. The result of this will be that once the system is pumped, atoms will accumulate in level E_2, which is called the metastable level. Hence, unlike the two-level system, atoms in level E_2 are immune to getting stimulated by photons of energy $h\nu_{13}$. Thus we can increase the number of atoms in level E_2 at the expense of those in level E_1 by increasing the intensity of the exciting radiation at $h\nu_{13}$. This means that we can make $N_2 > N_1$, i.e., achieve population inversion by pumping at frequency ν_{13}.

After population inversion has been achieved, if photons of energy $h\nu_{21}$ corresponding to the energy difference $E_2 - E_1$ (such as those produced by the spontaneous transition from E_2 to E_1) are released into this system, they will stimulate the downward transition from E_2 to E_1, producing more photons at energy $h\nu_{21}$ in the process. Thus the system acts as an optical amplifier.

It is obvious that in order to achieve population inversion, more than half the atoms from the heavily populated ground state must be excited by the pump. It is indeed a hard work for the pump to excite all these atoms. Let us now consider the four-level system shown in Fig. 7.19(b). Here, the system is pumped by photons of energy $h\nu_{14}$ corresponding to the energy difference between levels E_1 and E_4. The absorption of such photons excites atoms to E_4, from where they quickly relax (through non-radiative decay) to level E_3 (metastable or lasing state). The transition from E_3 to E_2 is radiative but slow and the transition from E_2 to E_1 is again fast and non-radiative. In this scheme, it is relatively easy to provide level E_3 with an inverted population over level E_2 because (i) E_2 is not well populated in the first place (by virtue of being above the ground state) and (ii) atoms do not accumulate there, as they quickly relax to the ground state. Hence it is quite easy to ensure that the population in level E_3 exceeds that in level E_2. The amplification at $h\nu_{32}$ is much more efficient, and hence a four-level system is a better amplifier.

Now let us assume that this assembly of atoms (two-level, three-level, or four-level atomic system) exists in a medium which we now call an active medium. Further, assume that this medium is in the form of a cylinder of length L whose axis is along the z-axis. Also assume that this system is appropriately (depending on the atomic system in the active medium) pumped so that population inversion has been achieved. Under this condition, if a beam of light corresponding to the difference of energy between the lasing levels ($E_2 - E_1$ for the two- and three-level systems and $E_3 - E_2$ for the four-level system) is allowed to pass through the medium in the z-direction, the power P in the beam will grow as it passes through the medium according to the relation

$$P_z = P_0 \exp(g_{21} z) \tag{7.82}$$

where P_0 is the incident power and P_z is the power at a distance of z along the axis. g_{21} is the gain coefficient (for the two- and three-level system). For a four-level system it can be represented as g_{32}.

This is laser action. In fact, the laser is more analogous to an oscillator than an amplifier, and hence it is necessary to provide some positive feedback to turn this optical amplifier into an optical oscillator. This is done by placing the active medium between a pair of mirrors, which reflect the amplified light back and forth to form an optical cavity as shown in Fig. 7.20. This is also called a Fabry–Perot resonator. It has a set of characteristic resonant frequencies. Therefore, the radiation is characteristic of these frequencies rather than the normal emission spectrum of the atomic system. Under equilibrium, the optical power loss (which includes the transmission loss at the mirrors) during one round trip through the active medium just balances the gain. Thus the self-oscillation will start only after the gain exceeds the losses.

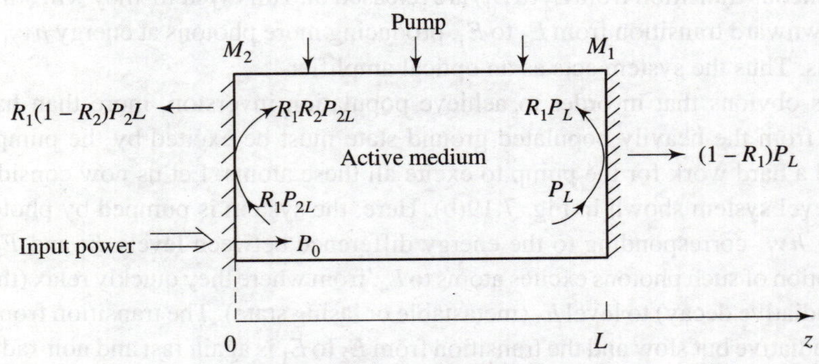

Fig. 7.20 An optical cavity

The total loss of optical power is due to a number of different processes, one of which is the transmission at the mirrors and forms the output beam of the laser. To simplify things, let us represent all the losses except the transmission losses at the mirrors by a single effective loss coefficient α_{eff}. This reduces the effective gain coefficient to $g_{21} - \alpha_{\text{eff}}$. We assume that the active medium fills the space between the mirrors M_1 and M_2, which have reflectivities R_1 and R_2 and a separation L. Then the optical power in the beam will vary with distance according to the following expression:

$$P_z = P_0 \exp[(g_{21} - \alpha_{\text{eff}})z] \tag{7.83}$$

Then, in travelling from M_2 (at $z = 0$) to M_1 (at $z = L$), the power in the beam will increase to

$$P_L = P_0 \exp[(g_{21} - \alpha_{\text{eff}})L] \tag{7.84}$$

At the mirror M_1, a fraction R_1 of the incident power is reflected, and hence the power in the reflected beam will be R_1P_L, and after a complete round trip (i.e., after traversing back to M_2 and suffering a reflection there), the power will be

$$R_1R_2P_{2L} = R_1R_2P_0 \exp[(g_{21} - \alpha_{\text{eff}})2L] \tag{7.85}$$

Therefore, the gain in power G in one round trip of the active medium will be

$$G = \frac{R_1 R_2 P_0 \exp[(g_{21} - \alpha_{\rm eff})\, 2L]}{P_0} = R_1 R_2 \exp[2(g_{21} - \alpha_{eff})L] \qquad (7.86)$$

For sustained oscillations, G must be greater than unity. We may, therefore, write the threshold condition for laser action as

$$G = R_1 R_2 \exp[2(g_{\rm th} - \alpha_{\rm eff})L] = 1 \qquad (7.87)$$

where $g_{\rm th}$ is the threshold gain coefficient. From this equation, we may arrive at an expression for the threshold gain coefficient as follows

$$g_{\rm th} = \alpha_{\rm eff} + (1/2L)\ln(1/R_1 R_2) \qquad (7.88)$$

Here, the first term represents the losses in the volume of the cavity and the second term gives the loss in the form of a useful output.

7.9.2 Laser Modes

The oscillations are sustained in the optical cavity over a narrow range of frequencies for which the gain is sufficient to overcome the net loss. Thus, the output of the cavity is not perfectly monochromatic (i.e., consisting of a single frequency) but is a narrow band of frequencies centred around that corresponding to the energy difference between the levels involved in stimulated emission as shown in Fig. 7.21(a). The

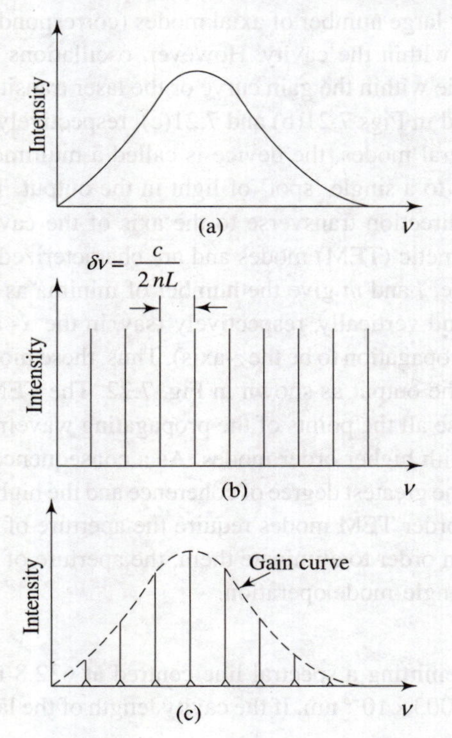

Fig. 7.21 (a) The gain curve or the broadened laser transition line, (b) possible axial modes, and (c) axial modes in the laser output

radiation in the form of electromagnetic waves emitted along the axis of the cavity forms a standing-wave pattern between the two mirrors. The condition for reinforcement of the waves, or resonance, is that the optical path length L between the mirrors must be an integral multiple of the half-wavelength of the waves in the active medium. Hence all the waves which satisfy the condition given below will form standing-wave patterns:

$$L = m\lambda/2n = mc/2n\nu \tag{7.89}$$

where λ is the wavelength in vacuum, n is the refractive index of the active medium, m is an integer, c is the speed of light, and ν is the frequency of the wave. Thus the resonant frequencies may be given by the expression

$$\nu = mc/2nL \tag{7.90}$$

These frequencies corresponding to different integer values of m and are also known as axial or longitudinal modes of the cavity.

From Eq. (7.90) we can obtain the frequency separation between the adjacent modes ($\delta m = 1$) as follows:

$$\delta\nu = c/2nL \tag{7.91}$$

Since $\delta\nu$ is independent of m, the mode separation is the same, irrespective of the actual mode frequencies.

Equation (7.90) indicates that a very large number of axial modes (corresponding to all values of m) may be generated within the cavity. However, oscillations are sustained only for those modes which lie within the gain curve or the laser transition line. These two situations are illustrated in Figs 7.21(b) and 7.21(c), respectively.

As the laser output consists of several modes, the device is called a multimode laser. All these axial modes contribute to a single 'spot' of light in the output. The resonant modes may be formed in a direction transverse to the axis of the cavity. These are called transverse electromagnetic (TEM) modes and are characterized by two integers l and m; e.g., TEM_{lm}. Here, l and m give the number of minima as the output beam is scanned horizontally and vertically, respectively (say, in the x- and y-direction, assuming the direction of propagation to be the z-axis). Thus, these modes may give rise to a pattern of spots in the output as shown in Fig. 7.22. The TEM_{00} mode is called a uniphase mode because all the points of the propagating wavefront are in phase. However, this is not so with higher order modes. As a consequence, a laser operating in the TEM_{00} mode has the greatest degree of coherence and the highest spectral purity. Oscillations of higher order TEM modes require the aperture of the cavity to be large enough. Therefore, in order to eliminate them, the aperture of the cavity is suitably narrowed down for single-mode operation.

Example 7.5 A typical gas laser is emitting a spectral line centred at 632.8 nm, whose gain curve has a half-width of 3.003×10^{-3} nm. If the cavity length of the laser

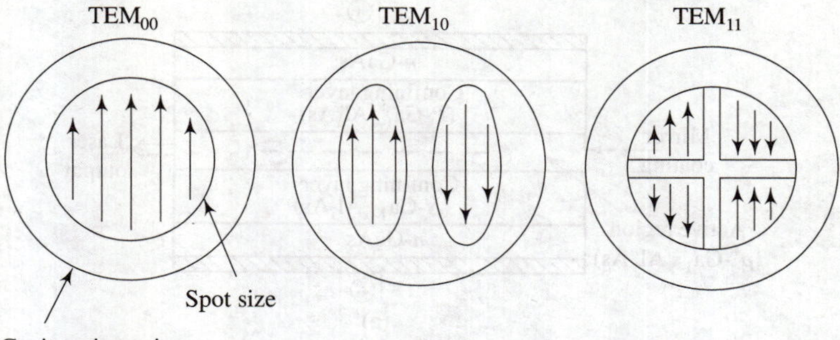

Fig. 7.22 Lower order TEM$_{lm}$ modes of a laser. Arrows represent the direction of the electric-field vectors within the light spot in the output beam

is 20 cm, calculate the number of longitudinal modes excited. Take the refractive index inside the gas medium to be 1.

Solution

Using Eq. (7.89), we have

$$m = \frac{2nL}{\lambda}$$

Therefore the separation $\Delta\lambda$ of the spectral lines between adjacent longitudinal modes can be calculated as follows:

$$\Delta m = 1 = \frac{1}{\lambda^2} 2nL\Delta\lambda$$

or $$\Delta\lambda = \frac{\lambda^2}{2nL}$$

Here, $\lambda = 632.8 \times 10^{-9}$ m, $n = 1$, and $L = 20 \times 10^{-2}$ m. Therefore,

$$\Delta\lambda = \frac{(632.8 \times 10^{-9})^2}{2 \times 1 \times 20 \times 10^{-2}}$$
$$= 1.001 \times 10^{-12} \text{ m}$$
$$= 1.001 \times 10^{-3} \text{ nm}$$

The separation $\Delta\lambda$ of the spectral lines is one-third of the half-width of the gain curve, and hence three longitudinal modes can be excited.

7.9.3 Laser Action in Semiconductors

Laser action in semiconductors may be achieved by forming an optical cavity in the active region of a DH, shown in Fig. 7.23(a). The configuration is analogous to a broad-area DH LED, with the difference that the end faces of the crystal forming the

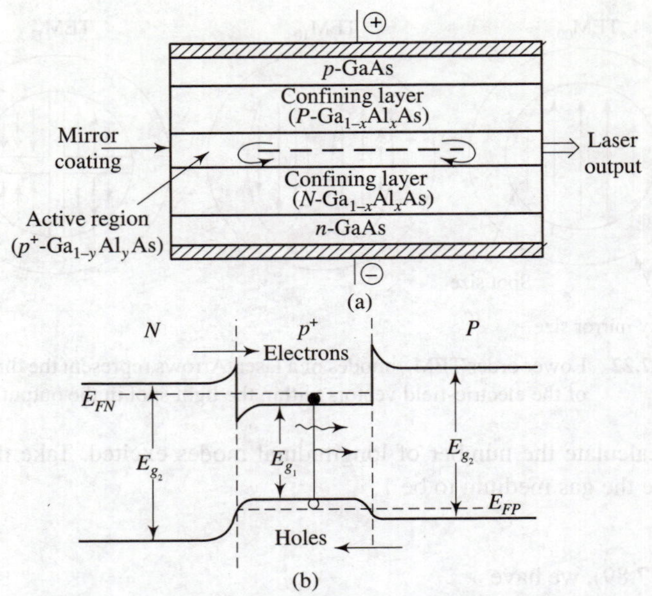

Fig. 7.23 (a) A DH laser diode. (b) Energy-level diagram of confining and active layers under heavy forward bias. p^+ denotes that the active region is a heavily doped p-type material.

active region along the longitudinal direction are cleaved so that they act as mirrors. In some cases, a mirror coating is deposited on one side to make the reflectivity nearly unity. When the device is strongly forward-biased, the energy levels of the different regions of the N-p^+-P DH typically take the form shown in Fig. 7.23(b). The electrons are injected from the N-region into the CB of the p-region and the holes are injected from the P-region into the VB of the p-region. As a result, population inversion occurs corresponding to the transition between those levels for which the photon energy is greater than the band gap E_{g_1} of the active region, but is less than the energy difference between the quasi-Fermi levels E_{FN} and E_{FP}. The refractive index of the active region is greater than that of the confining layer, and hence optical confinement is provided in the transverse direction, but optical feedback is provided in the longitudinal direction. Thus the laser action takes place along the longitudinal direction as in Fig. 7.23(a).

The gain coefficient g_{th} given by Eq. (7.88) for a two-level system will be modified in this case because only a fraction Γ of the optical power that lies within the active region can participate in stimulated emission. This parameter Γ is called the *confinement factor* and it causes the condition for the lasing threshold (i.e., when the gain just exceeds the total loss) to be given by

$$\Gamma g_{th} = \alpha_{eff} + (1/2L)\ln(1/R_1R_2) \tag{7.92}$$

where L is the length of the active region.

The external quantum efficiency of the laser diode is measured in terms of the differential quantum efficiency η_D, which is defined as the number of photons emitted per radiative electron–hole pair recombination above the lasing threshold. If the gain coefficient is assumed to be constant above the threshold, then η_D may be given by (Kressel & Butler 1977)

$$\eta_D = \eta_{\mathrm{int}}(g_{\mathrm{th}} - \alpha_{\mathrm{eff}})/g_{\mathrm{th}} \tag{7.93}$$

where η_{int} is the internal quantum efficiency of stimulated emission.

Experimentally, η_D is calculated from the slope of the curve for emitted optical power or the flux ϕ as a function of drive current I above the threshold current I_{th}. This is also sometimes called the *slope efficiency*. The curve is shown in Fig. 7.24.

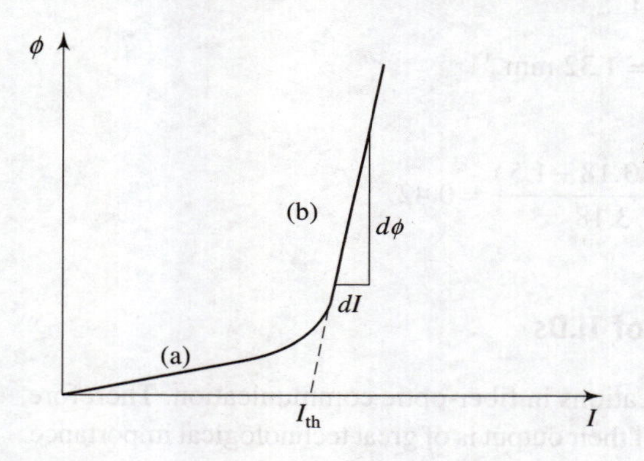

Fig. 7.24 Curve illustrating output optical power as a function of ILD drive current. (a) At low current, the optical output is a spontaneous LED-type emission, (b) above threshold, I_{th}, the output radiation is dominated by stimulated emission.

Thus,

$$\eta_D = (e/E_{g_1})\,(d\phi/dI) \tag{7.94}$$

where e is the electronic charge and E_{g_1} is the band gap of the semiconductor used in the active region. η_D of the order of 15–20 % is common in standard ILDs. Higher efficiencies are possible with improvement in designs.

Example 7.6 A DH GaAs/GaAlAs ILD has a cavity length of 0.5 mm, an effective loss coefficient α_{eff} of 1.5 mm^{-1}, a confinement factor Γ of 0.8, and uncoated facet reflectivities of 0.35. (a) Calculate the reduction that occurs in the threshold gain coefficient when the reflectivity of one of the facets is increased to 1. (b) In the latter case, if the internal quantum efficiency of stimulated emission is 0.80, calculate the differential quantum efficiency of the device.

Solution

(a) Using Eq. (7.92), we have

$$g_{\mathrm{th}} = \frac{1}{\Gamma}\left[\alpha_{\mathrm{eff}} + \frac{1}{2L}\ln\left(\frac{1}{R_1 R_2}\right)\right]$$

Thus, with $R_1 = R_2 = 0.35$,

$$(g_{th})_1 = \frac{1}{0.80}\left[1.5 + \frac{1}{2 \times 0.5}\, ln\left(\frac{1}{0.35 \times 0.35}\right)\right]$$

$$= 4.50 \text{ mm}^{-1}$$

and with $R_1 = 0.35$ and $R_2 = 1.0$

$$(g_{th})_2 = \frac{1}{0.80}\left[1.5 + \frac{1}{2 \times 0.5}\, ln\left(\frac{1}{0.35 \times 1}\right)\right]$$

$$= 3.18 \text{ mm}^{-1}$$

Therefore, $\Delta g_{th} = (g_{th})_1 - (g_{th})_2 = 1.32 \text{ mm}^{-1}$

(b) Using Eq. (7.93), we get

$$\eta_D = \frac{0.80(3.18 - 1.5)}{3.18} = 0.42$$

7.9.4 Modulation Response of ILDs

ILDs are increasingly finding applications in fiber-optic communication. Therefore, the study of high-speed modulation of their output is of great technological importance. They can be modulated either externally (using optoelectronic modulators, discussed in Chapter 9) or directly by modulating the excitation current. In this section, we discuss direct current modulation. The treatment given below follows that of Yariv (1997a) closely.

As discussed earlier, stimulated emission will dominate spontaneous emission only when population inversion has occurred. For semiconductor lasers, this condition is satisfied by doping *p*-type and *n*-type confining layers so heavily that the Fermi level separation exceeds the band gap of the active region under forward bias [see Fig. 7.23(b)]. When the injected carrier density in the active region exceeds a certain value (n_{tr}) called the *transparency value*, population inversion is achieved and the active region starts exhibiting optical gain. An input signal propagating through the active region would then amplify as exp(gz), where g is the gain coefficient. At transparency, $g = 0$. For semiconductors, g is normally calculated numerically. In general, the value of the peak gain coefficient g_p is approximated by the relation

$$g_p = B(n - n_{tr}) \tag{7.95}$$

where B is the gain constant. Typically, B is about 1.5×10^{-16} cm^2 for GaAs/GaAlAs lasers at 300 K. However, it increases with decrease in temperature. n_{tr} is typically around 1.55×10^{18} cm^{-3}.

The rate equations governing the change in the photon density P and injected electron (and hole) density n for an ILD can be written as follows:

$$\frac{dn}{dt} = \frac{I}{eV} - \frac{n}{\tau} - A(n - n_{tr})P \tag{7.96}$$

$$\frac{dP}{dt} = A(n - n_{tr})P\Gamma - \frac{P}{\tau_p} \tag{7.97}$$

where I is the total current, V is the volume of the active region, τ is the spontaneous recombination lifetime, and τ_p is the photon lifetime as limited by absorption in the bounding media, scattering, and coupling through output mirrors. The term $A(n - n_{tr})P$ is the net rate per unit volume of induced transitions, n_{tr} is the minimum inversion density needed to achieve transparency, and A is the temporal growth constant that is related to the constant B [defined by Eq. (7.95)] by the relation $A = (Bc)/n_s$, where c is the speed of light and n_s is the refractive index of the semiconductor. Γ is the confinement factor.

The confinement factor ensures that the total number [rather than the density variables used in Eqs (7.96) and (7.97)] of electrons undergoing stimulated transitions is equal to the number of photons emitted. The contribution of spontaneous emission to the photon density is neglected (since a very small fraction ~10^{-4} of the spontaneously emitted power enters the lasing mode).

Steady-state solutions n_0 and P_0 may be obtained by setting the LHSs of Eqs (7.96) and (7.97) equal to zero:

$$0 = \left(\frac{I_0}{eV}\right) - \frac{n_0}{\tau} - A(n_0 - n_{tr})P_0 \tag{7.98}$$

and

$$0 = A(n_0 - n_{tr})P_0\Gamma - \frac{P_0}{\tau_p} \tag{7.99}$$

We consider the case where the current is made up of dc and ac components

$$I = I_0 + I_m \exp(j\omega_m t) \tag{7.100}$$

and define the small-signal modulation responses n_m and P_m as

$$n = n_0 + n_m \exp(j\omega_m t) \tag{7.101}$$

and

$$P = P_0 + P_m \exp(j\omega_m t) \tag{7.102}$$

where n_0 and P_0 are the dc solutions given by Eqs (7.98) and (7.99). Using Eqs (7.101) and (7.102) and the result $A(n_0 - n_{tr}) = 1/(\tau_p\Gamma)$, from Eq. (7.99), in Eqs (7.96) and (7.97), we get

$$-j\omega_m n_m = \frac{I_m}{eV} + \left(\frac{1}{\tau} + AP_0\right) n_0 + \frac{1}{\tau_p \Gamma} P_m \tag{7.103}$$

and

$$j\omega P_m = AP_0 \Gamma n_m \tag{7.104}$$

From Eqs (7.103) and (7.104), we get the modulation response

$$\frac{P_m(\omega)}{I_m(\omega)} = \frac{-(1/eV)\,AP_0\,\Gamma}{\omega_m^2 - j\omega_m/\tau - j\omega_m AP_0 - AP_0/\tau_p} \tag{7.105}$$

The response curve remains flat at small frequencies and peaks at the *relaxation resonance frequency* ω_R, which can be obtained by minimizing the denominator of Eq. (7.105). Thus,

$$\omega_R = \sqrt{\left[\frac{AP_0}{\tau_p} - \frac{1}{2}\left(\frac{1}{\tau} + AP_0\right)^2\right]} \tag{7.106}$$

To very good accuracy, it may be approximated to

$$\omega_R\,(\text{rad/s}) = \sqrt{\frac{AP_0}{\tau_p}} \tag{7.107}$$

or

$$\nu_R\,(\text{Hz}) = \frac{\omega_R}{2\pi} = \frac{1}{2\pi}\sqrt{\frac{AP_0}{\tau_p}} \tag{7.108}$$

This result suggests that in order to increase ω_R and, thus, the useful linear region of the modulation response, we need to increase the optical gain coefficient A, decrease the photon lifetime τ_p, and operate the laser at an internal photon density P_0 as high as possible. It is also possible to write Eq. (7.107) as

$$\omega_R = \sqrt{\frac{1 + A\tau_p\,\Gamma n_{\text{tr}}}{\tau\tau_p}\left(\frac{I_0}{I_{\text{th}}}\right)} \tag{7.109}$$

7.9.5 ILD Structures

The simplest structure of an ILD is shown in Fig. 7.23(a). Here, a thin active layer is sandwiched between P and N-type cladding layers of higher band gap semiconductors. The resulting DH is forward-biased through metallic contacts. Such ILDs are known as broad-area lasers. Light is emitted in the form of an elliptic spot from the cleaved end facets of the active layer. The size of the spot depends on the size of the active region. The active layer behaves as a planar waveguide for the light generated within the layer because its refractive index is higher than that of the confining layers. Thus, apart from the longitudinal modes, it may also support a few transverse modes depending on the thickness of the active layer. Normally this layer is made so thin that it supports only one transverse mode. As there is no confinement in the lateral direction (i.e., in a direction parallel to the junction plane), the light spreads out over the entire width of the emitting side. The drawback of such a structure is that it requires high threshold current, and the spatial pattern is also highly elliptical with a shape that varies with the current. Therefore, this structure is not suitable for fiber-optic communication systems. Reduction in the threshold current and the spot size have been achieved by introducing a mechanism of optical confinement in the lateral direction. A few of these mechanisms are discussed here.

A simple scheme to provide the optical confinement in the horizontal plane is to limit the current injection over a narrow stripe, similar to that shown in Fig. 7.16. Such lasers are called stripe geometry lasers. Two configurations based on this scheme are shown in Fig. 7.25. In the first structure, shown in Fig. 7.25(a), a dielectric (SiO_2) layer is deposited on top of the *p*-layer, so that the current is injected through a narrow stripe. An alternative to the scheme (though not shown in the figure) is that a highly resistive region may be formed on two sides of the stripe in the *p*-layer itself by proton bombardment. In the second structure, shown in Fig. 7.25(b), an *n*-layer is deposited on the *P*-layer. Diffusion of Zn over the central stripe converts the *n*-region into a *p*-region, thus forming a *p-n* junction in the central region. Current flows only through this region and is blocked elsewhere because of the reverse bias in the other parts. The optical gain in both these cases peaks at the centre of the stripe and the light is also confined to the stripe region. Since the optical confinement is aided by gain, these devices are called gain-guided lasers.

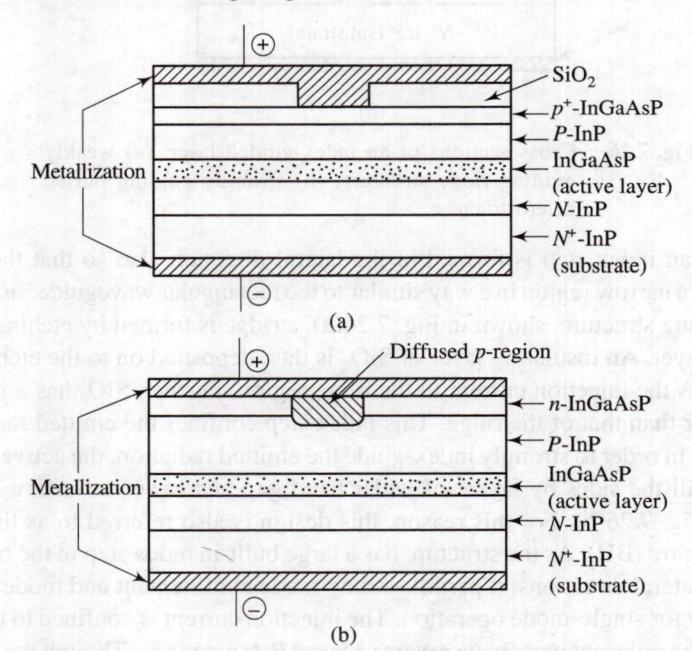

Fig. 7.25 Cross section of an InGaAsP/InP (a) oxide-isolated stripe geometry laser and (b) junction-isolated stripe geometry laser

A major drawback of gain-guided lasers is that the laser modes tend to be unstable as the injection current is increased. This drawback is overcome in structures called index-guided lasers, shown in Fig. 7.26.

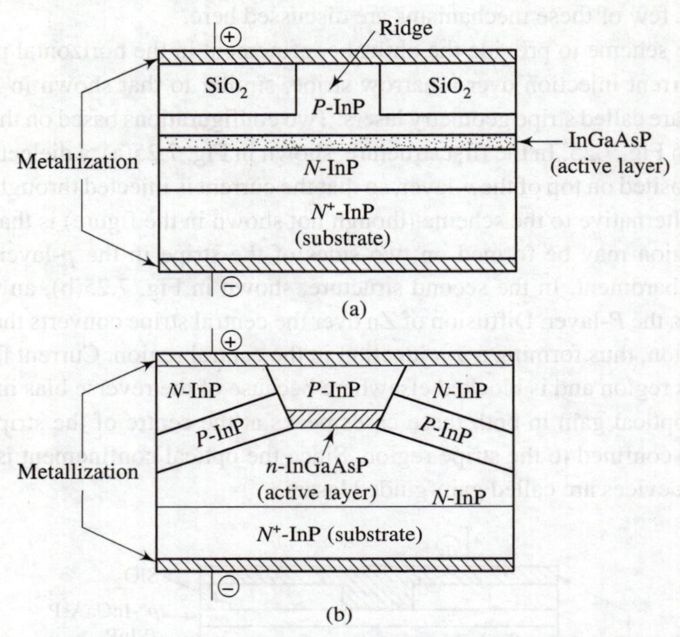

Fig. 7.26 Cross sections of an index-guided laser: (a) weakly guiding ridge structure, (b) strongly guiding buried heterostructure

Herein, an index step is formed in the lateral direction also so that the light is confined in a narrow region in a way similar to the rectangular waveguide. In a weakly index-guiding structure, shown in Fig. 7.26(a), a ridge is formed by etching parts of the top P-layer. An insulating layer of SiO_2 is then deposited on to the etched parts, which limits the injection current to the ridge region. Further, SiO_2 has a refractive index lower than that of the ridge. This index step confines the emitted radiation to this region. In order to strongly index-guide the emitted radiation, the active region is buried on all the sides by layers of wider band gap and lower refractive index, as shown in Fig. 7.26(b). For this reason, this design is also referred to as the *buried heterostructure* (BH). As the structure has a large built-in index step in the transverse as well as lateral directions, it permits strong mode confinement and mode stability, particularly for single-mode operation. The injection current is confined to the active region, as the adjacent parts have reverse-biased P-N junctions. Though the BH laser shown in Fig. 7.26(b) is based on InP/InGaAsP material, such devices may be fabricated using other materials as well. Further, many other configurations of the BH laser are also available, offering both multimode and single-mode operations.

These devices also have lower threshold currents (typically 10–20 mA) as compared to gain-guided or weakly index-guided lasers. Therefore, BH lasers are very suitable for fiber-optic communications.

High-speed, long-haul communications require highly powerful single-mode lasers which possess only one longitudinal and one transverse mode. One way to obtain a single longitudinal mode is to reduce the length L of the optical cavity so much that the mode separation is larger than the laser transition line width. However, such a small length makes the device difficult to handle and it also gives less power output. Therefore alternative structures have been developed for single-mode lasers.

An elegant way to obtain single-mode operation, which has found widespread application, is to introduce a distributed feedback (DFB) system in the structure of the ILD. In this scheme, the feedback is not localized at the facets but is distributed throughout the length of the optical cavity. This is achieved by etching a Bragg diffraction grating on to one of the cladding layers surrounding the active layer in the DH, as shown in Fig. 7.27(a). The grating provides periodic variation in the refractive index along the direction of wave propagation. The feedback is obtained through the phenomenon of Bragg diffraction, which couples the waves propagating in the forward and backward directions. Mode selectivity or coupling occurs only for those wavelengths λ_B which satisfy the Bragg condition given below:

$$\Lambda = m\lambda_B/2n_e \tag{7.110}$$

where Λ is the grating period, n_e is the effective refractive index of the waveguide for that particular mode wavelength (also called the mode index), and m is an integer representing the order of Bragg diffraction. The coupling is strongest for first-order diffraction ($m = 1$).

In another configuration, shown is Fig. 7.27(b), called the distributed Bragg reflector (DBR) laser, the gratings are etched near the cavity ends. These regions act as mirrors

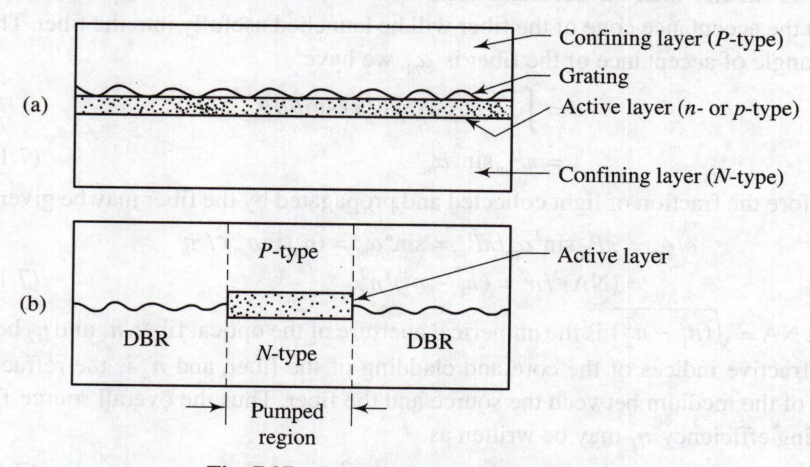

Fig. 7.27 (a) DFB and (b) DBR lasers

whose reflectivity is maximum for a wavelength λ_B satisfying the condition of Eq. (7.110). Here, the feedback does not take place in the active region. Therefore the loss is minimum for the longitudinal mode around λ_B and increases greatly for other longitudinal modes. This confinement has the advantage of separating the perturbed regions from the active region. Depending on the requirement, other structures are also possible for single-mode operation. We will take up the discussion of laser configurations again in Chapter 11.

7.10 SOURCE-FIBER COUPLING

In a fiber-optic communication link, the source is coupled to the fiber at the transmitter end and the detector is coupled to it at the receiver end. Thus, the system's performance depends on how effectively the source is coupled to the optical fiber. As there is a variety of sources and optical fibers, coupling efficiency is governed by many factors such as the size, radiance, and angular power distribution of the source, and the numerical aperture, core size, refractive index profile, etc. of the optical fiber.

A simplified calculation of the overall source-fiber coupling efficiency η_T may be done for a *p-n* homojunction SLED coupled to a step-index fiber. For this case, η_T may be defined as a ratio of the number of photons usefully launched into and propagated by the optical fiber to the number of carriers crossing the junction. Since SLED is a surface emitter, the flux emitted by it may be calculated using the equation

$$\phi_m = \int_0^{\pi/2} (P_m \cos\theta)(2\pi)\sin\theta d\theta = \pi P_m \tag{7.111}$$

where P_m is the power emitted by the top surface of the source per unit solid angle along the normal to the emitted surface. If we assume that the emitting area of the SLED is smaller than the core diameter, then out of the total flux ϕ_m, only ϕ lying within the acceptance cone of the fiber will be launched usefully into the fiber. Thus, if the angle of acceptance of the fiber is α_m, we have

$$\begin{aligned}\phi &= \int_0^{\alpha_m} (P_m \cos\theta)\,(2\pi)\sin\theta d\theta \\ &= \pi P_m \sin^2\alpha_m\end{aligned} \tag{7.112}$$

Therefore the fraction of light collected and propagated by the fiber may be given by

$$\begin{aligned}\phi/\phi_m &= \pi P_m \sin^2\alpha_m/\pi P_m = \sin^2\alpha_m = (n_a \sin\alpha_m)^2/n_a^2 \\ &= (\text{NA})^2/n_a^2 = (n_1^2 - n_2^2)/n_a^2\end{aligned} \tag{7.113}$$

where $\text{NA} = \sqrt{(n_1^2 - n_2^2)}$ is the numerical aperture of the optical fiber, n_1 and n_2 being the refractive indices of the core and cladding of the fiber, and n_a is the refractive index of the medium between the source and the fiber. Thus the overall source-fiber coupling efficiency η_T may be written as

$$\eta_T = \eta_{\text{ext}}\phi/\phi_m = \eta_{\text{ext}}(n_1^2 - n_2^2)/n_a^2 \tag{7.114}$$

where η_{ext} is given by Eq. (7.65). If we take our earlier figure of η_{ext} to be equal to 0.011 and the NA of the fiber to be 0.17, with air ($n_a = 1$) in between the source and the fiber, $\eta_T = 3.179 \times 10^{-4}$, that is, out of 10, 000 carriers crossing the *p-n* junction in the LED, only about three photons will be collected and propagated by the optical fiber. This oversimplified calculation simply indicates that η_T is normally poor. Thus, an attempt must be made to improve upon it.

Several schemes have been suggested and also implemented in practice for improving η_T. Some of these are discussed in brief as follows. First of all, the source that is employed for communication purposes is not a simple *p-n* homojunction LED, but is either a Burrus-type DH SLED or a stripe geometry ELED for short-haul links, or ILDs for long-haul application. Second, the source is normally coupled with the fiber in a way that optimizes the coupling efficiency. One method is to use an index-matching epoxy between the source and the fiber as shown in Fig. 7.15. The second method utilizes microlenses for coupling power into the fiber as shown in Fig. 7.28. Thus, the lens may be grown either on the surface of the LED or on the tip of the fiber. ILDs may be coupled to the fiber using two or more lenses.

Several other schemes have been suggested. Because of the Lambertian nature of the LED, the best coupling efficiency reported is of the order of 0.5%. However, with

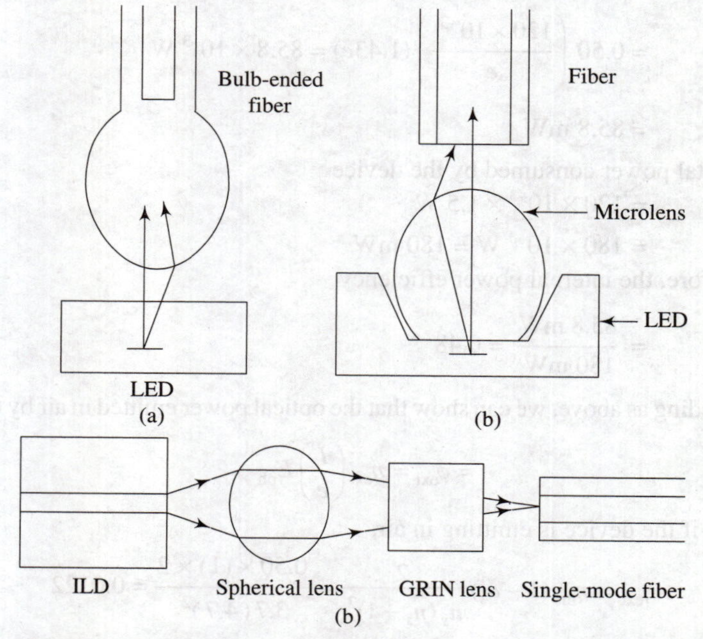

Fig. 7.28 Source-fiber coupling: (a) bulb-ended fiber coupled to SLED, (b) multimode fiber coupled to a Burrus-type SLED with a truncated microlens on it, and (c) ILD coupled to a single-mode fiber.

a single-mode ILD coupled to a single-mode fiber, the coupling efficiency may increase by a factor of 50 to 100.

Example 7.7 Assuming that the LED of Example 7.4 is forward-biased with a current of 120 mA and a voltage of 1.5 V, and the emitted photons possess energy $E_{ph} = 1.43$ eV, calculate (a) the internal power efficiency of the device, (b) the external power efficiency of the device, if it is emitting in air, and (c) the overall source-fiber power coupling efficiency (defined in terms of the ratio of the total power launched into and propagated by the fiber to the total power supplied to the device) and the optical loss (in dB). Assume that the refractive indices of the core and cladding of the optical fiber are 1.5 and 1.48, respectively.

Solution

(a) The optical power emitted within the LED

$$= \phi_{int} = \text{(rate of photon generation)} \times \text{(photon energy)}$$
$$= \eta_{int}\ \text{(rate of carriers crossing the junction)} \times E_{ph}$$
$$= \eta_{int}(I/e)E_{ph}$$
$$= 0.50\left(\frac{120 \times 10^{-3}}{e}\right)(1.43e) = 85.8 \times 10^{-3}\ \text{W}$$
$$= 85.8\ \text{mW}$$

The total power consumed by the device

$$= 120 \times 10^{-3} \times 1.5\ \text{W}$$
$$= 180 \times 10^{-3}\ \text{W} = 180\ \text{mW}$$

Therefore, the internal power efficiency

$$= \frac{85.8\ \text{mW}}{180\ \text{mW}} = 0.48$$

(b) Proceeding as above, we can show that the optical power emitted in air by the LED

$$= \phi_{ext} = \eta_{ext}\left(\frac{I}{e}\right)E_{ph}$$

Since, if the device is emitting in air,

$$\eta_{ext} = \eta_{int}\,(1 - a_s)\frac{2}{n_s\,(n_s + 1)^2} = \frac{0.50 \times (1) \times 2}{3.7\,(4.7)^2} = 0.0122$$

$$\phi_{ext} = 0.0122 \times \left(\frac{120 \times 10^{-3}}{e}\right)(1.43e)$$
$$= 2.100 \times 10^{-3}\ \text{W} = 2.1\ \text{mW}$$

Therefore, the external power efficiency (when the LED is emitting in air)

$$= \frac{2.1\ \text{mW}}{180\ \text{mW}} = 0.012$$

(c) From Eq. (7.114),

$$\eta_T = \eta_{\text{ext}}\left(\frac{n_1^2 - n_2^2}{n_a^2}\right)$$

$$= 0.0337\left[\frac{(1.5)^2 - (1.48)^2}{(1.5)^2}\right]$$

$$= 0.0337\left[\frac{(1.5)^2 - (1.48)^2}{(1.5)^2}\right]$$

$$= 8.93 \times 10^{-4}$$

Therefore, the optical power usefully launched into the fiber

$$= \phi_T = \eta_T\left(\frac{I}{e}\right)E_{\text{ph}} = 8.93 \times 10^{-4} \times \left(\frac{120 \times 10^{-3}}{e}\right)(1.43e)$$

$$= 153.2 \times 10^{-6}\ \text{W} = 153.2\ \mu\text{W}$$

Therefore, the overall source–fiber power coupling efficiency

$$= \frac{0.1532\ \text{mW}}{180\ \text{mW}} = 8.51 \times 10^{-4}$$

The optical loss (in dB)

$= -10 \log_{10}$ (source-fiber power coupling efficiency) = 30.7 dB

SUMMARY

- In a fiber-optic system, electrical signals at the transmitter end have to be converted into optical signals. This function is performed by an optoelectronic source. It appears that only semiconductor devices—light-emitting diodes (LEDs) or injection laser diodes (ILDs)—fulfil the requirements of an optoelectronic source.
- In order to understand the principle of operation, efficiency, and designs of an LED or ILD, it is essential to understand the properties of semiconductors, *p-n* homojunctions, and heterojunctions, light-emission processes, etc.
- An intrinsic semiconductor at absolute zero of temperature has no charge carriers, and behaves like an insulator; but at any other temperature the material develops equal number of two types of charge carriers, namely, (i) electrons (negative charge carriers) and (ii) holes (positive charge carriers). The intrinsic carrier

density, n_i, in an intrinsic semiconductor, can be related to temperature T by the relation

$$n_i = 2(2\pi kT/h^2)^{3/2}(m_e m_n)^{3/4} \exp(-E_g/2kT)$$

where the symbols have their usual meaning.

- An extrinsic semiconductor at room temperature has a greater density of one type of carrier than the other, which is normally accomplished by doping it with suitable impurities. Thus, an n-type semiconductor has electrons as majority charge carriers and holes as minority carriers, and a p-type semiconductor has a majority of holes and a minority of electrons. However, for moderate doping concentrations and at normal temperature, the product of electron and hole densities is almost independent of the dopant concentration.
- A p-n junction is a transition region between the two sides (one side doped p-type and the other n-type) of the same semiconductor. This transition region contains a potential V_D given by

$$V_D = \left(\frac{kT}{e}\right) ln\left(\frac{N_a N_d}{n_i^2}\right)$$

Upon forward-biasing, the potential barrier lowers, so that the excess electrons and holes cross the junction and reach the other side, where they recombine with opposite charge carriers either radiatively or non-radiatively.

The injection efficiency of a p^+-n junction (p^+ indicates that the p-side is more heavily doped as compared to the n-side) is given by

$$\eta_{\text{inj}} = \frac{1}{1 + J_e/J_n}$$

and that of a p-n^+ junction is given by

$$\eta_{\text{inj}} = \frac{1}{1 + J_h/J_e}$$

- The radiative recombination of injected carriers (e.g., electrons with holes) gives rise to the emission of light. This phenomenon is known as *injection luminescence*. The p-n junction diode exhibiting this phenomenon is known as a light-emitting diode. This process is more probable in direct band gap semiconductors, e.g., GaAs.
- The *internal quantum efficiency* η_{int} of an LED is defined as the ratio of the rate of photons generated within the semiconductor to the rate of carriers crossing the junction and is given by

$$\eta_{\text{int}} = \frac{1}{1 + \tau_{\text{rr}}/\tau_{\text{nr}}}$$

- Due to many factors, light emitted within a semiconductor cannot be fully collected at its surface. The ratio of the rate of photons emitted from the surface of a semiconductor to the rate of carriers crossing the junction is called *external quantum efficiency* η_{ext}, given by

$$\eta_{\text{ext}} = \eta_{\text{int}} \frac{(1-a_s)\,2\,n_a^3}{n_s\,(n_s+n_a)^2}$$

These efficiencies are quite low. Therefore heterojunctions are being used for fabricating LEDs. Either a surface-emitting double-hetero structure (DH) or an edge-emitting DH is used. These have better efficiencies.

- A more efficient source is an injection laser diode. Laser is an acronym for light amplification by stimulated emission of radiation. Therefore this device uses a specific structure (normally a DH), since upon heavy forward-biasing, a build-up of electrons in the conduction band and that of holes in the valence band of the active region may occur, their recombination giving rise to a large flux of monochromatic radiation. Depending on the size of the active region, the ILD can give either a single-mode or multimode output. The condition for the lasing threshold is given by

$$\Gamma g_{\text{th}} = \alpha_{\text{eff}} + \left(\frac{1}{2L}\right) ln\left(\frac{1}{R_1 R_2}\right)$$

where the symbols have their usual meaning. Several designs are possible.

- Finally, source-fiber coupling is very important. For LEDs, the overall source-fiber coupling efficiency is given by

$$\eta_T = n_{\text{ext}} \frac{(\text{NA})^2}{n_a^2}$$

This is of the order of 0.5%. For a single-mode ILD coupled to a single-mode fiber this may increase by a factor of 50 or more.

MULTIPLE CHOICE QUESTIONS

7.1 Which of the following materials is not suitable for making an LED?

(a) GaAs (b) Silicon
(c) InGaAsP (d) GaAlAs

7.2 In an LED, which of the following factors affects most severely the efficiency of the diode and cannot be eliminated even in principle?

(a) Fresnel reflection (b) Back emission
(c) Total internal reflection (d) Absorption

7.3 The densities of electrons and holes are the same in

(a) an intrinsic semiconductor.
(b) an extrinsic semiconductor.
(c) a *p-n* junction in equilibrium.
(d) a forward-biased *p-n* junction.

7.4 In a *p-n* homojunction, the majority carrier concentrations are almost equal to the dopant concentrations at
(a) absolute zero temperature. (b) normal temperature.
(c) high temperature. (d) all temperatures.

7.5 The material for making an efficient LED should be
(a) an indirect band gap type semiconductor.
(b) a direct band gap type semiconductor.
(c) a metal. (d) an insulator.

7.6 Which of the following pairs are suitable for making a heterojunction?
(a) Si and Ge (b) Si and GaAs
(c) GaAs and AlAs (d) GaAs and GaAlAs

7.7 A typical DH ILD has a cavity length of 0.6 mm, an effective loss coefficient of 1.0 mm^{-1}, a confinement factor of 0.90, and uncoated facet reflectivities of 0.33. What is the reduction in the threshold gain coefficient that occurs when the reflectivity of one of the facets is increased to 1?
(a) 3.16 mm^{-1} (b) 2.13 mm^{-1} (c) 1.03 mm^{-1} (d) 0.56 mm^{-1}

7.8 If the internal quantum efficiency of stimulated emission for the ILD of Question 7.7 is 75%, what is its approximate differential quantum efficiency?
(a) 10% (b) 20% (c) 30% (d) 40%

7.9 What should be the grating period Λ in a DFB laser to obtain single-mode operation at $\lambda_B = 1.55$ μm, assuming that the effective refractive index of the waveguide for λ_B is $n_e = 3.3$?
(a) 85 nm (b) 177 nm (c) 235 nm (d) 360 nm

7.10 An LED with an external quantum efficiency of 0.012 is coupled to an optical fiber of NA = 0.15 (with air between them). What is the overall source-fiber coupling efficiency?
(a) 1.8×10^{-4} (b) 2.7×10^{-4} (c) 3.2×10^{-4} (d) 7.8×10^{-3}

Answers

7.1	(b)	7.2	(c)	7.3	(a)	7.4	(b)	7.5	(b)
7.6	(d)	7.7	(c)	7.8	(d)	7.9	(c)	7.10	(b)

REVIEW QUESTIONS

7.1 What are direct band gap and indirect band gap type of semiconductors? Give at least two examples of each. Which of these are more suitable for fabricating LEDs? Give reasons.

7.2 Show schematically the variations of carrier concentration across an n^+-p homojunction (a) in equilibrium and (b) under forward bias.

7.3 Assume that an n^+-p homojunction is forward-biased and that the net rate of recombination per unit volume in the *p*-side is equal to $\Delta n/\tau_p$, where Δn is the local excess minority concentration in the *p*-side and τ_p is a constant. Prove that τ_p is the mean lifetime of the excess minority carriers.

7.4 Calculate the injection efficiency of a GaAs diode in which $N_a = 10^{21}$ m^{-3} and $N_d = 10^{23}$ m^{-3}. Assume that at room temperature of 300 K, $\mu_e = 0.85$ m^2 V^{-1} s^{-1}, $\mu_h = 0.04$ m^2 V^{-1} S^{-1}, and that $L_e \approx L_h$. The symbols have their usual meaning. In what manner does this diode differ from that of Example 7.3?

Ans: 0.9995

7.5 In a forward-biased *p-n* homojunction, the probability that an electron in the conduction band at energy E_2 will recombine with a hole in the valence band at energy E_1 is proportional to the concentration of electrons, $n(E_2)$, at E_2; and to the concentration of holes, $p(E_1)$, at E_1. Thus the probability of a photon of energy, E_{ph}, being radiated by the diode may be obtained by integrating the product $n(E_2)p(E_1)$ over all values of E_1 (or E_2), subject to the condition $E_2 - E_1 = E_{ph}$. Assume that

$$n(E_2) \approx A\exp[-(E_2 - E_c)/kT]$$

and

$$p(E_1) \approx B\exp[-(E_v - E_1)/kT]$$

where *A* and *B* are proportionality constants and the other symbols have their usual meaning. (a) Calculate the power *P* radiated by the *p-n* diode as a function of photon energy. (b) At what photon energy does the maximum power occur? What is the power at this photon energy?

Ans: (a) $P = \alpha(E_{ph} - E_g)\exp[-(E_{ph} - E_g)/kT]$ (where α is a constant)
(b) $(E_{ph})_{peak} = E_g + kT$; $P(\text{at peak } E_{ph}) = \alpha kT/e$

7.6 (a) What are homojunctions and heterojunctions?
(b) Discuss unique properties of the *P-n-N* double heterostructure LED and sketch (with proper labelling) the energy-level diagram of such a configuration.

7.7 A Burrus-type *p-n* GaAs LED is coupled to a step-index fiber (NA = 0.16) using an epoxy resin of refractive index similar to that of the fiber core ($n = 1.50$). The emitting surface of the LED is bloomed so that the transmission factor may be taken to be unity. The refractive index of GaAs is 3.7 and η_{int} of the source may be taken to be 0.40. (a) Calculate the fraction of total optical power that is able to escape from the semiconductor surface. (b) Assuming 10% self-absorption within the semiconductor, calculate the external quantum efficiency. (c) Calculate the fraction of incident radiation captured and propagated by the fiber. (d) Calculate the overall source-fiber coupling efficiency. (e) If the LED is emitting in air, calculate the fraction of radiation captured and propagated by the fiber.

Ans: (a) 0.0889 (b) 0.032 (c) 0.0105 (d) 3.36×10^{-4} (e) 0.0256

7.8 A Burrus-type *p-n* GaAs LED is coupled to a step-index fiber of core diameter larger than the emitting area of the LED, using transparent bonding cement. The diode is forward-biased with a current of 100 mA and a voltage of 1.5 V, and each emitted photon possesses energy of 1.42 eV. Relevant data are given as follows: for the optical fiber, core refractive index = 1.46, relative refractive index difference = 2%; for the bonding cement, refractive index = 1.5; for GaAs, refractive index = 3.7, τ_{rr} = 120 ns, τ_{nr} = 100 ns; and the absorption within GaAs = 10%. (a) Calculate the internal quantum efficiency of the LED, (b) the internal power efficiency of the device, (c) the external power efficiency if the diode is emitting in air, and (d) the overall source fiber coupling efficiency and optical loss (in dB).

Ans: (a) 0.4545 (b) 0.43 (c) 0.0095 (d) 9.86×10^{-4}, 30 dB

7.9 A GaAs LED is forward-biased with a current of 120 mA and a voltage of 1.5 V. Each emitted photon possesses an energy of 1.43 eV, and the refractive index of GaAs is 3.7. The configuration of the LED is such that we may neglect back emission and self-absorption within the semiconductor. Assuming the internal quantum efficiency of the LED to be 60%, calculate (a) the internal power efficiency of the device and (b) the external power efficiency of the device.

Ans: (a) 0.572 (b) 0.0276

7.10 (a) Derive the threshold condition for laser action.

(b) On what factors does the gain coefficient of a semiconductor laser depend?

7.11 The longitudinal modes of a DH GaAs/GaAlAs ILD operating at a wavelength of 850 nm are separated in frequency by 250 GHz. If the refractive index of the active region is 3.7, calculate the length of the optical cavity and the number of longitudinal modes emitted.

Ans: 162 μm, 1410

7.12 A DH GaAs/GaAlAs ILD operating at 850 nm has a cavity length of 500 μm, and the refractive index of the cavity is 3.7. How many longitudinal modes are emitted? What is the mode separation in terms of frequency (Hz) and wavelength? (Hint: $\delta\lambda = \lambda^2/2nL$)

Ans: 4353, 81 GHz, 0.195 nm

7.13 For a DH ILD, with strong carrier confinement, the threshold gain coefficient, g_{th}, to a good approximation, may be given by the relation $g_{th} = \beta J_{th}$, where β is a constant depending on the device configuration and J_{th} is the threshold current density for stimulated emission. Consider a GaAs ILD with a cavity of length 300 μm and width 100 μm. Assume that β = 0.02 cm A^{-1}, α_{eff} = 12 cm^{-1}, n = 3.7, and one of the facets has a reflectivity of 100%. Calculate the threshold current density and threshold current I_{th}, assuming the current flow is confined to the cavity region.

Ans: 1524 A cm^{-2}, 0.46 A

7.14 Calculate the maximum allowed length of the active region for the single-mode operation of a DH InGaAsP/InP ILD emitting at 1.3 μm. Assume that the refractive index of the active region is 3.5.

Ans: 0.185 μm

APPENDIX A7.1: LAMBERTIAN SOURCE OF RADIATION

A Lambertian source (named after the German scientist J. Lambert) is a reference model. Figure A7.1 shows the radiation pattern of a Lambertian source, in which the power radiated per unit solid angle in a direction at an angle θ to the normal to the emitting surface is given by

$$P(\theta) = P_0 \cos\theta \tag{A7.1}$$

where P_0 is the power radiated per unit solid angle normal to the surface.

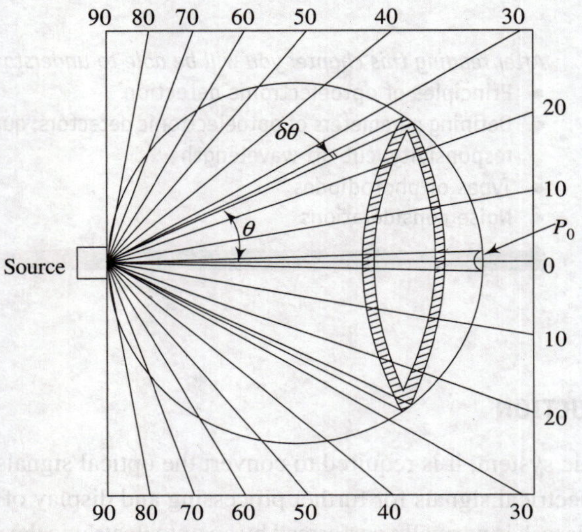

Fig. A7.1 The radiation pattern of a Lambertian source

Thus the power radiated into a small solid angle $\delta\Omega$ in a direction θ with respect to the normal to the surface will be $P(\theta)\delta\Omega = P_0\cos\theta\,\delta\Omega = P_0\cos\theta(2\pi\sin\theta\,\delta\theta)$. (Here the solid angle $\delta\Omega$ is equal to the solid angle subtended by the elementary annular ring whose radius subtends an angle θ and width subtends an angle $\delta\theta$ at the surface of the source. Thus $\delta\Omega$ will be equal to $2\pi\sin\theta\,\delta\theta$.) The total power ϕ_0 radiated by this source can be obtained by integrating $P(\theta)$ over all forward directions. Thus,

$$\phi_0 = \int_{\theta=0}^{\pi/2} (P_0\cos\theta)(2\pi\sin\theta)\,d\theta = \pi P_0 \tag{A7.2}$$

8 Optoelectronic Detectors

After reading this chapter you will be able to understand the following:

- Principles of optoelectronic detection
- Defining parameters of optoelectronic detectors: quantum efficiency, responsivity, cut-off wavelength
- Types of photodiodes
- Noise considerations

8.1 INTRODUCTION

In any fiber-optic system, it is required to convert the optical signals at the receiver end back into electrical signals for further processing and display of the transmitted information. This task is normally performed by an optoelectronic detector. Therefore, the overall system performance is governed by the performance of the detector. The detector requirements are very similar to those of optoelectronic sources; that is, they should have high sensitivity at operating wavelengths, high fidelity, fast response, high reliability, low noise, and low cost. Further, its size should be comparable with that of the core of fiber employed in the optical link. These requirements are easily met by detectors made of semiconducting materials, and hence we will discuss the detection principles and design of only such detectors.

8.2 THE BASIC PRINCIPLE OF OPTOELECTRONIC DETECTION

Figure 8.1 shows a reverse-biased *p-n* junction. When a photon of energy greater than the band gap of the semiconductor material (i.e., $h\nu \geq E_g$) is incident on or near the depletion region of the device, it excites an electron from the valence band into the conduction band. The vacancy of an electron creates a hole in the valence band. Electrons and holes so generated experience a strong electric field and drift rapidly towards the *n* and *p* sides, respectively. The resulting flow of current is proportional

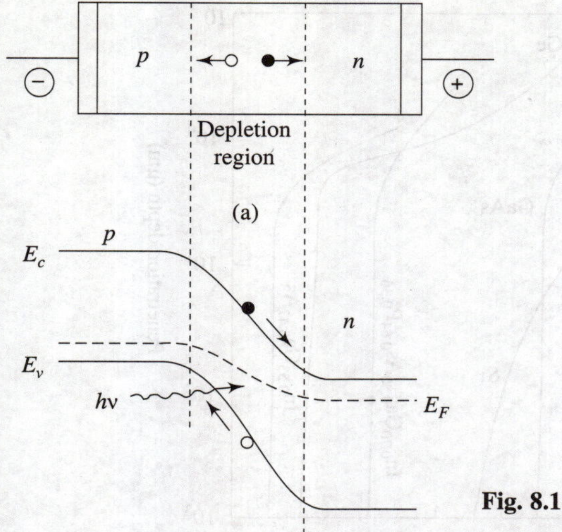

Fig. 8.1 (a) A reverse-biased *p-n* junction. (b) Energy band diagram showing carrier generation and their drift.

to the number of incident photons. Such a reverse-biased *p-n* junction, therefore, acts as a photodetector and is normally referred to as a *p-n* photodiode.

8.2.1 Optical Absorption Coefficient and Photocurrent

The absorption of photons of a specific wavelength in a photodiode to produce electron–hole pairs and thus a photocurrent depends on the absorption coefficient α of the semiconductor for that particular wavelength. If we assume that the total optical power incident on the photodiode is P_{in} and that the Fresnel reflection coefficient at the air–semiconductor interface is R, the optical power entering the semiconductor will be $P_{in}(1 - R)$. The power absorbed by the semiconductor is governed by Beer's law. Thus, if the width of the absorption region is d and α is the absorption coefficient of the semiconductor at the incident wavelength, the power absorbed by this width will be given by

$$P_{abs} = P_{in}(1 - R)\,[1 - \exp(-\alpha d)] \tag{8.1}$$

Let us assume that the incident light is monochromatic and the energy of each photon is $h\nu$, where h is Planck's constant and ν is the frequency of incident light. Then the rate of photon absorbtion will be given by

$$\frac{P_{abs}}{h\nu} = \frac{P_{in}(1-R)}{h\nu}[1 - \exp(-\alpha d)]$$

The dependence of α on wavelength for some commonly used semiconductors is shown in Fig. 8.2. It is clear from these curves that α depends strongly on the wavelength of incident light. Thus, a specific semiconductor material can be used

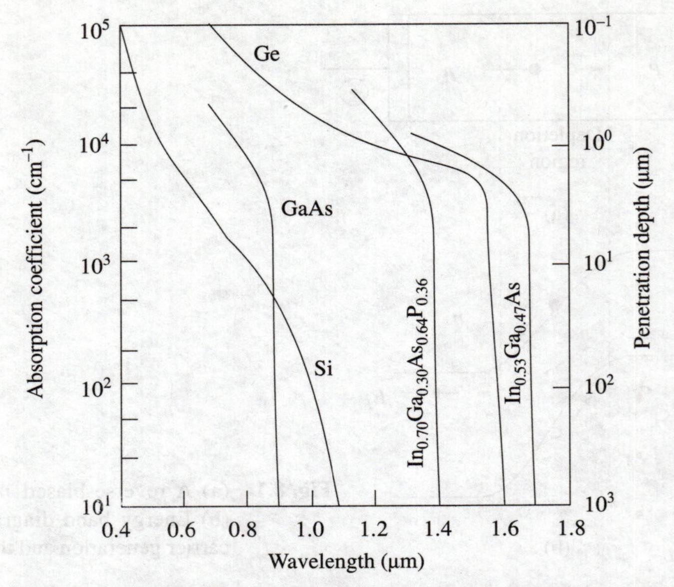

Fig. 8.2 Wavelength dependence of the absorption coefficient α for some semiconductors (Lee & Li 1979)

only in a specific range. There is also an upper wavelength limit (explained in Sec. 8.2.4) for each semiconductor material.

Further, if we assume that (i) the semiconductor is an intrinsic absorber (i.e., the absorption of photons excites the electrons from the valence band directly to the conduction band), (ii) each photon produces an electron–hole pair, and (iii) all the charge carriers are collected at the electrodes, then the photocurrent (rate of flow of charge carriers) I_p so produced will be given by

$$I_p = \frac{P_{\text{in}}(1-R)e}{h\nu}[1-\exp(-\alpha d)] \tag{8.2}$$

where e is the electronic charge.

8.2.2 Quantum Efficiency

Quantum efficiency η is defined as the ratio of the rate (r_e) of electrons collected at the detector terminals to the rate r_p of photons incident on the device. That is,

$$\eta = \frac{r_e}{r_p} \tag{8.3}$$

η may be increased if the Fresnel reflection coefficient R is decreased and the product αd in Eq. (8.2) is much greater than unity. It must be noted, however, that η is also a function of the photon wavelength.

8.2.3 Responsivity

The responsivity $\Re$ of a photodetector is defined as the output photocurrent per unit incident optical power. Thus, if I_p is the output photocurrent in amperes and P_{in} is the incident optical power in watts,

$$\Re = \frac{I_p}{P_{\text{in}}} \text{ (in A W}^{-1}\text{)} \tag{8.4}$$

The output photocurrent I_p may be written in terms of the rate, r_e, of electrons collected as follows:

$$I_p = er_e \tag{8.5}$$

where e is the electronic charge. Combining Eqs (8.3) and (8.5), we get

$$I_p = e\eta r_p \tag{8.6}$$

Now the rate of incident photons is given by

$$r_p = \frac{\text{Incident optical power}}{\text{Energy of the photon}} = \frac{P_{\text{in}}}{h\nu} \tag{8.7}$$

Thus $$I_p = \frac{\eta e P_{\text{in}}}{h\nu} \tag{8.8}$$

Substituting for I_p from Eq. (8.8) in Eq. (8.4), we get an expression for $\Re$ in terms of η as follows:

$$\Re = \frac{\eta e}{h\nu} = \frac{\eta e \lambda}{hc} \tag{8.9}$$

where λ and c are the wavelength and speed of the incident light in vacuum, respectively. Equation (8.9) shows that the responsivity is directly proportional to the quantum efficiency at a particular wavelength and in the ideal case, when $\eta = 1$, $\Re$ is directly proportional to λ. For a practical diode, as the wavelength of the incident photon becomes longer, its energy becomes smaller than that required for exciting the electron from the valence band to the conduction band. The responsivity thus falls off near the cut-off wavelength λ_c. This can be seen in Fig. 8.3(b).

8.2.4 Long-wavelength Cut-off

In an intrinsic semiconductor, the absorption of a photon is possible only when its energy is greater than or equal to the band gap energy E_g of the semiconductor used to fabricate the photodiode. That is, the photon energy $hc/\lambda \geq E_g$. Thus, there is a long-wavelength cut-off λ_c, above which photons are simply not absorbed by the semiconductor, given by

$$\lambda_c = \frac{hc}{E_g} \tag{8.10}$$

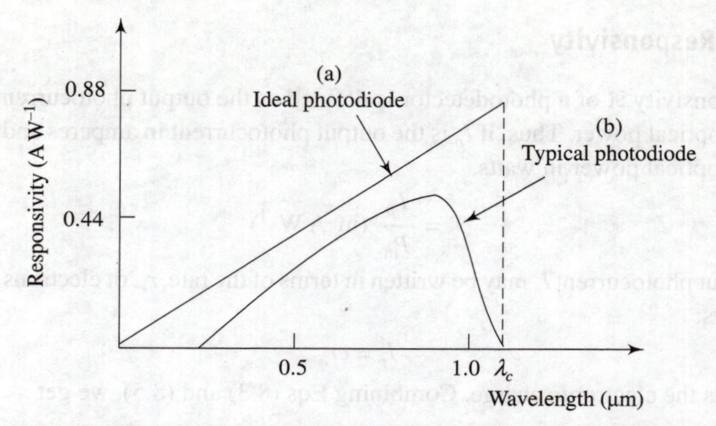

Fig. 8.3 Responsivity as a function of wavelength for (a) an ideal Si photodiode and (b) a practical Si diode

If E_g is expressed in eV, substituting the values of $h = 6.626 \times 10^{-34}$ J s, $c = 3 \times 10^8$ m s^{-1}, and 1 eV = 1.6×10^{-19} J, we get

$$\lambda_c\,(\mu\text{m}) \approx \frac{1.24}{E_g\,(\text{eV})} \tag{8.11}$$

This expression allows us to calculate λ_c for different semiconductors in order to select them for specific detection purposes.

Example 8.1 A *p-n* photodiode has a quantum efficiency of 70% for photons of energy 1.52×10^{-19} J. Calculate (a) the wavelength at which the diode is operating and (b) the optical power required to achieve a photocurrent of 3 μA when the wavelength of incident photons is that calculated in part (a).

Solution

(a) The photon energy

$$E = h\nu = \frac{hc}{\lambda}$$

Therefore $$\lambda = \frac{hc}{E} = \frac{6.626 \times 10^{-34} \times 3 \times 10^8}{1.52 \times 10^{-19}} = 1.30 \times 10^{-6}\,\text{m}$$

$$= 1.30\ \mu\text{m}$$

(b) $$\Re = \frac{\eta e}{h\nu} = \frac{0.70 \times 1.6 \times 10^{-19}}{1.52 \times 10^{-19}} = 0.736\ \text{A W}^{-1}$$

Since $$\Re = \frac{I_p}{P_{\text{in}}}$$

$$= 4.07 \times 10^{-6}\ \text{W}$$

or $$P_{\text{in}} = 4.07\ \mu\text{W}$$

Example 8.2 A *p-i-n* photodiode, on an average, generates one electron–hole pair per two incident photons at a wavelength of 0.85 μm. Assuming all the photo-generated electrons are collected, calculate (a) the quantum efficiency of the diode; (b) the maximum possible band gap energy (in eV) of the semiconductor, assuming the incident wavelength to be a long-wavelength cut-off; and (c) the mean output photocurrent when the incident optical power is 10 μW.

Solution

(a) $\eta = \frac{1}{2} = 0.5 = 50\%$

(b) $E_g = \frac{hc}{\lambda_c} = \frac{6.626 \times 10^{-34} \times 3 \times 10^8}{0.85 \times 10^{-6}} = 2.33 \times 10^{-19}$ J

$= 1.46$ eV

(c) $I_P = \Re P_{\text{in}} = \frac{\eta e}{h\nu} P_{\text{in}} = \frac{0.5 \times 1.6 \times 10^{-19}}{2.33 \times 10^{-19}} \times 10 \times 10^{-6} = 3.43 \times 10^{-6}$ A

$= 3.43$ μA

Example 8.3 Photons of wavelength 0.90 μm are incident on a *p-n* photodiode at a rate of 5×10^{10} s^{-1} and, on an average, the electrons are collected at the terminals of the diode at the rate of 2×10^{10} s^{-1}. Calculate (a) the quantum efficiency and (b) the responsivity of the diode at this wavelength.

Solution

(a) $\eta = \frac{2 \times 10^{10}}{5 \times 10^{10}} = 0.40$

(b) $\Re = \frac{\eta e \lambda}{hc} = \frac{0.40 \times 1.6 \times 10^{-19} \times 0.90 \times 10^{-6}}{6.626 \times 10^{-34} \times 3 \times 10^8} = 0.29$ A W^{-1}

8.3 TYPES OF PHOTODIODES

8.3.1 *p-n* Photodiode

The simplest structure is that of a *p-n* photodiode, shown in Fig. 8.4(a). Incident photons of energy, say *hν*, are absorbed not only inside the depletion region but also outside it, as shown in Fig. 8.4(b). As discussed in Sec. 8.2, the photons absorbed within the depletion region generate electron–hole pairs. Because of the built-in strong electric field [shown in Fig. 8.4(c)], electrons and holes generated inside this region

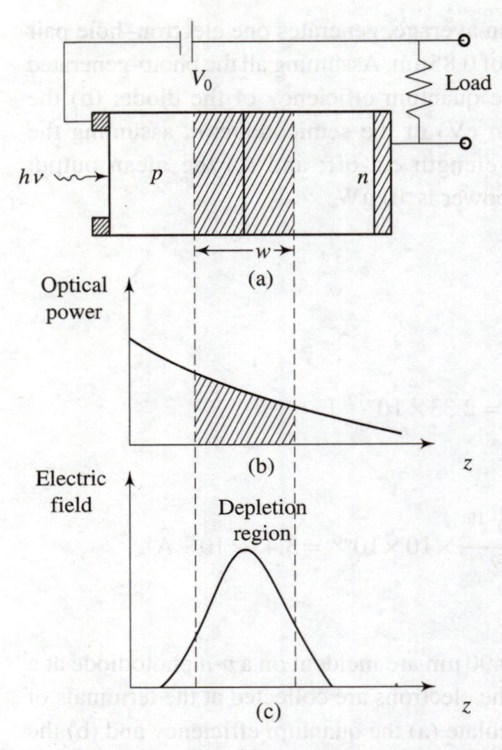

Fig. 8.4 (a) Structure of a *p-n* photodiode and the associated depletion region under reverse bias. (b) Variation of optical power within the diode. (c) Variation of electric field inside the diode.

get accelerated in opposite directions and thereby drift to the *n*-side and the *p*-side, respectively. The resulting flow of photocurrent constitutes the response of the photodiode to the incident optical power. The response time is governed by the transit time τ_{drift}, which is given by

$$\tau_{drift} = \frac{w}{v_{drift}} \tag{8.12}$$

where w is the width of the depletion region and v_{drift} is the average drift velocity. τ_{drift} is of the order of 100 ps, which is small enough for the photodiode to operate up to a bit rate of about 1 Gbit/s. In order to minimize τ_{drift}, both w and v_{drift} can be tailored. The depletion layer width w is given (Sze 1981) by

$$w = \left[\frac{2\varepsilon}{e}(V_{bi} + V_0)\left(\frac{1}{N_a} + \frac{1}{N_d}\right)\right]^{1/2} \tag{8.13}$$

where ε is the dielectric constant, e is the electronic charge, V_{bi} is the built-in voltage and depends on the semiconductor, V_0 is the applied bias voltage, and N_a and N_d are the acceptor and donor concentrations used to fabricate the *p-n* junction. The drift velocity v_{drift} depends on the bias voltage but attains a saturation value depending on the material of the diode.

As shown in Fig. 8.4(b), incident photons are absorbed outside the depletion region also. The electrons generated in the *p*-side have to diffuse to the depletion-region boundary before they can drift (under the built in electric field) to the *n*-side. In a similar fashion, the holes generated in the *n*-side have to diffuse to the depletion-region boundary for their drift towards the *p*-side. The diffusion process is inherently slow and hence the presence of a diffusive component may distort the temporal response of a photodiode, as shown in Fig. 8.5.

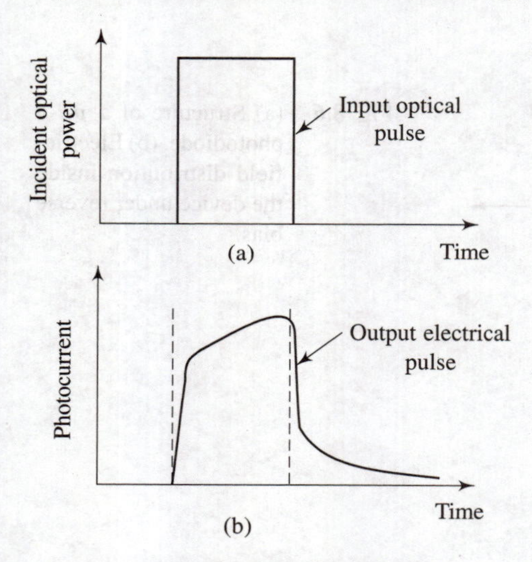

Fig. 8.5 Response of a typical *p-n* photodiode to a rectangular optical pulse when both drift and diffusion contribute to the photocurrent

8.3.2 *p-i-n*-Photodiode

The diffusion component of a *p-n* photodiode may be reduced by decreasing the widths of the *p*-side and *n*-side and increasing the width of the depletion region so that most of incident photons are absorbed inside it. To achieve this, a layer of semiconductor, so lightly doped that it may be considered intrinsic, is inserted at the *p-n* junction. Such a structure is called a *p-i-n* photodiode and is shown in Fig. 8.6 along with the electric-field distribution inside it under the reverse bias. As the middle layer is intrinsic in nature, it offers high resistance, and hence most of the voltage drop occurs across it. Thus, a strong electric field exists across the middle *i*-region. Such a configuration results in the drift component of the photocurrent dominating the diffusion component, as most of the incident photons are absorbed inside the *i*-region.

A double heterostructure, similar to that discussed in Chapter 7 for sources, improves the performance of the *p-i-n* photodiodes. Herein, the middle *i*-region of a material with lower band gap is sandwiched between *p*- and *n*-type materials of higher band gap, so that incident light is absorbed only within the *i*-region. Such a configuration is shown in Fig. 8.7. The band gap of InP is 1.35 eV and hence it is

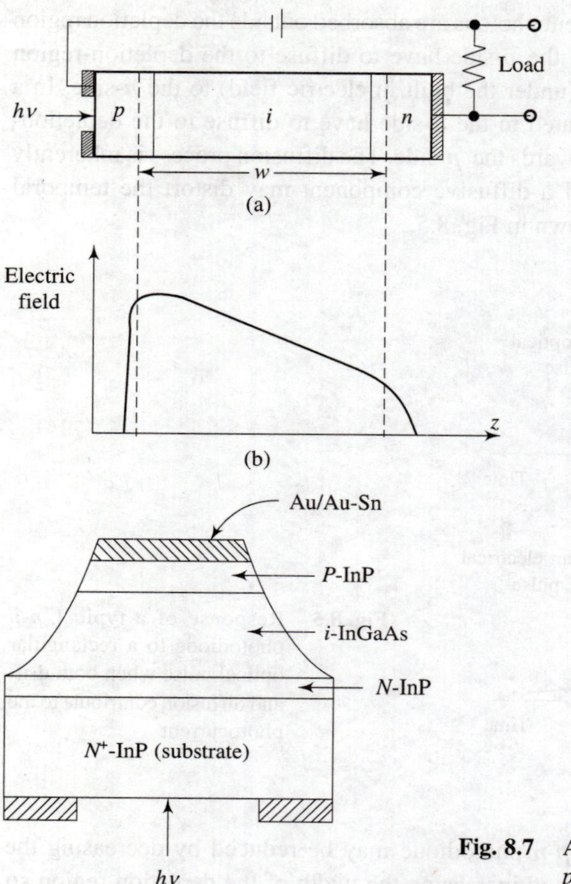

Fig. 8.6 (a) Structure of a *p-i-n* photodiode. (b) Electric-field distribution inside the device under reverse bias.

Fig. 8.7 A double heterostructure design of a *p-i-n* photodiode using InGaAs/InP

transparent for light of wavelength greater than 0.92 μm, whereas the band gap of lattice-matched InGaAs is about 0.75 eV, which corresponds to a λ_c of 1.65 μm. Thus the intrinsic layer of InGaAs absorbs strongly in the wavelength range 1.3–1.6 μm. The diffusive component of the photocurrent is completely eliminated in such a heterostructure simply because the incident photons are absorbed only within the depletion region. Such photodiodes are very useful for fiber-optic systems operating in the range 1.3–1.6 μm.

8.3.3 Avalanche Photodiode

Avalanche photodiodes (APDs) internally multiply the primary photocurrent through multiplication of carrier pairs. This increases receiver sensitivity because the photocurrent is multiplied before encountering the thermal noise associated with the receiver circuit. Through an appropriate structure of a photodiode, it is possible to create a high-field region within the device upon biasing. When the primary electron–hole pairs generated by incident photons pass through this region, they get accelerated and acquire so much kinetic energy that they ionize the bound electrons in the valence band upon collision and in the process create secondary electron–hole pairs. This phenomenon is known as impact ionization. If the field is high enough, the secondary carrier pairs may also gain sufficient energy to create new pairs. This is known as the avalanche effect. Thus the carriers get multiplied, all of which contribute to the photocurrent.

A commonly used configuration for achieving carrier multiplication with little excess noise is shown in Fig. 8.8 and is called a reach through avalanche photodiode (RAPD). It is composed of a lightly doped *p*-type intrinsic layer (called a π-layer) deposited on a p^+ (heavily doped *p*-type) substrate. A normal *p*-type diffusion is made in the intrinsic layer, which is followed by the construction of an n^+ (heavily doped *n*-type) layer. Such a configuration is called a p^+-π-p-n^+ reach through structure. When the applied reverse-bias voltage is low, most of the potential drop is across the p-n^+ junction. As the bias voltage is increased, the depletion layer widens and the latter just reaches through to the π-region when the electric field at the p-n^+ junction becomes sufficient for impact ionization. Normally, an RAPD is operated in a fully depleted mode. The photons enter the device through the p^+-layer and are absorbed in the intrinsic π-region. The absorbed photons create primary electron–hole pairs, which are separated by the electric field in this region. These carriers drift to the p-n^+ junction,

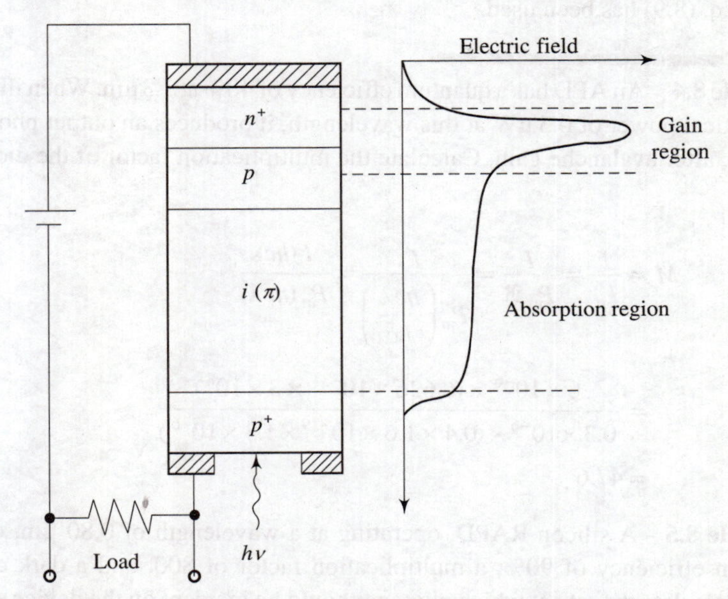

Fig. 8.8 Schematic configuration of RAPD and the variation of electric field in the depletion and multiplication regions

where a strong electric field exists. It is in this region that the carriers are multiplied first by impact ionization and then by the avalanche breakdown.

The average number of carrier pairs generated by an electron per unit length of its transversal is called the electron ionization rate α_e. Similarly, the hole ionization rate α_h may also be defined. The ratio $K = \alpha_h/\alpha_e$ is a measure of the performance of APDs. An APD made of a material in which one type of carrier dominates impact ionization exhibits low noise and high gain. It has been found that there is a significant difference between α_e and α_h only in silicon and hence it is normally used for making RAPDs for detection around 0.85 μm. For longer wavelengths, advanced structures of APDs are used.

The multiplication factor M of an APD is a measure of the internal gain provided by the device. It is defined as

$$M = \frac{I_M}{I_p} \tag{8.14}$$

where I_M is the average output current (after multiplication) and I_p is the primary photocurrent (before multiplication) defined by Eq. (8.2).

Analogous to *p-n* and *p-i-n* photodiodes, the performance of an APD is described by its responsivity $\Re_{\text{APD}}$, given by

$$\Re_{\text{APD}} = \frac{\eta e}{h\nu} M = M\Re \tag{8.15}$$

where Eq. (8.9) has been used.

Example 8.4 An APD has a quantum efficiency of 40% at 1.3 μm. When illuminated with optical power of 0.3 μW at this wavelength, it produces an output photocurrent of 6 μA, after avalanche gain. Calculate the multiplication factor of the diode.

Solution

$$M = \frac{I}{I_p} = \frac{I}{P_{\text{in}}\Re} = \frac{I}{P_{\text{in}}\left(\dfrac{\eta e\lambda}{hc}\right)} = \frac{I\,(hc)}{P_{\text{in}}\,(\eta e\lambda)}$$

$$= \frac{6\times10^{-6}\times(6.626\times10^{-34}\times3\times10^{8})}{0.3\times10^{-6}\times(0.4\times1.6\times10^{-19}\times1.3\times10^{-6})}$$

$$= 47.6$$

Example 8.5 A silicon RAPD, operating at a wavelength of 0.80 μm, exhibits a quantum efficiency of 90%, a multiplication factor of 800, and a dark current of 2 nA. Calculate the rate at which photons should be incident on the device so that the output current (after avalanche gain) is greater than the dark current.

Solution

$$I = I_P M = P_{in}\Re M = P_{in}\left(\frac{\eta e\lambda}{hc}\right)M$$

$$= \left(\frac{P_{in}}{(hc/\lambda)}\right)\eta e M$$

$$= [(\text{photon rate})e]\eta M$$

For $I = 2$ nA,

$$\text{Photon rate} = \frac{I}{e\eta M} = \frac{2\times10^{-9}}{1.6\times10^{-19}\times0.90\times800}$$

$$= 1.736\times10^{7}\ \text{s}^{-1}$$

For $I > 2$ nA,

$$\text{Photon rate} \approx 1.74\times10^{7}\ \text{s}^{-1}$$

8.4 PHOTOCONDUCTING DETECTORS

The basic detection process involved in a photoconducting detector is raising an electron from the valence band to the conduction band upon absorption of a photon by a semiconductor, provided the photon energy is greater than the band gap energy (i.e., $h\nu \ge E_g$). As long as an electron remains in the conduction band, it will contribute toward increasing the conductivity of the semiconductor. This phenomenon is known as photoconductivity.

The structure of a typical photoconducting detector designed for operation in the long-wavelength range (typically 1.1–1.6 μm) is shown in Fig. 8.9. The device comprises a thin conducting layer, 1–2 μm thick, of *n*-type InGaAs that can absorb photons in the wavelength range 1.1–1.6 μm. This layer is formed on the lattice-matched semi-insulating InP substrate. A low-resistance interdigital anode and cathode are made on the conducting layer.

In operation, the incident photons are absorbed by the conducting layer, thereby generating additional electron–hole pairs. These carriers are swept by the applied field towards respective electrodes, which results in an increased current in the external circuit. In the case of III-V alloys such as InGaAs, electron mobility is much higher than hole mobility. Thus whilst the faster electrons are collected at the anode, the corresponding holes are still proceeding towards the cathode. The process creates an absence of electrons and hence a net positive charge in the region. However, the excess charge is compensated for, almost immediately, by the injection of more electrons from the cathode into this region. In essence, this process leads to the generation of more electrons upon the absorption of a single photon. The overall

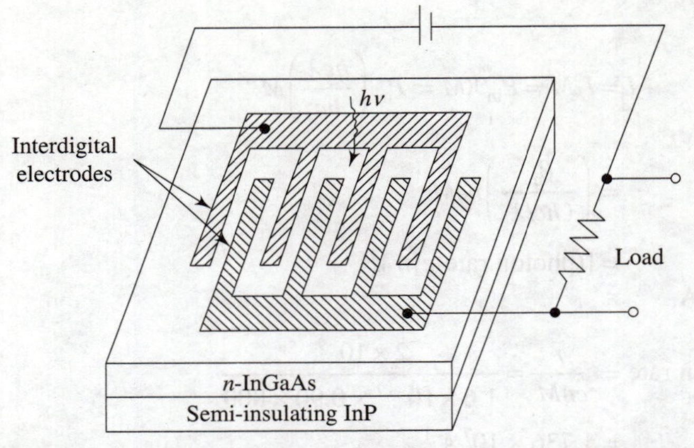

Fig. 8.9 A photoconducting detector for long-wavelength operation

effect is that it results in a photoconductive gain G, which may be defined as the ratio of transit time t_s for slow carriers to the transit time t_f for fast carriers. Thus,

$$G = \frac{t_s}{t_f} \tag{8.16}$$

The photocurrent I_g produced by the photoconductor can be written as

$$I_g = \frac{\eta P_{in} e}{h\nu} G = I_p G \tag{8.17}$$

where η is the quantum efficiency of the device, P_{in} is the incident optical power, and I_p is the photocurrent in the absence of any gain. Gain in the range 50–100 and a 3-dB bandwidth of about 500 MHz are currently achievable with the InGaAs photoconductors discussed above.

Example 8.6 The maximum 3-dB bandwidth permitted by an InGaAs photoconducting detector is 450 MHz when the electron transit time in the device is 6 ps. Calculate (a) the gain G and (b) the output photocurrent when an optical power of 5 μW at a wavelength of 1.30 μm is incident on it, assuming quantum efficiency of 75%.

Solution

The current response in the photoconductor decays exponentially with time once the incident optical pulse is removed. The time constant of this decay is equal to the slow carrier transit time t_s. Therefore, the maximum 3-dB bandwidth $(\Delta f)_m$ of the device will be given by

$$(\Delta f)_m = \frac{1}{2\pi t_s} = \frac{1}{2\pi t_f G}$$

where we have used Eq. (8.16).

(a) $G = \dfrac{1}{2\pi t_f (\Delta f)_m} = \dfrac{1}{2\pi \times 6 \times 10^{-12} \times 450 \times 10^6} = 58.94$

(b) $I = GI_P = \dfrac{G\eta P_{\text{in}} e\lambda}{hc} = \dfrac{58.94 \times 0.75 \times 5 \times 10^{-6} \times 1.6 \times 10^{-19} \times 1.3 \times 10^{-6}}{6.626 \times 10^{-34} \times 3 \times 10^{8}}$

$= 232.1\ \mu\text{A} = 2.321 \times 10^{-4}\ \text{A}$

8.5 NOISE CONSIDERATIONS

Optical signals at the receiver end in a fiber-optic communication system are quite weak. Therefore, even the simplest kind of receiver would require a good photodetector to be followed by an amplifier. Thus the signal power to noise ratio (S/N) at the output of the receiver would be given by

$$\frac{S}{N} = \frac{\text{Signal power from photocurrent}}{\text{Photodetector noise power + amplifier noise power}} \tag{8.18}$$

It is obvious from Eq. (8.18) that to achieve high S/N, (i) the photodetector should have high quantum efficiency and low noise so that it generates large signal power and (ii) the amplifier noise should be kept low. In fact the sensitivity of a detector is described in terms of the minimum detectable optical power, which is defined as the optical power necessary to generate a photocurrent equal in magnitude to the rms value of the total noise current. Therefore, in order to evaluate the performance of receivers, a thorough understanding of the various types of noises in a photodetector and their interrelationship is necessary. In order to see these interrelationships, let us analyse the simplest type of receiver model (Keiser 2000), shown in Fig. 8.10.

Herein, the photodiode has a negligible series resistance R_s, a load resistance R_L, and a total capacitance C_d. The amplifier has an input resistance R_a and capacitance C_a. There are three principal noises arising from the spontaneous fluctuation in a photodetector. These are (i) quantum or shot noise, (ii) dark current noise, and (iii) thermal noise. Thus, a current generated by a *p-n* or *p-i-n* photodiode in response to an instantaneous optical signal may be written as

$$I(t) = \langle i_p(t) \rangle + i_s(t) + i_d(t) + i_T(t) \tag{8.19}$$

where $\langle i_p(t) \rangle = I_p = \Re P_{\text{in}}$ is the average photocurrent, and $i_s(t)$, $i_d(t)$, and $i_T(t)$ are the current fluctuations related to shot noise, dark current noise, and thermal noise, respectively.

Quantum or shot noise arises because of the random arrival of photons at a photodetector and hence the random generation and collection of electrons. The mean square value of shot noise is given by

$$\langle i_s^2(t) \rangle = 2eI_p\, \Delta f\, M^2 F(M) \tag{8.20}$$

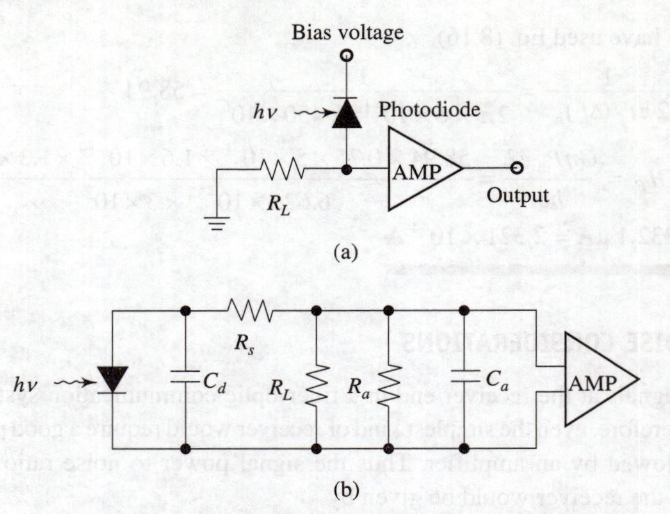

Fig. 8.10 (a) Simplest model of an optical receiver; (b) equivalent circuit

where $F(M)$ is a noise factor related to the random nature of an avalanche process and Δf is the effective noise bandwidth. Experimentally, it has been found that $F(M) \approx M^x$, where x depends on the material and varies from 0 to 1. For *p-n* and *p-i-n* photodiodes, $F(M)$ and M are unity.

Dark current is a reverse leakage current that continues to flow through the device when no light is incident on the photodetector. Normally, it arises from the electrons or holes which are thermally generated near the *p-n* junction of a photodiode. In an APD, these carriers get multiplied by the avalanche gain mechanism. Thus the mean square value of the dark current is given by

$$\langle i_d^2(t) \rangle = 2eI_d M^2 F(M)\Delta f \tag{8.21}$$

where I_d is the average primary dark current (before multiplication) of the detector.

Thermal (or Johnson) noise is a random fluctuation in current due to the thermally induced random motion of electrons in a conductor. The load resistance R_L adds such fluctuations to the current generated by the photodiode. The mean square value of this current is given by

$$\langle i_T^2 \rangle = \frac{4kT}{R_L}\Delta f \tag{8.22}$$

where k is Boltzmann's constant and T is the absolute temperature of the load resistor.

The noise power associated with the amplifier following the detector will depend on the active elements of the amplifier circuit. For the present discussion, let us assume that its mean square noise current is $\langle i_{\text{amp}}^2 \rangle$.

In general, therefore, the signal to noise ratio of an optical receiver may be written [using Eqs (8.18), (8.20), (8.21), and (8.22)] as

$$\frac{S}{N} = \frac{\langle i_p^2 \rangle M^2}{2e(I_p + I_d)\Delta f M^2 F(M) + \dfrac{4kT\Delta f}{R_L} + \langle i_{\text{amp}}^2 \rangle} \qquad (8.23)$$

When *p-n* and *p-i-n* photodiodes are used in the receiver, M and $F(M)$ become unity.

Example 8.7 An InGaAs *p-i-n* photodiode is operating at room temperature (300 K) at a wavelength of 1.3 μm. Its quantum efficiency is 70% and the incident optical power is 500 nW. Assume that the primary dark current I_d of the device is 5 nA, R_L is 1 kΩ, and the effective bandwidth is 25 MHz. Calculate (a) the rms values of shot noise current, dark current, and thermal noise current; and (b) S/N at the input end of an amplifier of the receiver.

Solution

(a) $$I_P = \Re P_{\text{in}} = \frac{\eta e \lambda}{hc} P_{\text{in}}$$

$$I_P = \frac{0.70 \times 1.6 \times 10^{-19} \times 1.3 \times 10^{-6}}{6.626 \times 10^{34} \times 3 \times 10^{8}} \times 500 \times 10^{-9} = 3.663 \times 10^{-7} \text{ A}$$

$$= 0.3662 \ \mu\text{A}$$

$$\langle i_s^2 \rangle = 2eI_p(\Delta f)M^2 F(M)$$

$$= 2 \times 1.6 \times 10^{-19} \times 3.662 \times 10^{-7} \times 25 \times 10^{6} \times 1 \times 1$$

$$= 293.03 \times 10^{-20} \text{ A}^2$$

$$\langle i_s^2 \rangle^{1/2} = 17.15 \times 10^{-10} \text{ A} = 1.715 \text{ nA}$$

$$\langle i_d^2 \rangle = 2eI_d(\Delta f) \qquad [M \text{ and } F(M) \text{ are unity}]$$

$$= 2 \times 1.6 \times 10^{-19} \times 5 \times 10^{-19} \times 25 \times 10^{6}$$

$$= 400 \times 10^{-22} \text{ A}^2 = 4 \times 10^{-20} \text{ A}^2$$

$$\langle i_d^2 \rangle^{1/2} = 20 \times 10^{-11} \text{ A} = 0.2 \text{ nA}$$

$$\langle i_T^2 \rangle = \frac{4kT(\Delta f)}{R_L} = \frac{4 \times 1.38 \times 10^{-23} \text{ (J/K)} \times 300 \text{ (K)}}{1{,}000} \times 25 \times 10^{6}$$

$$= 414 \times 10^{-18} \text{ A}^2$$

$$\langle i_T^2 \rangle^{1/2} = 20.34 \text{ nA}$$

(b) Sum of mean square noise currents $= 41{,}698.16 \times 10^{-20}$ A^2

$$= 4.17 \times 10^{-16} \text{ A}^2$$

and $$I_p^2 = 1.352 \times 10^{-13} \text{ A}^2$$

$$\frac{S}{N} = \frac{1.352 \times 10^{-13}}{4.17 \times 10^{-16}} = 0.324 \times 10^{3} = 324$$

SUMMARY

- In a fiber-optic system, it is required to convert the optical signals at the receiver end back into electrical signals. This task is performed by an optoelectronic detector.
- A reversed-biased *p-n* junction is used for this purpose. An incident photon of energy greater than the band gap of the semiconductor creates an electron–hole pair. The two charge carriers are swept in opposite directions by the applied bias, and photocurrent flows in the external circuit. The device is known as a *p-n* diode.
- The photocurrent depends on the absorption coefficient of the semiconductor for the incident wavelength.
- The ratio of the rate of electrons collected at the diode terminals to the rate of photons incident on it is called the quantum efficiency η of the device. It is related to the responsivity $\Re$ of the detector by the relation

$$\Re = \frac{\eta e}{h\nu} = \frac{\eta e\lambda}{hc}$$

- The absorption of a photon is possible only when its energy is greater than or equal to the energy gap of the semiconductor. Therefore, there is a long-wavelength cut-off λ_c, above which photons are not absorbed by the semiconductor. It is given by

$$\lambda_c = \frac{hc}{E_g}$$

- There are three types of photodiodes, namely, (i) *p-n*, (ii) *p-i-n*, and (iii) avalanche photodiodes. The first two produce current without gain while the third one produces current with gain. Photoconducting detectors can also be used for long-wavelength operations.
- There are three factors that may contribute to the noise at the detector end. These are (i) the random arrival of photons at the detector producing shot noise, (ii) thermally induced random motion of charge carriers giving rise to thermal noise, and (iii) even if there is no light, reverse leakage current known as dark current flowing through the circuit. Appropriate steps must be taken to improve the signal to noise ratio.

MULTIPLE CHOICE QUESTIONS

8.1 Practically, in order to create an electron–hole pair in a *p-n* diode, the energy of the incident photon should be

(a) less than E_g. (b) equal to E_g.
(c) greater than E_g. (d) much greater than E_g.

8.2 Given that germanium (Ge) has a band gap of 0.67 eV, what is the maximum wavelength that will be absorbed by it?
(a) 7,080 nm (b) 4,560 nm (c) 1,850 nm (d) 1,100 nm

8.3 The highest wavelength that silicon (Si) can absorb is 1.12 μm. What is the approximate band gap of Si?
(a) 1.1 eV (b) 1.4 eV (c) 1.74 eV (d) 2.3 eV

8.4 The following material is more suitable for making a *p-n* diode.
(a) A direct band gap semiconductor
(b) An indirect band gap semiconductor
(c) A metal
(d) An insulator

8.5 A *p-n* photodiode, on an average, generates one electron–hole pair per five incident photons at a wavelength of 0.90 μm. Assuming all the photogenerated electrons are collected, what is the quantum efficiency of the diode?
(a) 20% (b) 30% (c) 40% (d) 50%

8.6 Photons of wavelength 0.85 μm are incident on a *p-i-n* photodiode at the rate of 4×10^{10} s^{-1} and, on an average, electrons are collected at the terminals of the diode at the rate of 2×10^{10} s^{-1}. What is the responsivity of the diode at this wavelength?
(a) 0.15 A W^{-1} (b) 0.23 A W^{-1} (c) 0.34 A W^{-1} (d) 0.50 A W^{-1}

8.7 Which of the following detectors give amplified output?
(a) *p-n* photodiode
(b) *p-i-n* photodiode
(c) Avalanche photodiode
(d) Photovoltaic detector

8.8 The responsivity of a given *p-i-n* diode is 0.5 A W^{-1} for a wavelength of 1 μm. What is the output photocurrent when optical power of 0.2 μW at this wavelength is incident on it?
(a) 0.1 μA (b) 1 μA (c) 10 μA (d) 1 A

8.9 Which of the following is an inherent property of an optical signal and cannot be eliminated even in principle?
(a) Thermal noise (b) Shot noise
(c) Environmental noise (d) Background noise

8.10 A photoconducting detector can be constructed from
(a) an intrinsic semiconductor. (b) an extrinsic semiconductor.
(c) polycrystalline material. (d) all of these.

Answers

8.1	(c)	8.2	(c)	8.3	(a)	8.4	(b)	8.5	(a)
8.6	(c)	8.7	(c)	8.8	(a)	8.9	(b)	8.10	(d)

REVIEW QUESTIONS

8.1 Define the quantum efficiency and responsivity of a *p-n* diode. How are the two related to each other?

8.2 Distinguish between a *p-n* diode, a *p-i-n* diode, and an APD. Is it possible to make these three types of photodiodes using the same semiconductor?

8.3 A *p-n* photodiode has a quantum efficiency of 50% at $\lambda = 0.90$ μm. Calculate (a) its responsivity at this wavelength, (b) the received optical power if the mean photocurrent is 10^{-6} A, and (c) the corresponding number of received photons at this wavelength.
Ans: (a) 0.362 A W^{-1} (b) 2.76 μW (c) $r_p = 1.25 \times 10^{13}$ s^{-1}

8.4 An APD has a quantum efficiency of 50% at 1.3 μm. When illuminated with optical power of 0.4 μW at this wavelength, it produces an output photocurrent of 8 μA, after avalanche gain. Calculate the multiplication factor of the diode.
Ans: 38

8.5 Calculate the responsivity of an ideal *p-n* photodiode at the following wavelengths: (a) 0.85 μm, (b) 1.30 μm, and (c) 1.55 μm.
Ans: (a) 0.684 A W^{-1} (b) 1.046 A W^{-1} (c) 1.248 A W^{-1}

8.6 A typical photodiode has a responsivity of 0.40 A W^{-1} for a He–Ne laser source ($\lambda = 632.8$ nm). The active area of the photodiode is 2 mm^2. What will be the output photocurrent if the incident flux is 100 μW/mm^2?
Ans: 80 μA

8.7 (a) What are the factors responsible for making the responsivity versus wavelength curve (shown in Fig. 8.3) for a practical Si diode deviate from an ideal curve?
(b) How can the quantum efficiency of such a diode be improved?

8.8 The avalanche photodiode and the photoconducting detector both provide gain. Compare their merits for use in optical communication and other applications.

9 Optoelectronic Modulators

After reading this chapter you will be able to understand the following:

- The basic principles of optical polarization
- Electro-optic effect
- Longitudinal electro-optic modulator
- Transverse electro-optic modulator
- Acousto-optic effect
- Raman–Nath modulator
- Bragg modulator

9.1 INTRODUCTION

In a fiber-optic communication system, data are encoded in the form of the variation of some property of the optical output of a light-emitting diode (LED) or an injection laser diode (ILD). This property may be the amplitude or the phase of the optical signal or the width of the pulses being transmitted. There are two ways of modulating an optical signal from a LED or ILD. The first scheme involves an internal or direct modulation, in which a circuit is designed to modulate the current injected into the device. As the output of the source (LED/ILD) is controlled by the injected current, one can achieve the desired modulation. Direct modulation is simple but has several disadvantages: the upper modulation frequencies are limited to about 40 GHz; the emission frequency may change as the drive current is changed; and only amplitude modulation is possible with ease (for phase or frequency modulation, additional care is required to design the drivers).

The second scheme involves external modulation, in which an optical signal from the source is passed through a material (or device) whose optical properties can be altered by external means. In this scheme, the device speed is controlled by the modulator property and may be quite fast; the emission frequency remains unaffected and both amplitude and phase modulation are possible with ease. The only disadvantage is that the modulator is normally large on the scale of microelectronic devices and hence cannot become a part of the same integrated circuit (IC). In this chapter, we will discuss external modulators only. The design of these modulators is based either on the electro-optic effect or the acousto-optic effect. Devices based on the electro-optic effect are most common in high-speed communications. The technology used in these devices is also becoming compatible with that of semiconductor devices.

9.2 REVIEW OF BASIC PRINCIPLES

In order to appreciate a modulator design, it is essential to be familiar with certain phenomena of optics; e.g., polarization, birefringence or double refraction, etc. A detailed discussion on these topics may be found in any textbook on optics. Here, we will review them in brief.

9.2.1 Optical Polarization

An electromagnetic wave is said to be plane-polarized if the vibrations of its electric-field vector, say, **E**, are parallel to each other for all points in the wave as shown in Fig. 9.1. At all these points the vibrating **E**-vector and the direction of propagation form a plane called the *plane of vibration* or the *plane of polarization*. All such

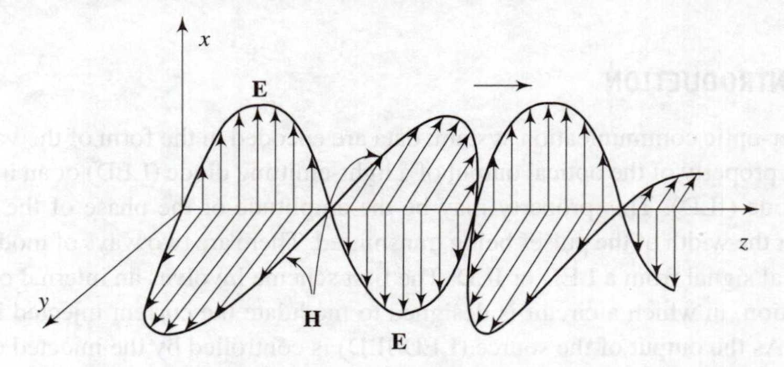

Fig. 9.1 An instantaneous snapshot of a plane-polarized transverse wave showing the electric- and magnetic-field vectors **E** and **H,** with the wave moving to the right

planes are parallel in a plane-polarized wave. In general, light produced by a radiation source and propagated in a given direction consists of such independent wave trains whose planes of vibration are randomly oriented about the direction of propagation, as shown in Fig. 9.2. This light is still transverse in nature but is said to be unpolarized. However, the resultant electric-field vector can be resolved into two components, and one may think of unpolarized light as comprising two orthogonal plane-polarized components with a random phase difference [see Fig. 9.2(c)]. It is possible to separate

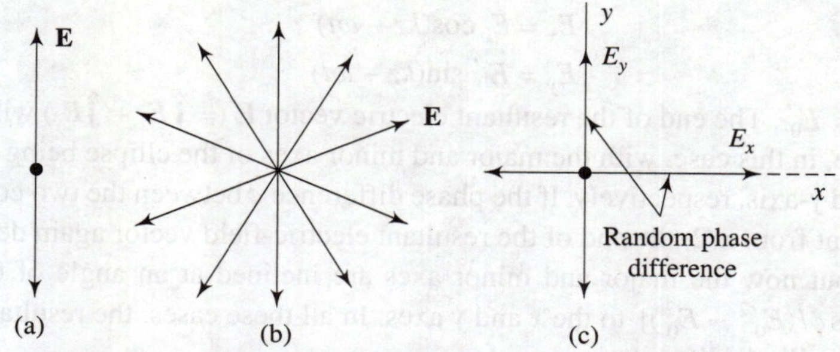

Fig. 9.2 (a) A plane-polarized transverse wave, shown only by the electric vector **E**. The direction of propagation is shown by the point of an arrow coming out of the page and perpendicular to it. (b) An unpolarized transverse wave depicted as comprising many randomly oriented plane-polarized transverse wave trains. (c) An equivalent description of an unpolarized transverse wave. Here the unpolarized wave has been depicted as consisting of two plane-polarized waves which have a random phase difference. The orientation of the x and y axes about the direction of propagation is also completely arbitary.

these components from each other by either selective reflection or absorption, and get a plane-polarized beam.

Let us consider a special case where the two components of a resultant electric-field vector have the same amplitudes, E_0, but a phase difference of $\pi/2$. Assume that the resultant wave is travelling in the z-direction and the orthogonal components are along x and y directions. We may then write the expressions for the component electric fields as

$$E_x = E_0 \cos(kz - \omega t) \tag{9.1}$$

$$E_y = E_0 \sin(kz - \omega t) \tag{9.2}$$

where k $(=2\pi/\lambda)$ is the propagation constant, ω is the angular frequency, and t is time. The resultant electric field is the vector sum of the two components:

$$\mathbf{E} = \hat{\mathbf{i}} E_x + \hat{\mathbf{j}} E_y \tag{9.3}$$

or

$$\mathbf{E} = E_0 [\hat{\mathbf{i}} \cos(kz - \omega t) + \hat{\mathbf{j}} \sin(kz - \omega t) \tag{9.4}$$

Here $\hat{\mathbf{i}}$ and $\hat{\mathbf{j}}$ are unit vectors in the x and y directions, respectively. The resultant field vector **E**, given by Eq. (9.4), at a given point in space is constant in amplitude but rotates with an angular frequency ω. Such an electromagnetic wave is said to be circularly polarized.

If the amplitude of the component electric-field vectors is not same but the phase difference between them is still $\pi/2$, the resultant electric-field vector at any point in space rotates with angular frequency ω and also changes in magnitude. Let us assume that the components are represented by the following equations:

$$E_x = E_0 \cos(kz - \omega t) \tag{9.5}$$

and

$$E_y = E_0' \sin(kz - \omega t) \tag{9.6}$$

with $E_0 \neq E_0'$. The end of the resultant electric vector $\mathbf{E}\ (= \hat{\mathbf{i}}\, E_x + \hat{\mathbf{j}}\, E_y)$ will describe an ellipse, in this case, with the major and minor axes of the ellipse being parallel to the x- and y-axis, respectively. If the phase difference ϕ between the two components is different from $\pi/2$, the end of the resultant electric-field vector again describes an ellipse, but now the major and minor axes are inclined at an angle of $(1/2)\tan^{-1}[E_0 E_0' \cos\phi / (E_0'^2 - E_0^2)]$ to the x and y axes. In all these cases, the resultant wave is said to be elliptically polarized.

9.2.2 Birefringence

In earlier chapters, we have assumed that the speed of light and hence the refractive index of a material is independent of the direction of propagation in the medium and of the state of polarization of the light. The materials that show this behaviour are called optically isotropic. Ordinary glass, for example, is isotropic in its behaviour. But, there are certain crystalline materials, such as calcite, quartz, KDP (potassium dihydrogen phosphate) etc., that are optically anisotropic, that is, the velocity of propagation in such materials, in general, depends on the direction of propagation and also the state of polarization of light. In other words, the refractive index of these crystals varies with direction, and such crystals are called birefringent or doubly refracting.

This nomenclature has been derived from the fact that a single beam of unpolarized light falling on such a crystal (e.g., calcite), in general, splits into two beams at the crystal surface. These two beams propagate in two different directions with different velocities and have mutually orthogonal planes of polarization, as shown in Fig. 9.3. This phenomenon is known as double refraction or birefringence. One of the beams, shown in Fig. 9.3 and represented by the ordinary ray or *o*-ray, follows Snell's law of refraction at the crystal surface, but the other beam labelled the extraordinary ray or *e*-ray does not.

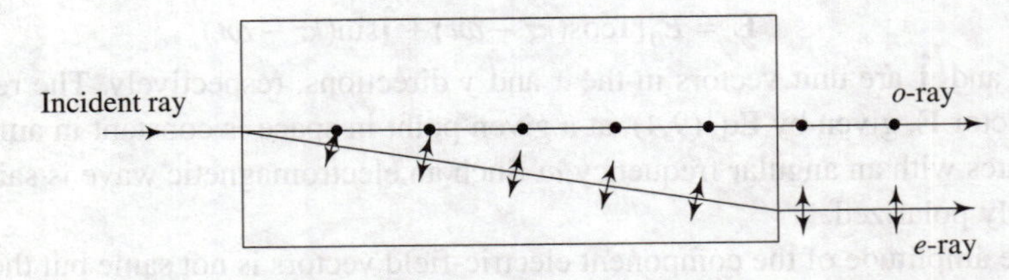

Fig. 9.3 Phenomenon of double refraction exhibited by a birefringent crystal. The electric-field vectors of the *o*-ray and the *e*-ray are shown to vibrate perpendicular and parallel to the plane of the figure, respectively.

This difference in the behaviour of the *o*-ray and the *e*-ray may be explained as follows. The *o*-ray propagates inside the crystal with the same speed v_o in all the directions and hence the refractive index n_o (defined by c/v_o, where *c* is the speed of light in vacuum) of this crystal for this wave is constant. On the other hand, the *e*-wave propagates inside the crystal with a speed that varies with direction from v_o to v_e ($v_e > v_o$ for calcite), which means that the refractive index of the crystal, for the wave, varies with direction from n_o to n_e ($n_e < n_o$ for calcite). However, there exists one direction along which both the *o*- and *e*-ray travel with the same velocity. This direction is referred to as the optic axis of the crystal. The quantities n_o and n_e are referred to as the principal refractive indices for the crystal. Such crystals, e.g., calcite, which are described by two principal refractive indices and one optic axis, are called uniaxial crystals. However, some birefringent crystals (e.g., mica, topaz, lead oxide, etc.) are more complex as compared to calcite and, therefore, require three principal refractive indices and two optic axes for a complete description of their optical behaviour. Such crystals are referred to as biaxial crystals. It should be noted that in a biaxial crystal, the wave velocities of the *o*- and *e*-wave (and not the ray velocities) are equal along the two optic axes. However, in a uniaxial crystal, both the wave and ray velocities for the *o*- and *e*-wave are equal along the optic axis.

The propagation of light waves in uniaxial doubly refracting crystals may be understood with the aid of Huygen's principle[1]. It states that all the points on a wavefront can be considered as point sources of secondary wavelets, and the position of the wavefront at a later time will be the surface of tangency to these secondary wavelets. Let us consider an imaginary point source of light, *S*, embedded in a uniaxial crystal. This source will generate two wavefronts—a spherical one corresponding to the *o*-ray and an ellipsoid of revolution about the optic axis corresponding to the *e*-ray, as shown in Figs 9.4(a) and 9.4(b).

The two wave surfaces represent light having two different polarization states. Considering only the rays lying in the plane of Fig. 9.4, we see that the plane of polarization of the *o*-rays is perpendicular to the plane of the figure (as shown by dots) and that of the *e*-rays coincides with the plane of the figure (as shown by lines). Another aspect must be noted regarding the wave surfaces of Figs 9.4(a) and 9.4(b): In a crystal, if $v_o > v_e$ (or $n_o < n_e$) in all the directions (except, of course, the optic axis), then the spherical wavefront would be completely outside the ellipsoidal wavefront, as shown in Fig. 9.4(b). The two surfaces will touch at two diametrically opposite points on the optic axis, along which the two wavefronts travel with equal velocities. Such a crystal is called a positive uniaxial crystal. Quartz and rutile (titanium oxide) are examples of this kind. On the other hand, if $v_o \le v_e$ (or $n_o \ge n_e$), the spherical wavefront lies completely inside the ellipsoidal wavefront, as shown in Fig. 9.4(a),

[1]A detailed discussion on Huygen's principle may be found in any textbook on geometrical optics.

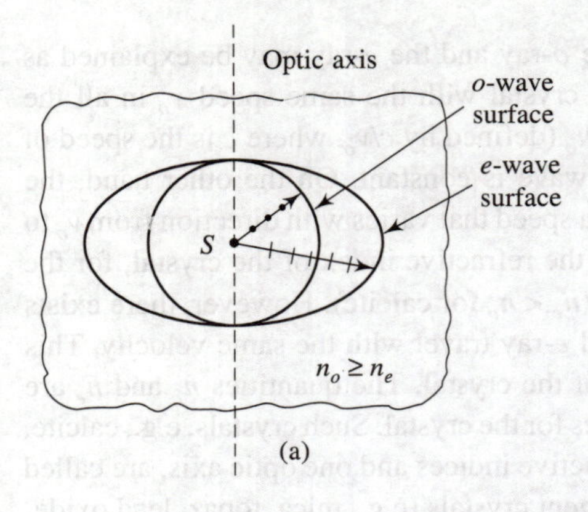

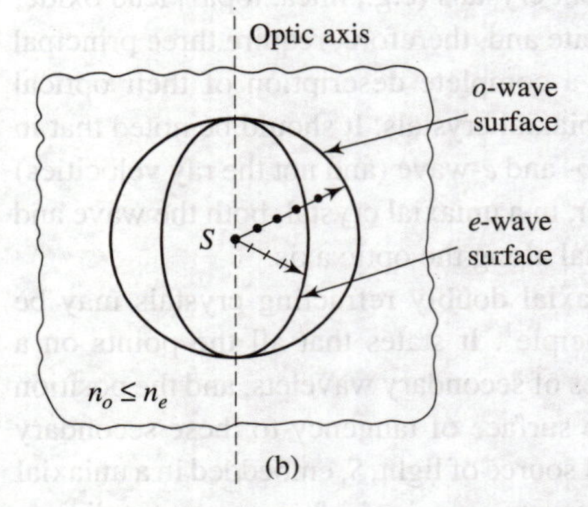

Fig. 9.4 Huygen's wave surfaces generated by a point source S embedded in a (a) negative and (b) positive uniaxial crystal. The polarization states for the *o*- and *e*-ray are shown by dots and lines, respectively.

and the crystal is referred to as a negative uniaxial crystal. Calcite, KDP, etc. are examples of this type of crystals.

Figure 9.5 shows four cases in which unpolarized light is incident normally on the surface of a negative uniaxial crystal. In the first case [Fig. 9.5(a)], the crystal slab is cut in such a way that the optic axis is normal to the surface. Let us consider a wavefront that coincides with the crystal surface at time $t = 0$. According to Huygen's principle, we may let any point on this surface serve as a point source for a double set of Huygen's *o*- and *e*-wavelet. The plane of tangency to these wavelets gives the new position of this wavefront at a later time t. As the speed of propagation for the two wavelets is same, v_o, along the optic axis, the beam emerging from the crystal slab will have the same polarization as that of the incident beam.

Figures 9.5(b) and 9.5(c) show another special case where the unpolarized incident beam falls normally on the surface of the crystal cut so that the optic axis is parallel to the surface. In this case too the incident beam is propagated without deviation. However, one can identify the *o*- and *e*-wavefront that propagate through the crystal with different speeds v_o and v_e, respectively. These waves are polarized in orthogonal directions. The *o*-ray is polarized in a direction that is perpendicular to the plane of

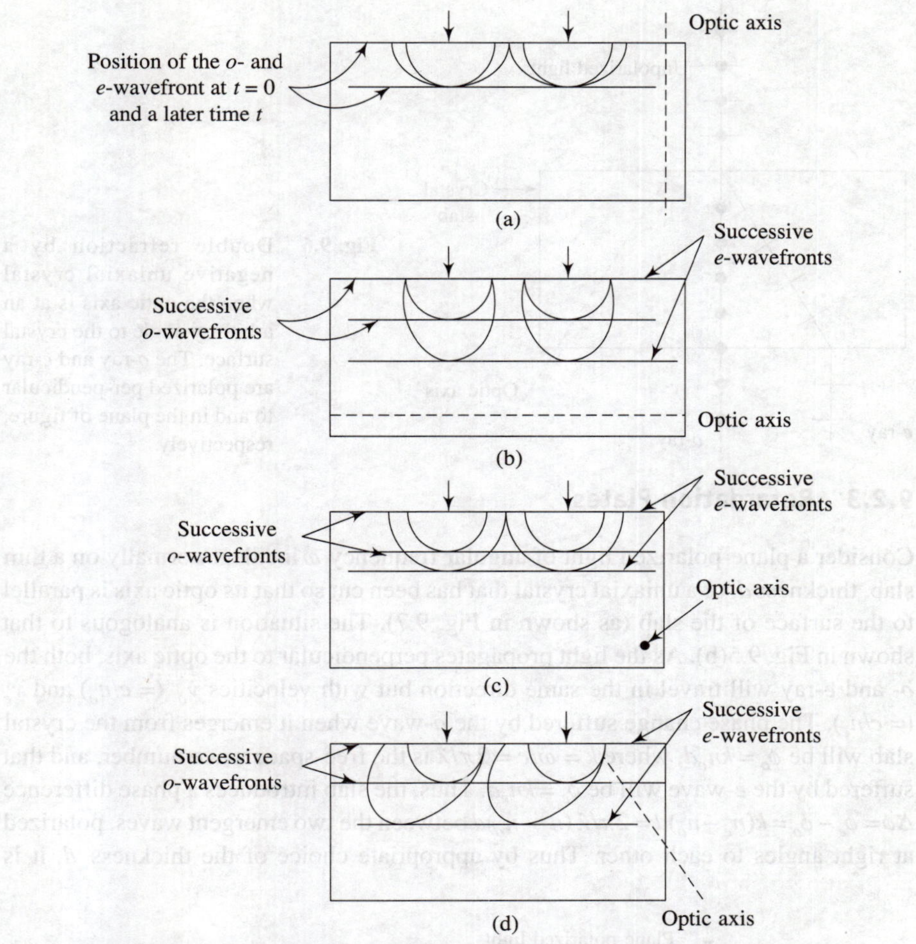

Fig. 9.5 Representation of Huygen's wavefronts for the *o*- and *e*-wave in a negative uniaxial crystal when the optic axis is (a) perpendicular, (b) and (c) parallel, and (d) at an oblique angle to the crystal surface.

the figure and the *e*-ray is polarized in the plane of the figure. The two different velocities of the *o*-ray and the *e*-ray produce a phase difference between the two states of polarization as they propagate. This phenomenon is used in making retardation plates, discussed in the next subsection. Figure 9.5(d) shows unpolarized light incident normally on the surface of a crystal that has been cut so that its optic axis makes an arbitrary angle with the surface. Herein, the *o*- and *e*-ray travel through the crystal at different speeds, the *o*-ray with speed v_o and the *e*-ray with a speed in between v_o and v_e (as a consequence of the fact that its speed varies with direction). In the process, two spatially separated beams are produced at the other surface of the crystal as shown in Fig. 9.6. These two beams are polarized at right angles to each other.

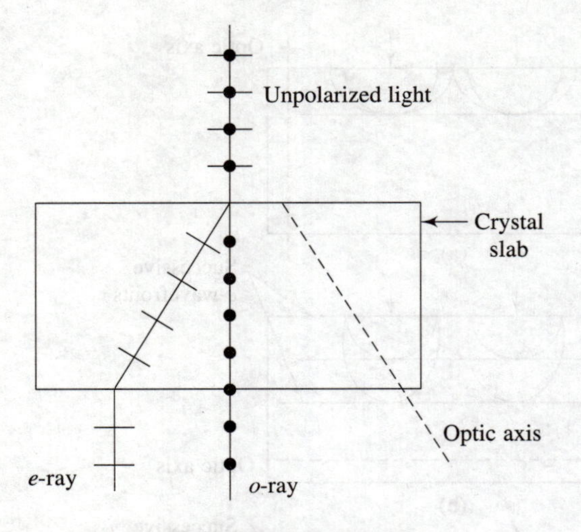

Fig. 9.6 Double refraction by a negative uniaxial crystal when the optic axis is at an arbitrary angle to the crystal surface. The *o*-ray and *e*-ray are polarized per-pendicular to and in the plane of figure, respectively.

9.2.3 Retardation Plates

Consider a plane-polarized light of angular frequency ω incident normally on a thin slab, thickness d, of a uniaxial crystal that has been cut so that its optic axis is parallel to the surface of the slab (as shown in Fig. 9.7). The situation is analogous to that shown in Fig. 9.5(b). As the light propagates perpendicular to the optic axis, both the *o*- and *e*-ray will travel in the same direction but with velocities v_o $(= c/n_o)$ and v_e $(= c/n_e)$. The phase change suffered by the *o*-wave when it emerges from the crystal slab will be $\phi_o = kn_od$, where $k = \omega/c = 2\pi/\lambda$ is the free-space wave number, and that suffered by the *e*-wave will be $\phi_e = kn_ed$. Thus, the slab introduces a phase difference $\Delta\phi = \phi_e - \phi_o = k(n_e - n_o)d = 2\pi/\lambda\,(n_e - n_o)d$ between the two emergent waves, polarized at right angles to each other. Thus by appropriate choice of the thickness d, it is

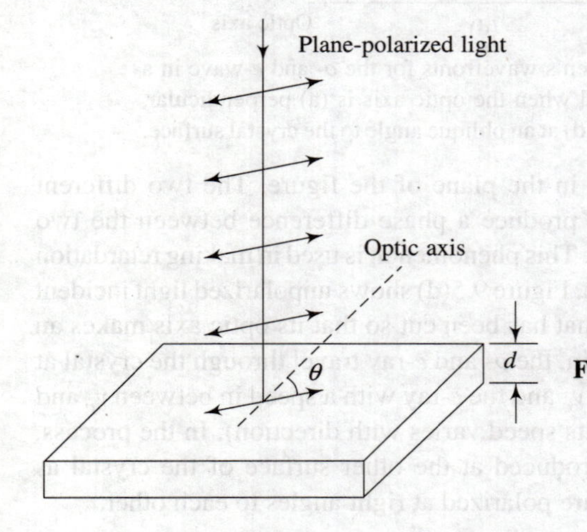

Fig. 9.7 Plane-polarized light incident normally on a uniaxial crystal slab of thickness d. The crystal is cut with its optic axis parallel to the surface.

possible to introduce a predetermined phase difference between the two components for a given wavelength of light. If the incident plane of vibration is at an angle θ to the optic axis, the output beam, in general, will be elliptically polarized. If the thickness of the slab chosen is such that the two emerging plane-polarized waves have a phase difference $\Delta\phi$ of $\pi/2$, the slab is called a quarter-wave plate (as a phase difference of $\pi/2$ corresponds to a path difference of $\lambda/4$). The required thickness may be obtained by the following relation:

$$\Delta\phi = k(n_e - n_o)d = \frac{1}{2}\pi$$

or

$$\frac{2\pi}{\lambda}(n_e - n_o)\,d = \frac{\pi}{2}$$

or

$$d = \frac{\lambda}{4(n_e - n_o)} \tag{9.7}$$

Such a quarter-wave plate is useful for converting plane-polarized light into circularly polarized light and vice versa. Thus, if the input beam of Fig. 9.7 is polarized at an angle $\theta = 45°$ with respect to the optic axis, the two polarization components (parallel and perpendicular to the optic axis) will have equal amplitudes and phases, and upon emerging from the slab they will have a phase difference of $\pi/2$. Thus the resulting output beam will be circularly polarized. In such a device, the directions parallel and perpendicular to the optic axis are called the axes of the quarter-wave plate. For a positive uniaxial crystal, such as quartz, $n_e > n_o$, the o-wave travels faster than the e-wave ($v_o > v_e$). The directions parallel and perpendicular to the optic axis are called slow and fast axes. The situation reverses for the negative uniaxial crystal.

If it is desirable to introduce a phase difference of π, i.e., a path difference of $\lambda/2$, between the two emerging beams, then it can be shown that the required thickness of the slab would be

$$d = \frac{\lambda}{2(n_e - n_o)} \tag{9.8}$$

Such a device is known as a half-wave plate and is useful for rotating the plane of vibration of plane-polarized beams.

Example 9.1 Calculate the thickness of a quarter-wave plate made of quartz and to be used with sodium light, $\lambda = 589.3$ nm. It is given that the principal refractive indices n_e and n_o for quartz are 1.553 and 1.544, respectively

Solution

$$x = \frac{\lambda}{4(n_e - n_o)} = \frac{589.3 \times 10^{-9}\ \text{m}}{4\,(1.553 - 1.544)} = 0.0164\ \text{mm}$$

Example 9.2 A calcite half-wave plate is to be used with sodium light ($\lambda = 589.3$ nm). What should its thickness be? It is given that n_o and n_e for calcite are 1.658 and 1.486, respectively.

Solution

$$x = \frac{\lambda}{4(n_o - n_e)} = \frac{589.3 \times 10^{-9} \text{ m}}{2(1.658 - 1.486)} = 0.0017 \text{ mm}$$

9.3 ELECTRO-OPTIC MODULATORS

9.3.1 Electro-optic Effect

The application of an electric field across a crystal may change its refractive indices. Thus, the field may induce birefringence in an otherwise isotropic crystal or change the birefringent property of a doubly refracting crystal. This is known as the electro-optic effect. If the refractive index varies linearly with the applied electric field, it is known as the Pockels effect, and if the variation in refractive index is proportional to the square of the applied field, it is referred to as the Kerr effect.

In general, the change in the refractive index n as a function of the applied electric field E can be given by an equation of the form

$$\Delta\left(\frac{1}{n^2}\right) = rE + PE^2 \tag{9.9}$$

where r is the linear electro-optic coefficient and P is the quadratic electro-optic coefficient.

In this section, we will study the Pockels effect in detail and see how such an effect may be used to modulate a light beam in accordance with the applied field. Such modulators are very useful in fiber-optic communications.

9.3.2 Longitudinal Electro-optic Modulator

The precise effects of an applied electric field across a crystal showing the Pockels or linear electro-optic effect depend on the crystal structure and symmetry. Let us consider the example of KDP, which is one of the most widely used electro-optic crystals. KDP is a uniaxial birefringent crystal. It possesses a fourfold axis of symmetry (that is, the rotation of the crystal structure about this axis by an angle $2\pi/4$ leaves it invariant), which is chosen as the optic axis of the crystal, normally designated the z-axis. It also possesses two mutually orthogonal twofold axes of symmetry, designated the x-axis and y-axis.

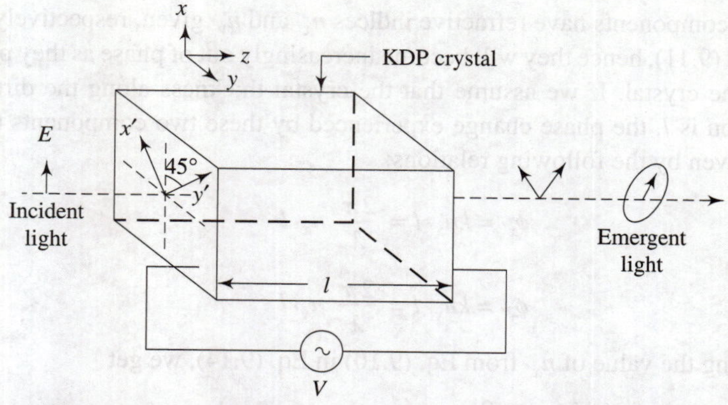

Fig. 9.8 A beam of plane-polarized light propagating along the *z*-axis (optic axis) of a KDP crystal subject to an external electric field applied in the same direction

Let us consider a longitudinal configuration (Fig. 9.8). Herein, a plane-polarized light is propagating along the optic axis (designated the *z*-axis) of a KDP crystal which is being acted upon by an external electric field directed along the *z*-axis. In the absence of external field, the incident wave polarized normal to the *z*-axis (i.e., in the *x-y* plane) will propagate as a principal wave with an ordinary refracting index n_o, because KDP is a uniaxial crystal and the optic axis is along the *z*-axis. Upon the application of electric field E_z along the *z*-axis, the crystal no longer remains uniaxial but becomes biaxial. The principal *x*-axis and *y*-axis of the crystal are rotated through 45° into new principal axes x' and y'. It can be shown that the refractive index $n_{x'}$ for a wave propagating along the *z*-direction and polarized along the x'-direction is given by

$$n_{x'} = n_o + \frac{1}{2} n_o^3 r_{63} E_z \tag{9.10}$$

where r_{63} is an appropriate electro-optic coefficient. Similarly the refractive index $n_{y'}$ for a wave polarized in the y'-direction is given by

$$n_{y'} = n_o - \frac{1}{2} n_o^3 r_{63} E_z \tag{9.11}$$

If the incident wave is represented by the equation

$$E = E_o \cos(\omega t - kz)$$

the components along the x'- and y'-direction will be, respectively, given by

$$E_{x'} = \frac{E_o}{\sqrt{2}} \cos(\omega t - kz) \tag{9.12}$$

and

$$E_{y'} = \frac{E_o}{\sqrt{2}} \cos(\omega t - kz) \tag{9.13}$$

But these components have refractive indices $n_{x'}$ and $n_{y'}$ given, respectively, by Eqs (9.10) and (9.11), hence they will become increasingly out of phase as they propagate through the crystal. If we assume that the crystal thickness along the direction of propagation is l, the phase change experienced by these two components (at $z = l$) may be given by the following relations:

$$\phi_{x'} = kn_{x'}\, l = \frac{2\pi}{\lambda}\, n_{x'}\, l \tag{9.14}$$

and

$$\phi_{y'} = kn_{y'}\, l = \frac{2\pi}{\lambda}\, n_{y'}\, l \tag{9.15}$$

Substituting the value of $n_{x'}$ from Eq. (9.10) in Eq. (9.14), we get

$$\phi_{x'} = \frac{2\pi}{\lambda}\, ln_o\left(1 + \frac{1}{2}\, r_{63}\, n_o^2\, E_z\right) \tag{9.16a}$$

Assuming $2\pi ln_o/\lambda = \phi_o$ and $\pi ln_o^3 r_{63} E_z/\lambda = \Delta\phi$, we have

$$\phi_{x'} = \phi_o + \Delta\phi \tag{9.16b}$$

Similarly, substituting for n_y from Eq. (9.11) in Eq. (9.15), we get

$$\phi_{y'} = \frac{2\pi}{\lambda}\, ln_o\left(1 - \frac{1}{2}\, r_{63}\, n_o^2\, E_z\right) \tag{9.17a}$$

or

$$\phi_{y'} = \phi_o - \Delta\phi \tag{9.17b}$$

where

$$\Delta\phi = \frac{\pi}{\lambda}\, lr_{63}\, n_o^3\, E_z = \frac{\pi}{\lambda}\, r_{63}\, n_o^3\, V \tag{9.18}$$

and $V = E_z l$ is the applied voltage. We see that an extra phase shift $\Delta\phi$ (due to the application of the electric field) for each component is directly proportional to the applied voltage V. Thus if V is made to oscillate with frequency ω_m, that is, if $V = V_0 \sin \omega_m t$, the phase shift $\Delta\phi$ will also vary sinusoidally and the peak value will be $\pi r_{63} n_o^3 V_0/\lambda$. Thus the electro-optic effect may be used for phase modulation. The net phase shift or total retardation (at $z = l$) between the two waves polarized in the x'- and y'-direction as a result of the application of voltage V to the crystal will, therefore, be given by

$$\Phi = \phi_{x'} - \phi_{y'} = 2\Delta\phi = \frac{2\pi}{\lambda}\, r_{63}\, n_o^3\, V \tag{9.19}$$

We know that, in general, the superposition of two plane-polarized waves that are perpendicular to each other produces an elliptically polarized wave. Thus, inspecting Eq. (9.19) it appears in the present case that, in general, the wave emerging at $z = l$ will be elliptically polarized. However, if the superposition gives a phase difference which is an integral multiple of π, the emergent beam will be plane-polarized, and if the phase difference is an odd integer multiple of $\pi/2$, the emergent beam will be circularly polarized.

The voltage $V = V_\pi$ required to introduce a phase shift of π between the two polarization components is called half-wave voltage and may be obtained as follows:

$$\Phi = \pi = \frac{2\pi}{\lambda} r_{63}\, n_o^3\, V_\pi$$

or

$$V_\pi = \frac{\lambda}{2 n_o^3\, r_{63}} \tag{9.20}$$

The half-wave voltage is one of the important parameters of an electro-optic modulator.

Using Eqs (9.12), (9.13), (9.16b), and (9.17b), the equations for the components of the wave emerging from the crystal polarized in the x'- and y'-direction may be written (omitting common phase factors, ϕ_o) as

$$E_{x'(z=\ell)} = \frac{E_o}{\sqrt{2}} \cos(\omega t + \Delta\phi) \tag{9.21}$$

and

$$E_{y'(z=\ell)} = \frac{E_o}{\sqrt{2}} \cos(\omega t - \Delta\phi) \tag{9.22}$$

Now suppose we put a plane polarizer at the output end of the KDP crystal and orient it at right angles to the polarizer producing the original plane-polarized beam as shown in Fig. 9.9.

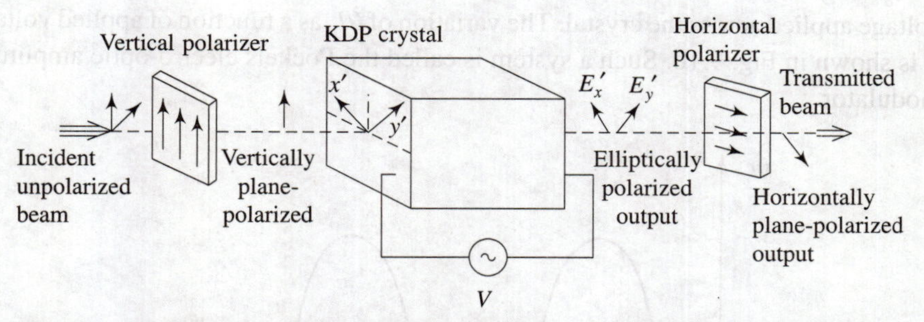

Fig. 9.9 A schematic diagram of a longitudinal electro-optic amplitude modulator using KDP

Then, the transmitted electric-field components will be given by $-E_{x'}/\sqrt{2}$ and $E_{y'}/\sqrt{2}$. Thus, using Eqs (9.21) and (9.22), we can write the following expression for the transmitted electric field:

$$E = \frac{E_o}{2} [-\cos(\omega t + \Delta\phi) + \cos(\omega t - \Delta\phi)]$$

or

$$E = E_o \sin\Delta\phi \sin\omega t \tag{9.23}$$

The intensity of the transmitted beam may be obtained by averaging E^2 [E being given by Eq. (9.23)], over a complete period $T = 2\pi/\omega$. Thus the intensity

$$I = \frac{1}{T}\int_0^T E^2\, dt = \frac{\omega}{2\pi}\int_{t=0}^{2\pi/\omega} E_o^2\,(\sin^2\Delta\phi)(\sin^2\omega t)\, dt$$

$$= \frac{E_o^2}{2}\sin^2\Delta\phi$$

or
$$I = I_o \sin^2\Delta\phi = I_o \sin^2\left(\frac{\Phi}{2}\right) \tag{9.24}$$

where $I_o = E_o^2/2$ is the amplitude of the intensity of the incident beam. Substituting for $\Delta\phi$ from Eq. (9.18) in Eq. (9.24), the transmittance I/I_o of the modulator, shown in Fig. 9.9, will be given by

$$\frac{I}{I_o} = \sin^2\left(\frac{\pi}{\lambda} r_{63}\, n_o^3\, V\right) \tag{9.25}$$

Using Eq. (9.20), this expression may be written as

$$\frac{I}{I_o} = \sin^2\left(\frac{\pi}{2}\frac{V}{V_\pi}\right) \tag{9.26}$$

Thus we can also define V_π as the voltage required for maximum transmission, i.e., $I = I_o$. In general, the transmittance of the modulator can be altered by changing the voltage applied across the crystal. The variation of I/I_o as a function of applied voltage V is shown in Fig. 9.10. Such a system is called the Pockels electro-optic amplitude modulator.

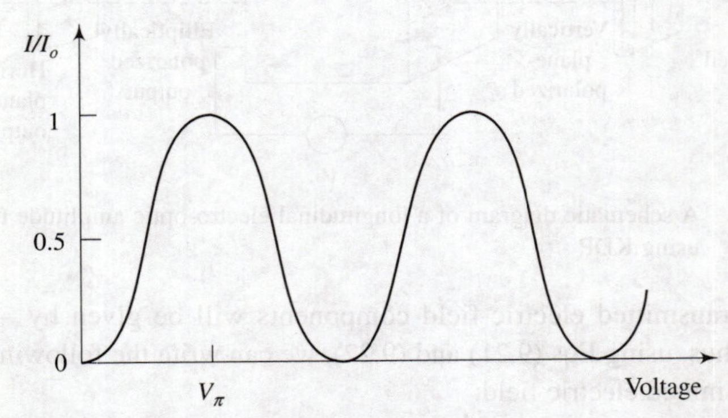

Fig. 9.10 The variation of the transmittance of an amplitude modulator as a function of applied voltage V.

It is clear that if such a modulator is operated around $V = 0$, the output intensity of the modulated beam does not vary linearly with the input signal. In fact, from Eq. (9.26) one can see that for $V \ll V_\pi$, the transmitted intensity is proportional to V^2.

In order to overcome this problem, a common practice is to introduce an external bias, so that with no signal, the transmittance of this modulator is 1/2. In general, it is more convenient to bias the modulator optically to the 50% transmittance point Q, by introducing a quarter-wave plate with its fast and slow axes parallel to the x' and y' axes of the modulator crystal, respectively, as shown in Fig. 9.11. This retarder plate introduces a phase difference of $\pi/2$ between $E_{x'}$ and $E_{y'}$.

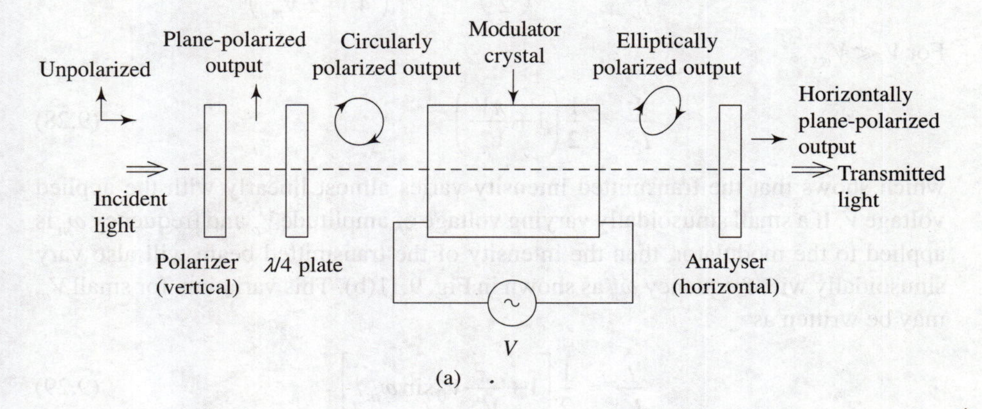

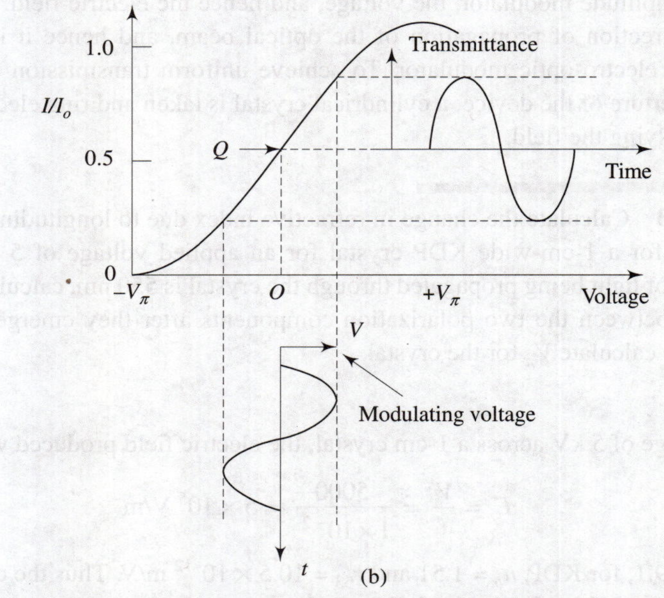

Fig. 9.11 (a) Pockels electro-optic amplitude modulator biased with a quarter-wave plate. (b) An almost linear variation of transmittance with applied voltage, when the modulator is operated about Q.

With this bias, the net retardation Φ between the two polarized components becomes

$$\Phi = \frac{\pi}{2} + 2\Delta\phi = \frac{\pi}{2} + \pi \frac{V}{V_\pi} \tag{9.27}$$

Substituting the value of Φ in Eq. (9.24), we have

$$\frac{I}{I_o} = \sin^2\left(\frac{\Phi}{2}\right) = \sin^2\left(\frac{\pi}{4} + \frac{\pi}{2}\frac{V}{V_\pi}\right)$$

For $V \ll V_\pi$,

$$\frac{I}{I_o} \approx \frac{1}{2}\left(1 + \frac{\pi V}{V_\pi}\right) \tag{9.28}$$

which shows that the transmitted intensity varies almost linearly with the applied voltage V. If a small sinusoidally varying voltage of amplitude V_o and frequency ω_m is applied to the modulator, then the intensity of the transmitted beam will also vary sinusoidally with frequency ω_m as shown in Fig. 9.11(b). This variation, for small V_o, may be written as

$$\frac{I}{I_o} \approx \frac{1}{2}\left[1 + \frac{\pi}{V_\pi} V_0 \sin \omega_m t\right] \tag{9.29}$$

In this amplitude modulator, the voltage, and hence the electric field, is applied along the direction of propagation of the optical beam, and hence it is called a longitudinal electro-optic modulator. To achieve uniform transmission across the effective aperture of the device, a cylindrical crystal is taken and ring electrodes are used for applying the field.

Example 9.3 Calculate the change in refractive index due to longitudinal electro-optic effect for a 1-cm-wide KDP crystal for an applied voltage of 5 kV. If the wavelength of light being propagated through the crystal is 550 nm, calculate the net phase shift between the two polarization components after they emerge from the crystal. Also calculate V_π for the crystal.

Solution

With a voltage of 5 kV across a 1-cm crystal, the electric field produced will be

$$E_z = \frac{V}{\ell} = \frac{5000}{1 \times 10^{-2}} = 5 \times 10^5 \text{ V/m}$$

From Table 9.1, for KDP, $n_o = 1.51$ and $r_{63} = 10.5 \times 10^{-12}$ m/V. Thus the change Δn in the refractive index (that is, the difference between n_o and $n_{x'}$ or n_o and $n_{y'}$) will be

$$\Delta n = \frac{1}{2} n_o^3 r_{63} E_z = \frac{1}{2} \times (1.51)^3 \times 10.5 \times 10^{12} \times 5 \times 10^5$$

or $$\Delta n = 9.04 \times 10^{-6}$$

Therefore, $$\Delta\phi = \frac{2\pi}{\lambda}\,\Delta n\ell$$

$$= \frac{2\pi}{550 \times 10^{-9}} \times 9.04 \times 10^{-6} \times 1 \times 10^{-2}$$

$$\simeq 0.33\pi$$

Thus the net phase shift suffered by the two polarization components [see Eq. (9.19)] will be

$$\Phi = 2\Delta\phi = 0.66\pi$$

Now, the half-wave voltage V_π may be calculated from Eq. (9.20) as

$$V_\pi = \frac{\lambda}{2n_o^3\, r_{63}} = \frac{550 \times 10^{-9}}{2 \times (1.51)^3 \times 10.5 \times 10^{-12}}$$

$$= 7606 \text{ V} \approx 7.6 \text{ kV}$$

9.3.2 Transverse Electro-optic Modulator

Figure 9.12 shows an electro-optic modulator in the transverse mode of operation, where the direction of propagation of light is perpendicular to the direction of the applied field. The advantages of this configuration are that electrodes do not obstruct the beam as in the case of the longitudinal modulator and the retardation (or phase difference), which is proportional to the electric field and the crystal length, can be increased by using longer crystals. It is worth mentioning here that retardation in the

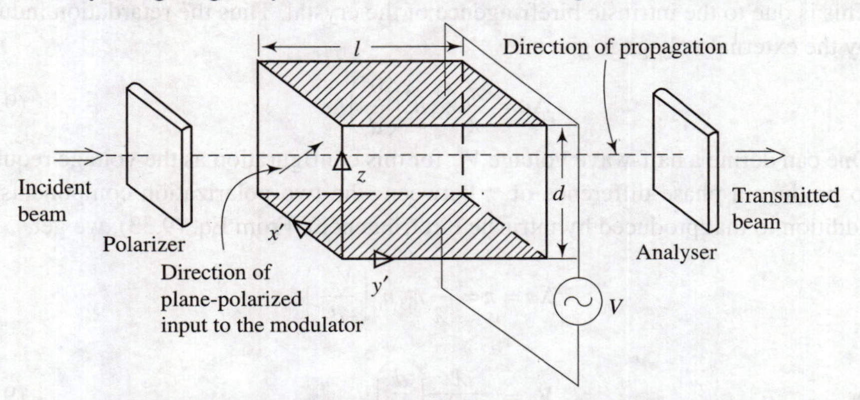

Fig. 9.12 Transverse electro-optic modulator. Herein, the input wave is plane-polarized at an angle of 45° to the x'-direction (in the x'-z plane) and propagated in the y'-direction. The electric field is applied along the z-direction. The analyser is placed with its pass axis normal to that of the polarizer.

case of longitudinal modulators is independent of crystal length. Assume that an electric field is applied along the z-direction, and the direction of propagation is the y' induced principal axis, as shown in Fig. 9.12. Further, assume that the incident light, after passing through the polarizer, is plane-polarized in the x'-z plane at 45° to the x'-principal axis. In the presence of electric field E_z in the z-direction, the refractive indices for a wave propagating along the y'-direction and polarized along the x'- and z-direction are, respectively, given by

$$n_{x'} = n_o + \frac{1}{2} n_o^3 r_{63} E_z \tag{9.30}$$

and

$$n_z = n_e \tag{9.31}$$

Thus the phase difference between the emergent field components along the x'- and y'-direction after traversing a length l of the crystal will be given by

$$\Delta\phi = \phi_{x'} - \phi_z = \frac{2\pi}{\lambda} l(n_{x'} - n_z)$$

$$= \frac{2\pi}{\lambda} l \left[n_o + \frac{1}{2} n_o^3 r_{63} E_z - n_e \right]$$

$$= \frac{2\pi}{\lambda} l(n_o - n_e) + \frac{\pi}{\lambda} r_{63} n_o^3 \left(\frac{V}{d} \right) l \tag{9.32}$$

where V is the voltage applied across the width d of the crystal. It is important to note that even when the applied voltage $V = 0$, there is finite retardation given by

$$(\Delta\phi)_{v=0} = \frac{2\pi}{\lambda} \ell(n_o - n_e)$$

This is due to the intrinsic birefringence of the crystal. Thus the retardation induced by the external voltage is given by

$$\Delta\phi = \frac{\pi}{\lambda} r_{63} n_o^3 \left(\frac{V}{d} \right) \ell \tag{9.33}$$

One can define a half-wave voltage V_π for this configuration as the voltage required to produce a phase difference of π between the two polarization components (in addition to that produced by intrinsic birefringence). From Eq. (9.33), we get

$$\Delta\phi = \pi = \frac{\pi}{\lambda} r_{63} n_o^3 \left(\frac{V_\pi}{d} \right) \ell$$

or

$$V_\pi = \frac{\lambda}{n_o^3 r_{63}} \left(\frac{d}{\ell} \right) \tag{9.34}$$

Contrary to the longitudinal modulator, V_π in this case is not independent of the length l of the modulator crystal, but depends on the ratio d/l. Thus the half-wave voltage may be reduced by employing long, thin crystals.

Some materials that are useful for making electro-optic modulators are listed in Table 9.1.

Table 9.1 Physical parameters of some electro-optic crystals used in Pockels modulators (the values quoted below are near $\lambda = 550$ nm)

Material	Refractive index		Relevant electro-optic
	n_o	n_e	coefficient r (10^{-12} m/V)
KDP (KH_2PO_4)	1.51	1.47	$r_{63} = 10.5$
KD*P (KD_2PO_4)	1.51	1.47	$r_{63} = 26.4$
ADP ($NH_4H_2PO_4$)	1.52	1.48	$r_{63} = 8.5$
Lithium niobate ($LiNbO_3$)	2.29	2.20	$r_{33} = 30.8$
Lithium tantalate (LiTaO3)	2.175	2.18	$r_{33} = 30.3$
Gallium arsenide (GaAs)	3.6	—	$r_{41} = 1.6$

Example 9.4 A transverse electro-optic modulator with a KD*P crystal is operating at a wavelength $\lambda = 550$ nm. The crystal has length $l = 3$ cm and width $d = 0.25$ cm. The optical constants of the crystal may be taken from Table 9.1. Calculate (a) the phase difference between the emergent field components with applied voltage $V = 0$, (b) the additional phase difference between the emergent field components with $V = 2$ V, and (c) the half-wave voltage V_π for the crystal.

Solution

(a) $\Delta\phi$ (due to intrinsic birefringence) $= \dfrac{2\pi l}{\lambda}(n_o - n_e)$

$$= \frac{2 \times 3 \times 10^{-2}}{550 \times 10^{-9}} (1.51 - 1.47)\pi = 4.363 \times 10^3 \pi$$

(b) $\Delta\phi$ (due to external field) $= \dfrac{\pi}{\lambda} r_{63} n_o^3 \left(\dfrac{V}{d}\right) l$

$$= \frac{26.4 \times 10^{-12} \times (1.51)^3}{550 \times 10^{-9}} \times \frac{200 \times 3 \times 10^{-2}}{0.25 \times 10^{-2}} \pi = 0.396\,\pi$$

(c) $$V_\pi = \frac{\lambda}{n_o^3 r_{63}}\left(\frac{d}{l}\right) = \frac{550 \times 10^{-9}}{(1.51)^3 \times 26.4 \times 10^{-12}} \times \left(\frac{25 \times 10^{-2}}{3 \times 10^{-2}}\right) = 504 \text{ V}$$

9.4 ACOUSTO-OPTIC MODULATORS

9.4.1 Acousto-optic Effect

The change in the refractive index of a medium caused by the mechanical strain produced due to the passage of an acoustic wave through the medium is referred to as the acousto-optic effect.

In general, the refractive index of a medium varies with mechanical strain in a complicated manner. Nevertheless, we may consider a simple case, shown in Fig. 9.13, where a monochromatic light of wavelength λ is incident normally on an acousto-optic medium, in which the periodic strain associated with an acoustic wave (wavelength Λ) has produced periodic variations in the refractive index of the medium. As the light enters the medium, the portion of the incident wavefront near the acoustic wave crests (or pressure maxima) encounter higher refractive index and hence advance with a lower velocity than those portions of the wavefront that encounter acoustic wave troughs (or pressure minima). As a consequence, the wavefront in the medium soon acquires a wave-like appearance, as shown in Fig. 9.13. As the velocity of the acoustic wave is much lesser than that of the optical wave, the refractive index in the medium may be considered to have almost no variation. In effect, the acoustic wave sets up a refractive index grating within the medium, so that when a light beam falls on it, either multiple-order or single-order diffraction takes place. The first one is called Raman–Nath diffraction, normally observed at low acoustic frequencies, and the second one is referred to as Bragg diffraction, usually observed at high acoustic frequencies. Acousto-optic modulators operating in the Raman–Nath and Bragg regimes are discussed in the following sections.

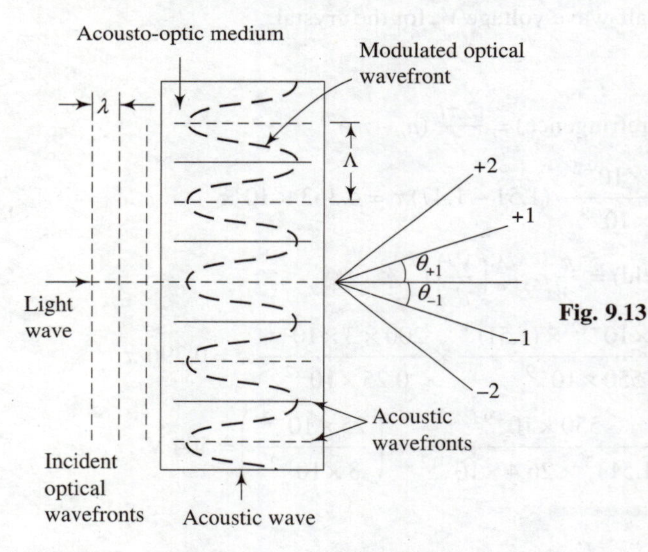

Fig. 9.13 Simplified illustration of acousto-optic modulation. Crests (pressure maxima) and troughs (pressure minima) in the acoustic wave are represented by solid and dashed horizontal lines, respectively.

9.4.2 Raman–Nath Modulator

In the Raman–Nath regime, the acousto-optic diffraction grating is so thin that it behaves almost like a plane transmission grating. In general, the *m*th-order diffracted wave propagates along a direction making an angle θ_m with the direction of the incident beam, given by

$$\sin \theta_m = m\left(\frac{\lambda}{n_0 \Lambda}\right) \tag{9.35}$$

where n_0 is the refractive index of the medium in the absence of the acoustic wave and

$$m = 0, \pm 1, \pm 2, \pm 3, \pm 4, \ldots$$

is the order number.

In this configuration, shown in Fig. 9.14, the signal carrying the information modulates the amplitude of the acoustic wave propagating through the medium. The light beam incident on the acousto-optic medium gets diffracted and the zeroth-order beam of the diffracted output is blocked using a stop. For small acoustic powers, the relative intensity in the first order is given by

$$\eta \approx \frac{\pi^2 (\Delta n)^2 L^2}{\lambda^2} \tag{9.36}$$

where Δn is the peak change in refractive index of the medium due to the acoustic wave and L is width of the acoustic beam, normally equal to the length of the medium. It can be shown that $(\Delta n)^2$ is proportional to the acoustic power. Thus, if the acoustic wave is amplitude-modulated, the first-order diffracted beam (corresponding to $m = \pm 1$) will be intensity-modulated.

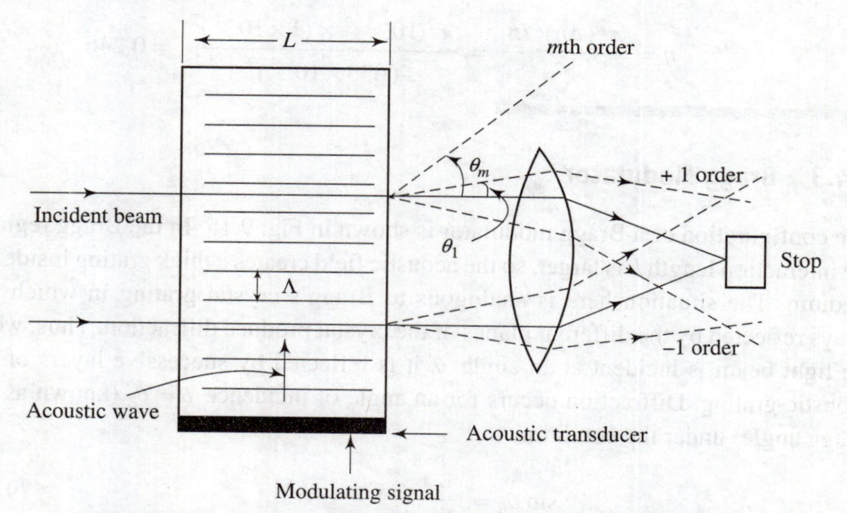

Fig. 9.14 Acousto-optic modulator based on Raman–Nath diffraction

Example 9.5 A typical acousto-optic cell of a Raman–Nath modulator contains water. A piezoelectric crystal (an acoustic transducer of Fig. 9.14) bonded to the cell generates an acoustic wave of frequency 5 MHz in water. The velocity of the acoustic wave in water is 1500 m s^{-1} and the thickness of the cell is 1 cm. If a He–Ne laser beam ($\lambda = 633$ nm) is incident on the cell, calculate the (a) angle between the first-order diffracted beam and the direct beam and (b) relative intensity of the diffracted beam in the first order if $\Delta n = 10^{-5}$.

Solution

(a) From Eq. (9.35), we have

$$\sin\theta_m = m\left(\frac{\lambda}{n_0 \Lambda}\right)$$

$m = 1$, $\lambda = 633$ nm, $n_0 = 1.33$ (for water).

$$\Lambda = \frac{\text{velocity of acoustic waves}}{\text{frequency (Hz)}} = \frac{1{,}500 \text{ m s}^{-1}}{5\times10^6 \text{ s}^{-1}} = 300\times10^{-6} \text{ m}$$

$$= 300\ \mu\text{m}$$

$$\theta_1 = \sin^{-1}\left[\frac{633\times10^{-9}}{1.33\times300\times10^{-6}}\right] = 0.09°$$

(b) From Eq. (9.36),

$$\eta = \frac{\pi^2(\Delta n)^2 L^2}{\lambda^2} = \frac{\pi^2(10^{-5})^2\times(1\times10^{-2})^2}{(633\times10^{-9})^2} = 0.246$$

9.4.3 Bragg Modulator

The configuration of a Bragg modulator is shown in Fig. 9.15. In the Bragg regime, the interaction length L is larger, so the acoustic field creates a thick grating inside the medium. The situation here is analogous to Bragg's crystal grating in which the x-rays reflected by the different planes of the crystal produce diffraction. Thus, when the light beam is incident at an angle θ, it is reflected by successive layers of the acoustic grating. Diffraction occurs for an angle of incidence $\theta = \theta_B$ (known as the Bragg angle) under the condition

$$\sin\theta_B = \frac{\lambda}{2n_0\Lambda} \tag{9.37}$$

For small acoustic power P_a, the diffraction efficiency for an angle of incidence θ_B may be given by

$$\eta \approx \frac{\pi^2 M}{2\lambda^2\cos^2\theta_B}\left(\frac{L}{H}\right)P_a \tag{9.38}$$

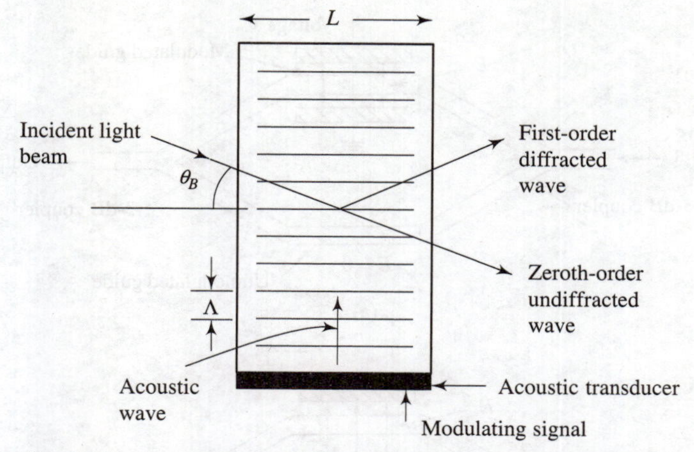

Fig. 9.15 An acousto-optic modulator based on Bragg diffraction

where M is the figure of merit of the acousto-optic device, and L and H are the length and height, respectively, of the acoustic transducer. Thus the intensity of the diffracted beam is directly proportional to the acoustic power and hence modulation of the acoustic power will lead to corresponding modulation of the diffracted beam.

9.5 APPLICATION AREAS OF OPTOELECTRONIC MODULATORS

In fiber-optic systems, electro-optic effect is used to convert phase modulation into intensity variation. Typical structures for achieving this objective are shown schematically in Fig. 9.16. These consist of two planar single-mode waveguides made from a similar electro-optic material. The input beam is split equally between two waveguides using a 3-dB coupler. The split beams after travelling through the waveguides are recombined through a second 3-dB coupler. In one scheme, shown in Fig. 9.16(a), the modulation voltage is applied to one waveguide but not to the other. If the two waveguides are equal in length and the voltage is zero, the optical path lengths of the two arms are equal and the two waves arrive at the second coupler in phase and constructively interfere (producing a 'high' signal). However, if we apply the voltage needed to delay the phase by π, the two waves get out of phase when they recombine. They interfere destructively (producing a 'low' signal). In the second scheme, shown in Fig. 9.16(b), voltages are applied across both the waveguides, but with opposite polarities, so that the voltage increases the refractive index in one arm and decreases it in the other. This differential change in the refractive index may be used to achieve the same modulation as in the first scheme but with lower voltage. Lithium niobate modulators are now widely used for this purpose; modulation at 10 Gbits/s is possible with these.

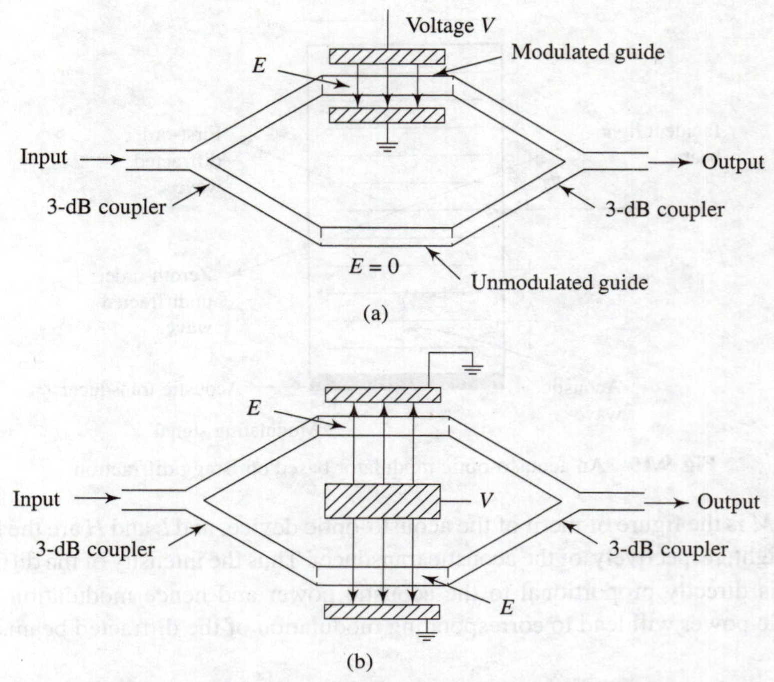

Fig. 9.16 Application of electro-optic planar waveguides in digital modulation

The acousto-optic effect may be used in the spectral analysis of radio-frequency (rf) signals. A typical integrated optic spectrum analyzer consists of an antenna that picks up the rf signal and sends it to an amplifier that drives an acousto-optic planar waveguide to periodically constrict or expand. The spatial period of the acoustic wave helps in getting the signature of the rf signal.

Optoelectronic modulators can also be used as switches or routers in optical networks.

SUMMARY

- Modulation of an optical signal at the source can be achieved either internally or externally. Internal or direction modulation has its limitation, whereas external modulation is faster and allows both amplitude and phase modulations with ease.
- An electromagnetic wave is said to be plane-polarized if the vibrations of its electric-field vector are parallel to each other for all points in the wave. In unpolarized light, the electric-field vectors are randomly oriented. In this case, however, the resultant electric-field vector may be resolved into two components

which have a random phase difference. If the wave is propagating along the z-axis and the two components are along the x- and y-axis and have same amplitudes with a phase difference of $\pi/2$, the wave is said to be circularly polarized. If the amplitudes of the two components are not same, but their phase difference is still $\pi/2$, the resultant wave will be elliptically polarized.

- In an optically isotropic material, such as glass, the velocity of light and, hence, the refractive index of the material do not change with the direction of propagation. However, in optically anisotropic materials, such as calcite, KDP, etc., the velocity of propagation depends on the direction as well as the state of polarization. Thus there are two principal refractive indices: one corresponding to the ordinary ray (which follows Snell's law) and the other corresponding to the extraordinary ray (which does not follow Snell's law). Crystals, for example, calcite and quartz, which have two principal refractive indices and one optic axis (the direction along which the velocity of both the rays is same) are called uniaxial crystals.
- When plane-polarized light is incident normally on a thin slab, of thickness d, of an uniaxial crystal that has been cut such that its optic axis is parallel to the surface of the slab, the ordinary and extraordinary waves emerging from the other side will have a phase difference of $\pi/2$ if

$$d = \frac{\lambda}{4(n_e - n_o)}$$

Such a slab is called a quarter-wave plate and is useful for converting plane-polarized light into circularly polarized light. However, if it is desirable to introduce a phase difference of π between the two emerging beams, the thickness of the slab should be

$$d = \frac{\lambda}{2(n_e - n_o)}$$

- The application of an electric field across a birefringent crystal, e.g., KDP, may change its refractive indices. If this change is linearly proportional to the electrical field, it is called the Pockels electro-optic effect, which may be used for phase modulation. When a voltage V is applied along the optic axis of such a crystal, the incident plane-polarized light splits into two components and the net phase shift between them is given by

$$\Phi = \frac{2\pi}{\lambda} r_{63} n_o^3 V$$

Thus the phase Φ can be changed by changing V. For a longitudinal modulator, the voltage required to introduce a phase shift of π is known as half-wave voltage V_π. Thus,

$$V_\pi = \frac{\lambda}{2 n_o^3 r_{63}}$$

For a transverse modulator,

$$V_\pi = \frac{\lambda}{n_o^3 r_{63}}\left(\frac{d}{l}\right)$$

- The refractive index of a crystal can also be changed by passing an acoustic wave through it. This is called the acousto-optic effect. It produces a grating within the crystal, so that when a light beam falls on it, either multiple-order (Raman–Nath regime) or single-order (Bragg regime) diffraction takes place. Based on this effect, acousto-optic modulators operating in the two regions can be made.

MULTIPLE CHOICE QUESTIONS

9.1 Consider an electromagnetic wave traveling in the *z*-direction. The orthogonal components of its resultant **E**-vector are along the *x*- and *y*-direction. Assume that the two components have same amplitudes but a phase difference of $\pi/2$. The resultant **E**-vector at any point in space

(a) is constant in amplitude but rotates with an angular frequency ω.
(b) changes in amplitude but rotates with an angular frequency ω.
(c) remains stationary.
(d) varies randomly.

9.2 The electromagnetic wave of Question 9.1 is said to be

(a) unpolarized. (b) plane-polarized.
(c) circularly polarized. (d) elliptically polarized.

9.3 In a birefringent crystal,

(a) the *o*-ray follows Snell's law but the *e*-ray does not.
(b) the *e*-ray follows Snell's law but the *o*-ray does not.
(c) both the *o*-ray and *e*-ray follow Snell's law.
(d) both the *o*-ray and *e*-ray do not follow Snell's law.

9.4 In a doubly refracting crystal, the optic axis is the direction in which

(a) v_o is greater than v_e. (b) v_o is equal to v_e.
(c) v_o is less than v_e. (d) v_o and v_e vary randomly.

9.5 A uniaxial crystal has

(a) one principal refractive index and no optic axis.
(b) one principal refractive index and one optic axis.
(c) two principal refractive indices and one optic axis.
(d) three principal refractive indices and two optic axis.

9.6 A quarter-wave plate will introduce a phase difference (between two emerging beams) of

(a) $\pi/4$. (b) $\pi/2$. (c) π. (d) $3\pi/2$.

9.7 A half-wave plate will introduce a path difference (between two emerging beams) of

(a) $\lambda/4$. (b) $\lambda/2$. (c) $3\lambda/4$. (d) λ.

9.8 In a longitudinal electro-optic modulator, half-wave voltage is that voltage which introduces the following phase shift between two polarization components:

(a) $\pi/4$ (b) $\pi/2$ (c) π (d) 2π

9.9 In a transverse electro-optic modulator

(a) V_π is independent of the length l and width d of the modulator crystal.
(b) V_π is dependent on the length l but not on the width d of the crystal.
(c) V_π is dependent on the width d but not on the length l of the crystal.
(d) V_π is dependent on the ratio d/l.

9.10 In a Raman–Nath modulator, the acousto-optic grating is

(a) so thin that it behaves almost like a plane transmission grating.
(b) so thick that it behaves almost like Bragg's crystal grating.
(c) analogous to a concave Rowland's grating.
(d) quite complicated.

Answers

9.1	(a)	9.2	(c)	9.3	(a)	9.4	(b)	9.5	(c)
9.6	(b)	9.7	(b)	9.8	(c)	9.9	(d)	9.10	(a)

REVIEW QUESTIONS

9.1 (a) Distinguish between plane-polarized, circularly polarized, and elliptically polarized light.
(b) How can you produce circularly polarized light?

9.2 Calculate the thickness of a quarter-wave plate made of calcite and to be used with sodium light ($\lambda = 589.3$ nm). It is given that the principal refractive indices n_o and n_e for calcite are 1.658 and 1.486, respectively.
Ans: 0.856 μm

9.3 Calculate the thickness of a half-wave plate made of quartz and to be used with sodium light ($\lambda = 589.3$ nm). It is given that the principal refractive indices n_e and n_o for quartz are 1.553 and 1.544, respectively.
Ans: 0.0327 mm

9.4 (a) What is the electro-optic effect?
(b) How can this effect be used for modulating the phase of an optical signal? How can the amplitude of the optical signal be modulated?

9.5 (a) Calculate the change in refractive index due to the longitudinal electro-optic effect for a 5-mm-long crystal of lithium niobate for an applied voltage of 100 V. If the wavelength of light propagating through the crystal is 550 nm, calculate the net phase shift between the two polarization components after they emerge from the crystal.

(b) For the crystal of part (a), calculate V_π. (Refer to Table 9.1 for constants.)

Ans: (a) 3.698×10^{-6}, 0.134π (b) 743.5 V

9.6 A typical transverse electro-optic modulator uses a lithium niobate crystal and operates at 550 nm. (The optical constants of the crystal are given in Table 9.1.)

(a) Calculate the length of the crystal required to produce a phase difference of $\pi/2$ between the emergent field components with zero applied voltage.

(b) Calculate the width of the crystal required to produce an additional phase difference of $\pi/2$ between these components with an applied voltage of 20 V.

(c) Calculate the half-wave voltage V_π for the crystal.

Ans: (a) 1.528 μm (b) 0.041 μm (c) 40 V

9.7 (a) What is the acousto-optic effect?

(b) Distinguish between modulators based on this effect and operating in the Raman–Nath and Bragg regimes.

9.8 A Raman–Nath modulator employs a cell that contains water. An acoustic transducer bonded to the cell generates a wave of frequency 4 MHz in water. The velocity of the acoustic wave in water is 1,500 m s^{-1} and the thickness of the cell is 1.2 cm. Consider a He–Ne laser beam (λ = 633 nm) incident on the cell. At what angle is the first-order diffracted beam observed? (Take n_o = 1.33 for water.)

Ans: 0.0727°

10 Optical Amplifiers

After reading this chapter you will be able to understand the following:

- Need for optical amplification
- Semiconductor optical amplifiers
- Erbium-doped fiber amplifiers
- Fiber Raman amplifiers
- Application areas of optical amplifiers

10.1 INTRODUCTION

In a fiber-optic communication system, digital or analog signals from a transmitter are coupled into the optical fiber. As they propagate along the length of the fiber, these signals get attenuated due to absorption, scattering, etc. and broadened due to dispersion. After a certain length, the cumulative effect of attenuation and dispersion causes the signals to become too weak and indistinguishable to be detected reliably. Before this happens, the strength and shape of the signals must be restored. This can be done by using either a regenerator or an optical amplifier at an appropriate point along the length of the fiber.

A regenerator is an optoelectronic device. It amplifies and cleans up the optical signal in three steps. The first step is to convert an optical signal into an electrical signal and then amplify it electronically; the second step is to clean up the signal pulses using re-timing and pulse-shaping circuits; and the third step is to reconvert the amplified electrical signal into an optical signal. This signal is then coupled into the next segment of the optical fiber. Such regenerators are designed to operate at a specific data rate and format. Their use is usually limited to digital systems, where the digital structure of the pulses makes it possible to discriminate between the signal and the noise. However, such systems may not be useful in long-haul communication systems where noise and dispersion can accumulate to obscure the signal.

An optical amplifier operates solely in the optical domain, that is, it takes in a weak optical signal from one segment of the link, amplifies it optically to produce a strong optical signal (without recourse to photon-to-electron conversion and vice versa), and couples it to the next segment of the link. Such devices offer several advantages over regenerators, namely, (i) they are insensitive to data rate or signal format and (ii) they have large gain bandwidths. Hence a single optical amplifier can simultaneously amplify many wavelength-division multiplexed (WDM) signals propagating through the same fiber. In contrast, if the system employs regenerators, it would need a regenerator for each wavelength. A major disadvantage of the present optical amplifiers is that they cannot regenerate signals, that is, they cannot clean up noise or compensate for dispersion. However, if appropriate steps are taken to reduce noise and compensate for dispersion, such amplifiers are much simpler, less expensive, and widely applicable. Therefore, optical amplifiers have become essential components in high performance, long-haul, and multichannel fiber-optic communication systems.

There are two main classes of optical amplifiers, namely, (i) semiconductor optical amplifiers, which utilize stimulated emission from injected carriers, and (ii) fiber amplifiers, in which the gain is provided by either rare-earth dopants or stimulated Raman scattering. These are discussed in the following sections.

10.2 SEMICONDUCTOR OPTICAL AMPLIFIERS

10.2.1 Basic Configuration

The process of amplification in semiconducting materials through injection laser diodes (ILDs) has already been discussed in Sec. 7.8. In an ILD, the amplifying medium is combined with the optical feedback mechanism (through cleaved optical facets acting as mirrors) to create a resonant cavity in which light passes back and forth. In a semiconductor optical amplifier (SOA), the light signal passes through the amplifying medium only once. The configuration of an SOA is shown in Fig. 10.1. This is the familiar double-hetero structure (DH) . The material of the active layer is chosen such that it has a band gap lower than that of the confining layers. When a forward bias is applied to this DH, electrons from the n-type semiconductor and holes from the p-type semiconductor travel towards the active layer where they get trapped in a low-band-gap potential well. If the biasing current is large enough, large concentrations of electrons and holes build up in the active layer, leading to *population inversion*. Signal photons passing through the active layer can stimulate radiative recombination of electrons and holes, resulting in the amplification of signal power. This is the basic principle underlying the functioning of this structure as an optical amplifier. It is also possible that the carriers (electrons and holes) recombine spontaneously, leading to amplified spontaneous emission (ASE), or even decay non-radiatively.

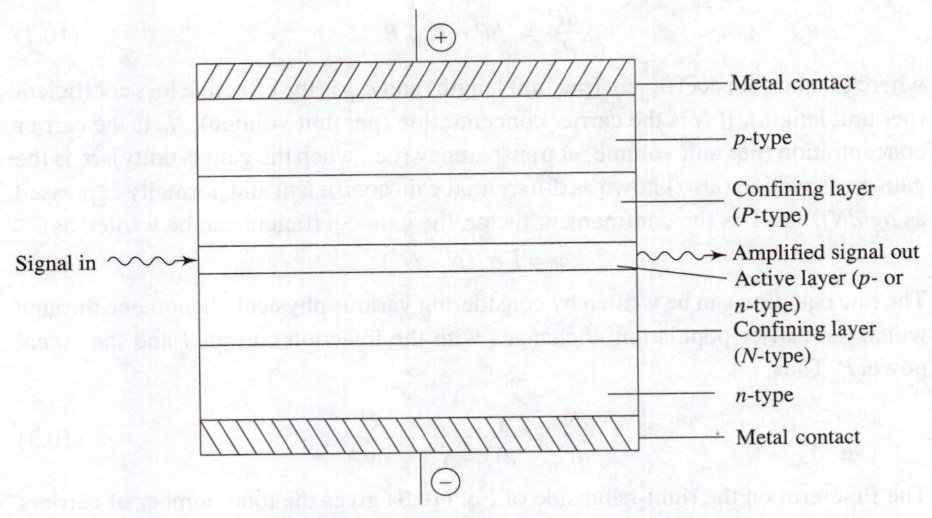

Fig. 10.1 Schematic diagram of an SOA

In order that a DH functions efficiently as an optical amplifier, several requirements must be met:

(i) The active layer should have a band gap lower than that of the surrounding layers so that the carriers are confined in this layer and population inversion is achieved. This implies that the lifetime of the carriers should also be sufficiently long.

(ii) The active layer should also confine the light passing through the structure. Its lower band gap with respect to the confining layers implies a larger refractive index of this layer, leading to waveguiding within this region.

(iii) The energy of signal photons should match with that of the inverted active layer in order to achieve optical gain. Further amplification should be independent of the polarization of the signal beam.

(iv) The signal beam must be coupled efficiently into and out of the SOA chip, usually to a single-mode optical fiber. This implies that the SOA should function as a single-mode waveguide with a circular beam waist matching the mode field diameter of the single-mode fiber.

(v) Finally, the optical feedback must be suppressed. This means that all measures must be taken to reduce optical reflections at the facets of the active layer to less than 0.01%.

10.2.2 Optical Gain

Let us consider the schematic structure of an SOA, shown in Fig. 10.1. The optical signal power P propagating through such an amplifier may be described by (Shimada & Ishio 1994)

$$\frac{dP}{dz} = gP - \alpha_{\text{eff}} P \tag{10.1}$$

where g is the gain coefficient (per unit length) and α_{eff} is the effective loss coefficient (per unit length). If N is the carrier concentration (per unit volume), N_{tr} is the carrier concentration (per unit volume) at transparency (i.e., when the gain is unity), σ_g is the gain cross section (also known as differential gain coefficient and normally expressed as dg/dN), and Γ is the confinement factor, the gain coefficient can be written as

$$g = \Gamma \sigma_g (N - N_{\text{tr}}) \tag{10.2}$$

The rate equation can be written by considering various physical phenomena through which the carrier population N changes with the injection current I and the signal power P. Thus,

$$\frac{dN}{dt} = \frac{I}{eV} - \frac{N}{\tau_c} - \frac{gP}{h\nu A} \tag{10.3}$$

The first term on the right-hand side of Eq. (10.3) gives the total number of carriers (per unit volume) pumped into the active region by the injection current I. Here, e is the electronic charge and V is the volume of the active region. The second term describes the carrier loss (per unit volume) through non-radiative processes, τ_c being the carrier lifetime. The third term gives the carrier loss (per unit volume) through the stimulated emission process. Here, $h\nu$ is the photon energy and A is the cross-sectional area of the active region.

Under steady-state conditions, $dN/dt = 0$ and the solution for N may be obtained. Setting the left-hand side (LHS) of Eq. (10.3) to be zero and solving for N, we get

$$N = \frac{I\tau_c}{eV} - \frac{gP\tau_c}{h\nu A} \tag{10.4}$$

Substituting this value of N in Eq. (10.2) gives

$$g = \frac{\Gamma \sigma_g \left[\dfrac{I\tau_c}{eV} - N_{tr} \right]}{\left[1 + \dfrac{P}{\left(\dfrac{h\nu A}{\Gamma \sigma_g \tau_c} \right)} \right]} \tag{10.5}$$

If the signal power P is small, the second term in the denominator of Eq. (10.5) may be neglected and hence a small-signal gain coefficient g_0 is obtained which is expressed by the following relation:

$$g_0 = \Gamma \sigma_g \left[\frac{I\tau_c}{eV} - N_{tr} \right]$$

The term $(h\nu A/\Gamma\sigma_g\tau_c)$ gives the saturation power P_{sat} of the amplifier. Thus, in terms of g_0 and P_{sat}, Eq. (10.5) may be written as

$$g = \frac{g_0}{(1 + P/P_{sat})} \tag{10.6}$$

Substituting the value of g from Eq. (10.6) in Eq. (10.1), we get

$$\frac{dP}{dz} = \frac{g_0}{\left(1 + \dfrac{P}{P_{sat}}\right)} P - \alpha_{eff} P \tag{10.7}$$

Neglecting α_{eff} and integrating Eq. (10.7), we get

$$\int_{P=P_{in}}^{P_{out}} \frac{dP}{\left[\dfrac{P}{1 + P/P_{sat}}\right]} = \int_{z=0}^{L} g_0\, dz$$

where P_{in} and P_{out} are the input and output signal powers, respectively, and L is the length of the active region. Solving this, we get

$$\int_{P_{in}}^{P_{out}} \frac{dP}{P} + \frac{1}{P_{sat}} \int_{P_{in}}^{P_{out}} dP = g_0 \int_0^L dz$$

or
$$[\ln P_{out} - \ln P_{in}] + \frac{1}{P_{sat}}[P_{out} - P_{in}] = g_0 L$$

or
$$\left[\ln \frac{P_{out}}{P_{in}}\right] = g_0 L - \left(\frac{P_{out} - P_{in}}{P_{sat}}\right) \tag{10.8}$$

Therefore, the amplifier gain G may be expressed as

$$G = \frac{P_{out}}{P_{in}} = \exp\left[g_0 L - \left(\frac{P_{out} - P_{in}}{P_{sat}}\right)\right] \tag{10.9}$$

If $P_{out} \gg P_{in}$ and the small-signal gain of the amplifier is expressed by $G_0 = \exp(g_0 L)$, we may write to a good approximation,

$$\ln G = \ln G_0 - \frac{P_{out}}{P_{sat}} \tag{10.10}$$

The 3-dB saturation power $(P_{sat})_{3\text{-dB}}$ is defined as the output power P_{out} at which the amplifier gain G has dropped to $G_0/2$. By putting $G = G_0/2$ and $P_{out} = (P_{sat})_{3\text{-dB}}$, we can determine the following from Eq. (10.10):

$$\ln(G_0/2) = \ln G_0 - \frac{(P_{sat})_{3\text{-dB}}}{P_{sat}}$$

or
$$(P_{\text{sat}})_{3\text{-dB}} = \ln(2)\,P_{\text{sat}} = \ln(2)\frac{h\nu A}{\Gamma \sigma_g \tau_c} \tag{10.11}$$

Thus amplifier saturation is governed by the material properties σ_g and τ_c.

10.2.3 Effect of Optical Reflections

Optical reflections at the facets of the active region can severely affect the performance of an amplifier, especially when the single-pass gain is high. At high signal powers, the reflections at the facets create a Fabry–Perot type of resonator, leading to oscillations in the amplifier gain versus wavelength curve. This is called *gain ripple*.

If one assumes that the reflections at the facets are independent of wavelength, the equation governing the transmitted optical power P_{out} relative to the input power P_{in} may be obtained, which leads to a well-known expression for a Fabry–Perot resonator with optical gain (O'Mahony 1988):

$$\frac{P_{\text{out}}}{P_{\text{in}}} = G = \frac{(1-R_1)(1-R_2)\,G_s}{(1-G_s\sqrt{R_1R_2})^2 + 4G_s\sqrt{R_1R_2}\,\sin^2(\phi)} \tag{10.12}$$

where R_1 and R_2 are input and output facet reflectivities, G is the real (i.e., measured) gain, G_s is the single-pass gain, and ϕ is the phase shift that the light wave undergoes on traversing the length L of the amplifier once (i.e., the single-pass phase shift).

$$\phi = \frac{2\pi n(\nu - \nu_0)L}{c} \tag{10.13}$$

where n is the refractive index of the active region material, ν is the incident signal frequency, ν_0 is the frequency of the resonant mode of the amplifier, and c is the speed of light in vacuum.

The 3-dB spectral bandwidth $\Delta\nu = 2(\nu - \nu_0)$ of a single longitudinal mode of a Fabry–Perot amplifier (FPA) may be calculated using Eqs (10.12) and (10.13):

$$\Delta\nu = \frac{c}{\pi n L}\sin^{-1}\left[\frac{1 - G_s\sqrt{R_1R_2}}{(4G_s\sqrt{R_1R_2})^{1/2}}\right] \tag{10.14}$$

It should be noted that the 3-dB spectral bandwidth of a single-pass amplifier (with $R_1 = R_2 = 0$), also known as a pure travelling-wave amplifier (TWA), is determined by the full gain width of the amplifier medium itself, as shown in Fig. 10.2. For near TWAs, however, the passband comprises peaks and troughs whose relative amplitudes are governed by R_1 and R_2, the single-pass gain and the input signal power. The peak-trough ratio of the passband ripple, ΔG, is defined as the difference between the resonant FPA and non-resonant TWA signal gain (see Fig. 10.2). It is given by the following expression:

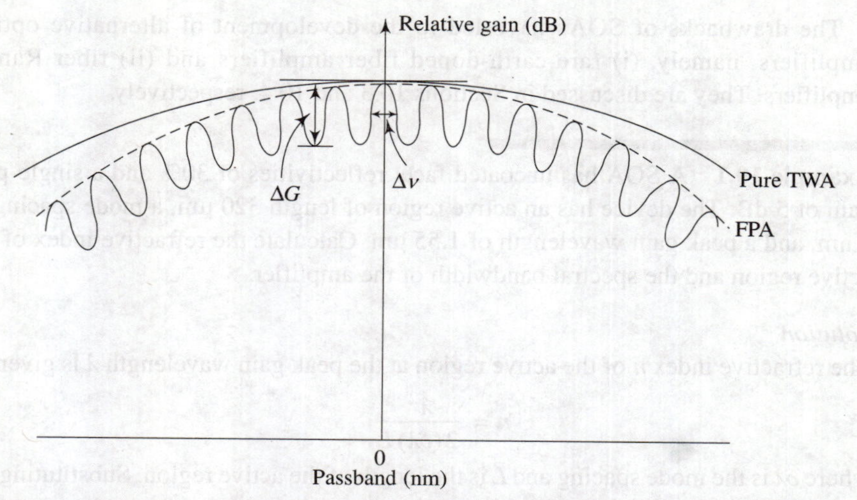

Fig. 10.2 Passband characteristics of pure TWA and FPA showing large ripples in the latter case (schematic)

$$\Delta G = \left(\frac{1 + G_s \sqrt{R_1 R_2}}{1 - G_s \sqrt{R_1 R_2}} \right)^2 \tag{10.15}$$

10.2.4 Limitations

The theory behind the SOA that we have discussed so far is valid under the assumption of continuous-wave (cw) operation. However, optical signals carrying information through optical fibers are modulated in time. If we assume that the signals are intensity-modulated, then according to Eq. (10.10), the gain will adjust itself to the changing signal intensity, but only as long as the carriers can respond to the time-varying signal. In the case of an SOA, the carrier lifetime is of the order of 0.1 ns, hence the gain recovery time is short with respect to the data rate of the order of gigahertz. Therefore, different levels of signal intensity (e.g., 0's and 1's in a digital signal) will experience different optical gains, leading to signal distortion. This effect becomes significant when the SOA is operating close to saturation conditions. This poses an upper limit on the maximum amplifier output power. Further, in many applications it is required that the gain be independent of the polarization of the input signal wave, but the semiconductor active layers are, in general, sensitive to polarization. In a multichannel operation, it is ideally expected that each channel should be amplified by the same amount and there is no crosstalk. In practice, however, the presence of several non-linear phenomena in SOAs leads to interchannel crosstalk, an undesirable feature. In spite of these drawbacks, SOAs have found application in wavelength conversion and fast-switching in WDM networks.

The drawbacks of SOAs have led to the development of alternative optical amplifiers, namely, (i) rare-earth-doped fiber amplifiers and (ii) fiber Raman amplifiers. They are discussed in Sections 10.3 and 10.4, respectively.

Example 10.1 A SOA has uncoated facet reflectivities of 30% and a single-pass gain of 5 dB. The device has an active region of length 320 μm, a mode spacing of 1 nm, and a peak gain wavelength of 1.55 μm. Calculate the refractive index of the active region and the spectral bandwidth of the amplifier.

Solution

The refractive index n of the active region at the peak gain wavelength λ is given by

$$n = \frac{\lambda^2}{2(\delta\lambda)L}$$

where $\delta\lambda$ is the mode spacing and L is the length of the active region. Substituting the values of λ, $\delta\lambda$, and L in the above equation, we get

$$n = \frac{(1.55 \times 10^{-6})^2}{2 \times 1 \times 10^{-9} \times 320 \times 10^{-6}} = 3.75$$

Equation (10.14) gives an expression for the 3-dB spectral bandwidth of the amplifier as follows:

$$\Delta\nu = \frac{c}{\pi n L} \sin^{-1}\left[\frac{1 - G_s\sqrt{R_1 R_2}}{(4G_s\sqrt{R_1 R_2})^{1/2}}\right]$$

$$= \frac{3 \times 10^8 \text{ (m s}^{-1})}{\pi \times 3.75 \times 320 \times 10^{-6} \text{ m}} \sin^{-1}\left[\frac{1 - 3.16\sqrt{0.30 \times 0.30}}{(4 \times 3.16\sqrt{0.30 \times 0.30})^{1/2}}\right]$$

(5-dB gain is equivalent to 3.16)

$$= 2.125 \times 10^9 \text{ Hz}$$

$$= 2.125 \text{ GHz}$$

10.3 ERBIUM-DOPED FIBER AMPLIFIERS

An optical fiber of suitable length (about 10–30 m) that has been doped with a rare-earth element such as erbium (Er), holmium (Ho), neodymium (Nd), samarium (Sm), or ytterbium (Yb) can also serve as an amplifier. A popular material for such an optical fiber is silica-rich glass doped with erbium ions. Therefore, these devices are called *erbium-doped fiber amplifiers* (EDFAs). The reason for their popularity is that they operate in the low-attenuation window around 1.55 μm. The basic configuration of an EDFA is shown in Fig. 10.3. The signal wave from one segment of the optical fiber link, at a wavelength λ_s, is coupled to a short length of an erbium-doped fiber

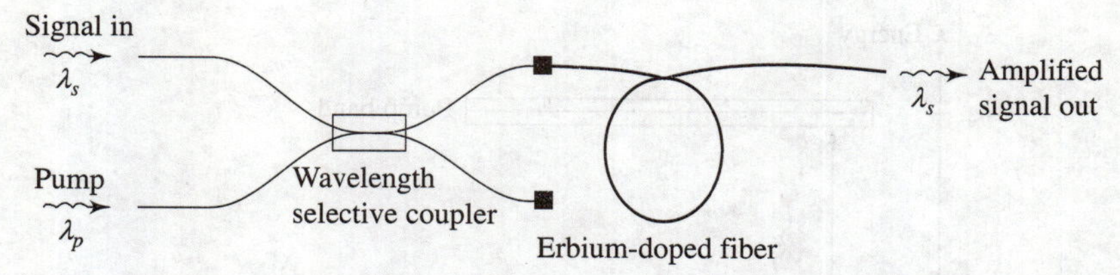

Fig. 10.3 Basic configuration of an EDFA

(EDF) along with the pump wave, usually from a diode laser, at a wavelength λ_p. Pump photons excite Er^{3+} ions and produce population inversion. The passage of the signal wave through the EDF triggers stimulated emission around λ_s and in this process gets amplified. The amplified signal is then coupled into the next segment of the communication link.

10.3.1 Operating Principle of EDFA

In order to understand how an EDFA works, let us look at the energy-level diagram of an erbium ion (Er^{3+}) in silica glass (see Fig. 10.4). Each level is labelled with the corresponding Russel–Saunders coupling term [$^{2S+1}L_J$], where S, L, and J denote the total spin, total orbital angular momentum, and total angular momentum, respectively. The quantum number L is denoted by the letters $S, P, D, F, G, H, I, \ldots$ for $L = 0, 1, 2, 3, 4, 5, 6, \ldots$, respectively. When an Er^{3+} ion is embedded in an amorphous host material such as silica, the individual energy levels are split into a number of sublevels and also get broadened to form energy bands. Only those bands that are important from the point of view of communication applications are shown in Fig. 10.4.

An EDFA uses the process of optical pumping. This requires at least three energy levels (the ground, metastable, and pump levels). The energy of the pumping photon, which corresponds to the difference between the ground and pump levels, is absorbed and the system is raised to the higher excited state (pump level). After reaching there, the electron rapidly loses part of its energy non-radiatively and falls to the metastable level (also known as the lasing level). If the pump power is high, the population in the lasing level may exceed that in the ground level. This is called *population inversion*. Under such a condition, if a signal photon (corresponding to the wavelength of light being transmitted through the optical fiber link, which is 1.55 μm in the present case) passes through this medium, it can trigger a stimulated emission from the lasing level to the ground level, thus producing a new photon that is identical to the signal photon. Therefore, this process requires the energy of the pumping photon to be greater than that of the signal photon. In other words, the pump wavelength should be shorter than the signal wavelength.

There are several ways to optically pump an erbium-doped optical fiber and achieve gain. An intense source of pumping, e.g., a laser emitting 0.98 μm can be used to excite Er^{3+} ions from the ground band $^4I_{15/2}$ to the pump band $^4I_{11/2}$ [as shown by

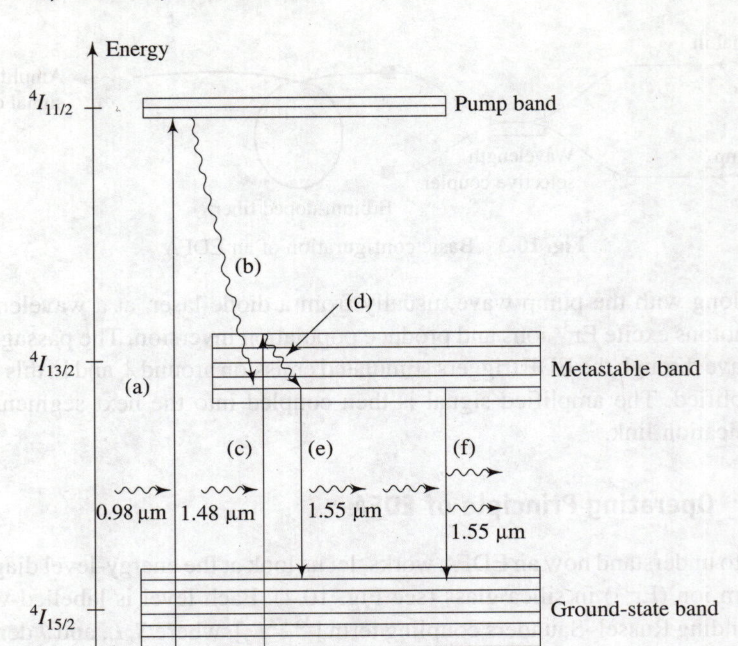

Fig. 10.4 Simplified energy-level diagram of Er^{3+} ion in silica fiber showing various possible transitions

transition (a) in Fig. 10.4]. The excited ions decay non-radiatively in about 1 μs from the pump band to the metastable band [as shown by transition (b) in Fig. 10.4]. Within this band, the electrons of the excited ions tend to populate the lower end of the band. The lifetime of spontaneous emission from this band to the ground-state band is very long (about 10 ms).

Similarly, $^4I_{15/2}$ to $^4I_{13/2}$ transition can be achieved using photons of wavelength 1.48 μm. The absorption of this pump photon excites an electron from the bottom of the ground band to the lightly populated top of the metastable band [as shown by transition (c) in Fig. 10.4]. These electrons then relax to the more populated lower end of the metastable band [transition (d)]. Some of these ions in the metastable state may get de-excited in the absence of any external photons and fall back randomly to ground state with the emission of photons of 1.55 μm. This is called *spontaneous emission* [transition (e)]. However, if a flux of signal photons of energies corresponding to the energy gap between the ground-state band and the metastable band passes through this medium (erbium-doped fiber), stimulated emission may occur, that is, the signal photon may trigger an excited ion to drop back to the ground state, thereby emitting a new photon with identical energy, wave vector, and polarization as the signal photon [transition (f)]; but this transition is possible when population inversion has occurred. Normally, stimulated emission occurs in the wavelength range 1.53–1.56 μm.

In practice, most EDFAs employ 0.98-μm pump lasers, as they are commercially available and can provide more than 100 mW of pump power. Sources of 1.48 μm are also available, but require larger fibers and higher pump powers.

10.3.2 A Simplified Model of an EDFA

The energy bands of an Er^{3+} ion in a silica matrix, shown in Fig. 10.4, can be approximately described as a non-degenerate three-level system, provided the transition is characterized by different absorption and emission cross sections (Desurvire 1994). Assume that the core of the erbium-doped silica fiber has erbium ion density of N_t and that the fiber is single-moded at both pump wavelength (λ_p = 0.98 μm) and signal wavelength (λ_s = 1.55 μm). Further, assume that the population density (number of ions per unit volume) of Er^{3+} ion in the ground state $^4I_{15/2}$ (with energy E_1) is N_1 and that in the upper amplifier level (metastable level) $^4I_{13/2}$ (with energy E_2) is N_2. Pumping this system by λ_p = 0.98 μm takes the ground-state Er^{3+} ions from E_1 to the pump level $^4I_{11/2}$ (with energy E_3), from which the ions rapidly relax to the metastable level (energy E_2). Since the relaxation rate of the pump level is very fast, we may assume that this top level (energy E_3) remains almost empty. Therefore, we may write

$$N_1 + N_2 = N_t \tag{10.16}$$

Let P_p and P_s represent the optical powers for the pump and signal waves, and σ_{pa}, σ_{sa}, and σ_{se} denote the absorption cross section at the pump frequency ($\nu_p = c/\lambda_p$), absorption cross section at the signal frequency ($\nu_s = c/\lambda_s$), and the emission cross section at the signal, respectively. Then the rate of change of population of the ground level (energy E_1) may be written as

$$\frac{dN_1}{dt} = -\frac{\sigma_{pa} P_p}{a_p h\nu_p} N_1 - \frac{\sigma_{sa} P_s}{a_s h\nu_s} N_1 + \frac{\sigma_{se} P_s}{a_s h\nu_s} N_2 + \frac{N_2}{\tau_{sp}} \tag{10.17}$$

where a_p and a_s are the cross-sectional areas of the fiber modes for λ_p and λ_s, and τ_{sp} is the spontaneous emission lifetime for the transition from E_2 to E_1. Further, $(\sigma_{pa}P_p/a_p h\nu_p)\,N_1$ is the rate of absorption per unit volume from the ground level E_1 to the pump level E_3 due to the pump at ν_p, $(\sigma_{sa}P_s/a_s h\nu_s)\,N_1$ is the rate of absorption per unit volume from level E_1 to the metastable level E_2 due to the signal at ν_s, $(\sigma_{se}P_s/a_s h\nu_s)\,N_2$ is the rate of stimulated emission (per unit volume) from level E_2 to level E_1 due to the signal at ν_s, and (N_2/τ_{sp}) is the rate of spontaneous emission (per unit volume) from level E_2 to level E_1.

Similarly, the rate equation for the upper amplifier level N_2 may be written as

$$\frac{dN_2}{dt} = \frac{\sigma_{pa} P_p}{a_p h\nu_p} N_1 + \frac{\sigma_{sa} P_s}{a_s h\nu_s} N_1 - \frac{\sigma_{se} P_s}{a_s h\nu_s} N_2 - \frac{N_2}{\tau_{sp}} \tag{10.18}$$

Equations (10.17) and (10.18) can be reduced to the following compact forms:

$$\frac{dN_1}{dt} = \frac{-\sigma_{\rm pa}\sigma_p}{a_p h\nu_p}N_1 + \frac{P_s}{a_s h\nu_s}[\sigma_{\rm se}N_2 - \sigma_{\rm sa}N_1] + \frac{N_2}{\tau_{\rm sp}} \tag{10.19}$$

and

$$\frac{dN_2}{dt} = \frac{\sigma_{\rm pa}P_p}{a_p h\nu_p}N_1 + \frac{P_s}{a_s h\nu_s}[\sigma_{\rm sa}N_1 - \sigma_{\rm se}N_2] - \frac{N_2}{\tau_{\rm sp}} \tag{10.20}$$

The pump power P_p and the signal power P_s vary along the length of the amplifier due to absorption, stimulated emission, and spontaneous emission. If we neglect the contribution of spontaneous emission, the variation of P_s and P_p along the amplifier length z will be given by

$$\frac{dP_s}{dz} = \Gamma_s\,(\sigma_{\rm se}N_2 - \sigma_{\rm sa}N_1)P_s - \alpha P_s \tag{10.21}$$

$$\pm\frac{dP_p}{dz} = \Gamma_p\,(-\sigma_{\rm pa}N_1)P_p - \alpha' P_p \tag{10.22}$$

where α and α' take into account fiber losses at the signal and pump wavelengths, respectively. Such losses can be neglected for small amplifier lengths.

The confinement factors Γ_s and Γ_p take into account the fact that the doped region within the core provides the gain for the entire fiber mode. The ± sign in the LHS of Eq. (10.22) indicates the direction of propagation of the pump wave (positive for the forward direction and negative for the backward direction).

For lumped amplifiers, the fiber length is small (10–30 m) and hence both the absorption coefficients α and α' can be assumed to be zero. Because N_1 and N_2 are related through Eq. (10.16), we only need to solve either Eq. (10.19) or Eq. (10.20). Let us take Eq. (10.20). Under steady-state conditions

$$\frac{dN_2}{dt} = 0$$

Therefore

$$N_2(z) = -\frac{\tau_{\rm sp}}{a_d h\nu_s}\frac{dP_s}{dz} - \frac{\tau_{\rm sp}}{a_d h\nu_p}\frac{dP_p}{dz} \tag{10.23}$$

assuming pump propagation in the forward direction. Hence $a_d = \Gamma_s a_s = \Gamma_p a_p$ is the cross-sectional area of the doped portion of the fiber core. Substituting $N_2(z)$ from Eq. (10.23) into Eqs (10.21) and (10.22) and integrating over the fiber length, we can get the pump power P_p and signal power P_s in the analytical form at the output end of the doped fiber.

The total gain G for an EDFA of length L can be obtained using the expression

$$G = \Gamma_s \exp\left[\int_0^L (\sigma_{\rm se}N_2 - \sigma_{\rm sa}N_1)\,dz\right] \tag{10.24}$$

where $N_1 = N_t - N_2$ and N_2 is given by Eq. (10.23). Figure 10.5 illustrates how the small-signal gain varies with the doped fiber length L and the pump power P_p.

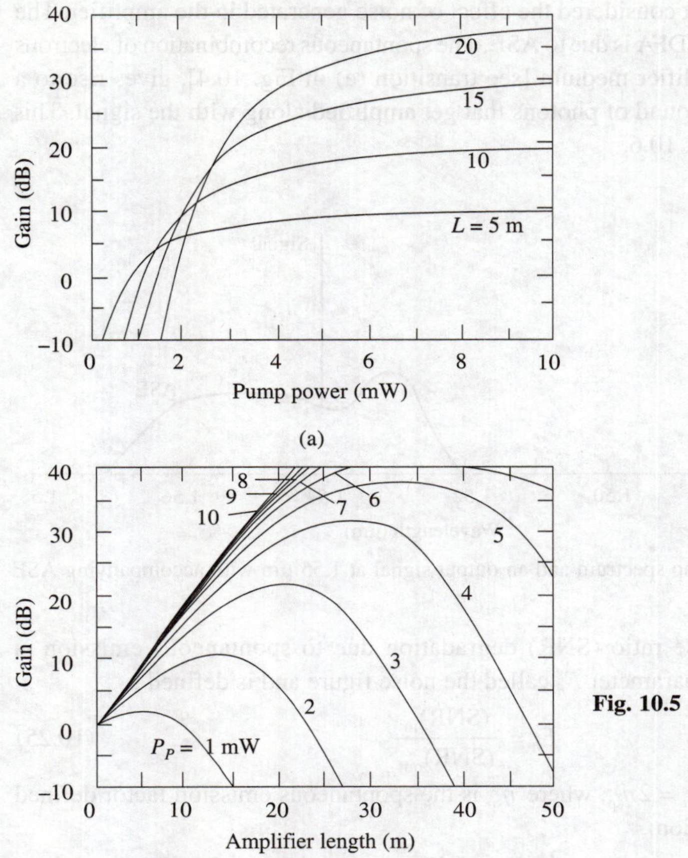

Fig. 10.5 Dependence of EDFA small-signal gain on doped fiber length and pump power for a 1.48-μm pump and 1.55-μm signal (Giles & Desurvire 1991)

For low pump powers, as the EDFA length increases, the gain first increases and becomes maximum at a specific value of L and then drops sharply with further increase in L. The reason for this behaviour is that the pump does not have enough energy to create complete population inversion in the lower portion of the doped fiber. The unpumped section of this fiber, therefore, absorbs the signal, resulting in signal loss rather than gain. As the optimum value of L depends on the pump power P_p, it is essential to choose L and P_p appropriately. The qualitative features shown in Fig. 10.5 are commonly observed in nearly all EDFAs.

The above analysis assumes that the pump and signal beams are continuous waves. However, in practice, the EDFA is pumped by cw ILD and the signal is in the form of pulses of 1's and 0's, whose duration is inversely proportional to the bit rate. Fortunately, owing to a relatively longer lifetime of the excited state (≈ 10 ms) for Er^{3+} ions, the gain does not vary from pulse to pulse.

So far we have not considered the effect of noise generated in the amplifier. The dominant noise in an EDFA is due to ASE. The spontaneous recombination of electrons and holes in the amplifier medium [see transition (e) in Fig. 10.4], gives rise to a broad spectral background of photons that get amplified along with the signal. This effect is shown in Fig. 10.6.

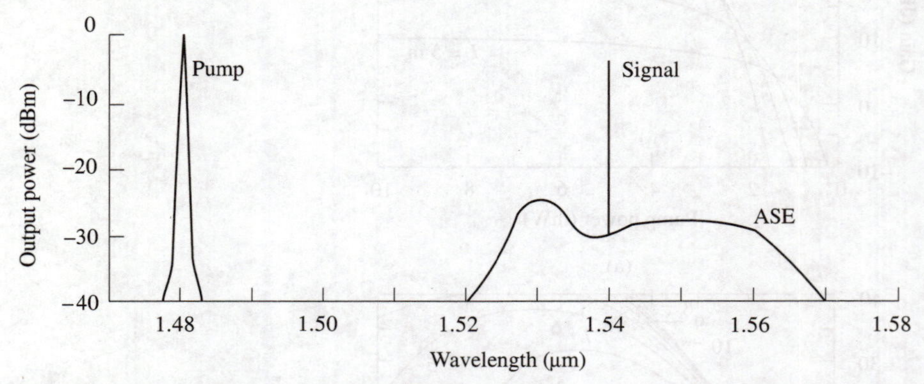

Fig. 10.6 1.48-μm pump spectrum and an output signal at 1.55 μm with accompanying ASE noise

The signal-to-noise ratio (SNR) degradation due to spontaneous emission is quantified through a parameter F_n called the noise figure and is defined as

$$F_n = \frac{(\mathrm{SNR})_{\mathrm{in}}}{(\mathrm{SNR})_{\mathrm{out}}} \tag{10.25}$$

In the present case, $F_n = 2\eta_{\mathrm{sp}}$, where η_{sp} is the spontaneous emission factor defined by the following relation:

$$\eta_{\mathrm{sp}} = \frac{N_2}{(N_2 - N_1)} \tag{10.26}$$

Here, N_1 and N_2 are the relative populations per unit volume in the ground and excited states. η_{sp} denotes the extent of population inversion between the two energy levels. From Eq. (10.26), we can infer that $\eta_{\mathrm{sp}} \geq 1$, with equality holding for an ideal EDFA when the population inversion is complete. Typical values of η_{sp} range from 1.3 to 3.5 depending on the signal wavelength and the pump.

Example 10.2 Consider a typical erbium-doped fiber amplifier with the following parameters: doping concentration = 6×10^{24} m^{-3}, signal wavelength λ_s = 1.536 μm, absorption cross section at λ_s, $\sigma_{\mathrm{sa}} = 4.644 \times 10^{-25}$ m^2, emission cross section at λ_s, $\sigma_{\mathrm{se}} = 4.644 \times 10^{-25}$ m^2, lifetime for spontaneous emission, $\tau_{\mathrm{sp}} = 1.2 \times 10^{-2}$ s, length of the doped fiber = L = 7 m, and Γ_s = 0.80. Assume that (N_2/N_t) is nearly constant over the length of the EDFA and is equal to 0.70. (In actual practice N_1 and N_2 vary with z.) Calculate the small-signal gain of EDFA and the maximum possible achievable gain.

Solution

The total gain G for a lumped EDFA of length L can be obtained using Eq. (10.24). With the above assumptions, simple mathematical manipulation of this equation gives

$$G = \Gamma_s \exp\left[\sigma_{sa} N_t \left\{\left(\frac{\sigma_{se}}{\sigma_{sa}} + 1\right)\frac{N_2}{N_t} - 1\right\} L\right]$$

Here N_1 has been replaced by $N_t - N_2$. Thus,

$$G = 0.80 \exp\left[(4.644 \times 10^{-25}\ \text{m}^2)(6 \times 10^{24}\ \text{m}^{-3}) \times \left\{\left(\frac{4.644 \times 10^{-25}}{4.644 \times 10^{-25}} + 1\right) \times 0.70 - 1\right\} \times 7\ \text{m}\right]$$

or

$$G = 0.80 \exp(7.80) = 1956$$
$$= 32.9\ \text{dB}$$

In the above expression for G, if we substitute $N_2 = N_t$ and $\Gamma_s = 1$, we get the expression for maximum possible achievable gain,

$$G_{\max} = \exp(\sigma_{se} N_t L)$$

Substituting the values of relevant parameters, we get for the present case,

$$G_{\max} = \exp[(4.644 \times 10^{-25}\ \text{m}^2) \times (6 \times 10^{24}\ \text{m}^{-3}) \times (7\ \text{m})]$$
$$= 2.9568 \times 10^8$$
$$= 84.70\ \text{dB}$$

Indeed this value of $G_{\max}$ is quite high. A more realistic estimate of $G_{\max}$ can be obtained using the principle of conservation of energy. Thus if $P_{s,\text{in}}$ and $P_{s,\text{out}}$ are the signal powers at the input and output ends of the erbium-doped fiber at the signal wavelength λ_s, and $P_{p,\text{in}}$ is the input pump power at wavelength λ_p, the following inequality should hold true:

$$P_{s,\text{out}} \le P_{s,\text{in}} + \frac{\lambda_p}{\lambda_s} P_{p,\text{in}}$$

Assuming there is no spontaneous emission, the gain may be written as

$$G = \frac{P_{s,\text{out}}}{P_{s,\text{in}}} \le 1 + \frac{\lambda_p}{\lambda_s}\frac{P_{p,\text{in}}}{P_{s,\text{in}}}$$

This gives

$$G_{\max} = 1 + \frac{\lambda_p}{\lambda_s}\frac{P_{p,\text{in}}}{P_{s,\text{in}}}$$

10.4 FIBER RAMAN AMPLIFIERS

In order to understand the principle of operation of fiber Raman amplifiers (FRAs), let us first visualize the Raman effect. Consider the interaction of light quantum of energy $h\nu$ and wavelength $\lambda = c/\nu$ with a molecule as a collision satisfying the law of conservation of energy. In this encounter, the light quantum suffers a loss of energy (new energy $h\nu'$; $h\nu' < h\nu$) and hence appears in the spectrum as a radiation of increased wavelength λ' $(= c/\nu')$. This is called *Stokes' shift*. The molecule, which takes up the energy, is transported to a higher level of rotation or vibration. This phenomenon is called normal *Raman scattering*.

When giant pulses of short duration and high peak power are incident on a scattering medium such as silica glass, non-linear phenomena are observed. One such process is stimulated Raman scattering (SRS). Herein, the incident light wave of frequency ν induces a gain in the scattering medium (e.g., silica) at another frequency $\nu' = \nu - \nu_r$, where ν_r is the frequency of some Raman-active vibration. If the incident power is above the threshold value, the gain can exceed losses and the scattered beam with frequency ν' gets amplified. The power of the stimulated Raman line has been found to be much greater than that of spontaneous emission. Further, this stimulated emission, unlike the normal Raman effect, is coherent. This phenomenon of SRS has been used in making a fiber Raman amplifier.

The basic configuration of an FRA is shown in Fig. 10.7. Both the pump beam at a frequency ν_p and the input signal beam at frequency ν_s are injected into a specific optical fiber serving as an optical amplifier, through an optical coupler. The pump wavelength λ_p $(= c/\nu_p)$ is converted into a signal wavelength λ_s $(= c/\nu_s)$ by SRS, thereby increasing the power at λ_s. In other words, if a suitable optical fiber is optically pumped by an appropriate source, the signal beam will get amplified as the two beams

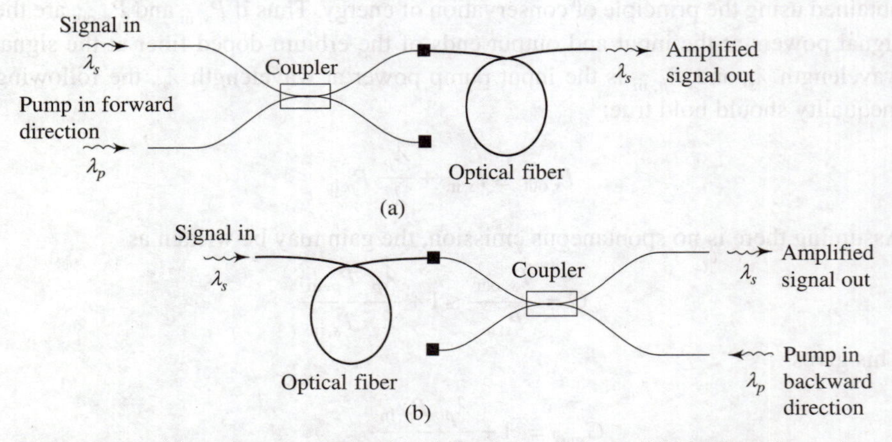

Fig. 10.7 Configuration of an FRA with (a) forward pumping and (b) backward pumping

co-propagate along the fiber. In practice, both forward pumping (i.e., the pump beam in the direction of propagation of the signal beam) and backward pumping (i.e., the pump beam in the direction opposite to that of the signal beam) are possible. Since SRS is not a resonant phenomenon, it does not require population inversion.

In the case of forward pumping, the variations of pump and signal powers along the FRA for small-signal amplification may be studied by solving the following two equations (Agrawal 2002):

$$\frac{dP_s}{dz} = -\alpha_s P_s + \left(\frac{g_R}{a_p}\right) P_p P_s \tag{10.27}$$

and

$$\frac{dP_p}{dz} \approx -\alpha_p P_p \tag{10.28}$$

where α_s and α_p represent fiber losses (per unit length) at the signal and pump frequencies ν_s and ν_p, respectively, P_s and P_p are the signal and pump powers, respectively, that vary along the length z of the fiber, g_R is the Raman gain coefficient, and a_p is the cross-sectional area of the pump beam inside the fiber.

Solving Eq. (10.28), we get the expression for pump power $P_p(z)$ at any point z along the length of the fiber. Thus

$$\int_{P_{p,\text{in}}}^{P_{p(z)}} \frac{dP_p}{P_p} = -\alpha_p \int_0^z dz$$

or

$$P_p(z) = P_{p,\text{in}} \exp(-\alpha_p z), \tag{10.29}$$

where $P_{p,\text{in}}$ is the input pump power (at $z = 0$). Substituting for P_p in Eq. (10.27) from Eq. (10.29), we get

$$\frac{dP_s}{dz} = -\alpha_s P_s + \left(\frac{g_R}{a_p}\right) P_s P_{p,\text{in}} \exp(-\alpha_p z)$$

$$= \left[-\alpha_s + \left(\frac{g_R}{a_p}\right) P_{p,\text{in}} \exp(-\alpha_p z)\right] P_s \tag{10.30}$$

If we assume that the signal power at the input end of the FRA is $P_{s,\text{in}}$ and that at the output end of total fiber length L is $P_s(L)$, solving Eq. (10.30), we have

$$\int_{P_{s,\text{in}}}^{P_s(L)} \frac{dP_s}{P_s} = \int_0^L \left[-\alpha_s + \left(\frac{g_R}{a_p}\right) P_{p,\text{in}} \exp(-\alpha_p z)\right] dz$$

or

$$\ln\left[\frac{P_s(L)}{P_{s,\text{in}}}\right] = \left[\frac{g_R}{a_p} P_{p,\text{in}} \frac{(1 - e^{-\alpha_p L})}{\alpha_p} - \alpha_s L\right]$$

or $$P_s(L) = P_{s,\text{in}} \exp\left[\frac{g_R}{a_p} P_{p,\text{in}} \frac{(1-e^{-\alpha_p L})}{\alpha_p} - \alpha_s L\right]$$

$$= P_{s,\text{in}} \exp\left[\frac{g_R}{a_p} P_{p,\text{in}} L_{\text{eff}} - \alpha_s L\right] \tag{10.31}$$

where L_{eff} is called the effective length of the fiber and is given by

$$L_{\text{eff}} = \left[\frac{1-\exp(-\alpha_p L)}{\alpha_p}\right] \tag{10.32}$$

If $\alpha_p L \gg 1$, $L_{\text{eff}} \approx 1/\alpha_p$. Thus the overall net gain, for small-signal amplification, will be given by

$$G_{\text{FRA}} = \frac{P_s(L)}{P_{s,\text{in}}} = \exp\left[\frac{g_R}{a_p} P_{p,\text{in}} L_{\text{eff}} - \alpha_s L\right] \tag{10.33}$$

This can also be written as

$$G_{\text{FRA}} = \exp\left[(g_0 - \alpha_s)L\right] \tag{10.34}$$

where $$g_0 = \frac{g_R P_{p,\text{in}} L_{\text{eff}}}{a_p L} \tag{10.35}$$

Gain (in dB) may be obtained as follows:

$$\text{Gain (dB)} = 10\log_{10}(G_{\text{FRA}}) = 4.343[(g_0 - \alpha_s)L] \tag{10.36}$$

In this case of backward pumping, for small-signal amplification, Eq. (10.27) for signal variation will be modified. Therein, dP_s/dz will be replaced by $-dP_s/dz$. Other things will remain the same. It is left as an exercise for the student to show that in this case the gain of the amplifier will be given by

$$G_{\text{FRA}} = \frac{P_{s,\text{in}}}{P_s(L)} = \exp[(g_0 - \alpha_s)L] \tag{10.37}$$

where the symbols have their usual meaning.

For fiber Raman amplifiers used either in the forward or backward configuration, gains exceeding 20 dB have been achieved experimentally in a silica fiber, which in principle exhibits a broad spectral bandwidth of up to 50 nm with suitable doping. Such a broad bandwidth is attractive for WDM-based system applications. The main drawback with the FRA is that it requires high power lasers for pumping.

Example 10.3 A fiber Raman amplifier has a length of 2 km. The attenuation coefficients α_s and α_p for signal and pump wavelengths for this fiber are 0.15 and 0.20 dB/km, respectively. Assume that $a_p = 60\ \mu\text{m}^2$ and $g_R = 5 \times 10^{-14}$ m/W. The amplifier is pumped by a laser of 1W power. If the input signal power is 1 μW, calculate (a) the output signal power for forward pumping and (b) the overall gain in dB.

Solution

In the case of forward pumping, Eq. (10.31) may be used. Thus

$$P_s(L) = P_{s,\text{in}} \exp\left[\frac{g_R}{a_p} P_{p,\text{in}} \left\{\frac{1-\exp(-\alpha_p L)}{\alpha_p}\right\} - \alpha_s L\right]$$

$$P_{s,\text{in}} = 1\ \mu\text{W} = 1 \times 10^{-6}\ \text{W}, \quad P_{p,\text{in}} = 1\ \text{W}$$

$$g_R = 5 \times 10^{-14}\ \text{mW}^{-1}, \quad a_p = 60\ \mu\text{m}^2 = 60 \times 10^{-12}\ \text{m}^2$$

$$L = 2\ \text{km} = 2000\ \text{m}, \quad \alpha_s = 0.15\ \text{dB/km}\ (= 3.39 \times 10^{-5}\ \text{m}^{-1})$$

$$\alpha_p = 0.20\ \text{dB/km}\ (= 4.50 \times 10^{-5}\ \text{m}^{-1})$$

$$P_s(L) = 1 \times 10^{-6}\ (\text{W}) \exp\left[\frac{5 \times 10^{-14}\ \text{mW}^{-1}}{6 \times 10^{-1}\ \text{m}^2} \times 1\ (\text{W})\right.$$

$$\left. \times \left\{\frac{1 - \exp[-4.50 \times 10^{-5}\ (\text{m}^{-1}) \times 2000\ (\text{m})]}{4.5 \times 10^{-5}\ \text{m}^{-1}}\right\} - 3.39 \times 10^{-5}\ (\text{m}^{-1}) \times 2000\ (\text{m})\right]$$

$$= 4.582 \times 10^{-6}\ \text{W}$$

$$= 4.582\ \mu\text{W}$$

Therefore the net gain of the amplifier will be

$$G_{\text{FRA}} = \frac{4.582\ \mu\text{W}}{1\ \mu\text{W}} = 4.582$$

$$\text{Gain (in dB)} = 10 \log_{10} G_{\text{FRA}} = 10 \log_{10} (4.582) = 6.61\ \text{dB}$$

10.5 APPLICATION AREAS OF OPTICAL AMPLIFIERS

Optical amplifiers serve a variety of purposes in the design of fiber-optic communication systems. Four of their general applications are illustrated in Fig. 10.8. These are discussed below.

In a fiber-optic communication link employing a single-mode fiber as a transmission medium, the effect of material dispersion may be very small. In such a case, the criterion for repeater spacing is mainly fiber attenuation. Therefore, an optical amplifier may be used to offset the loss and increase the transmission distance. In fact, the use of optical amplifiers is quite attractive for multichannel light wave systems, provided the bandwidth of the multichannel signal is smaller than the amplifier bandwidth. For all the three types of amplifiers discussed in this chapter, this bandwidth

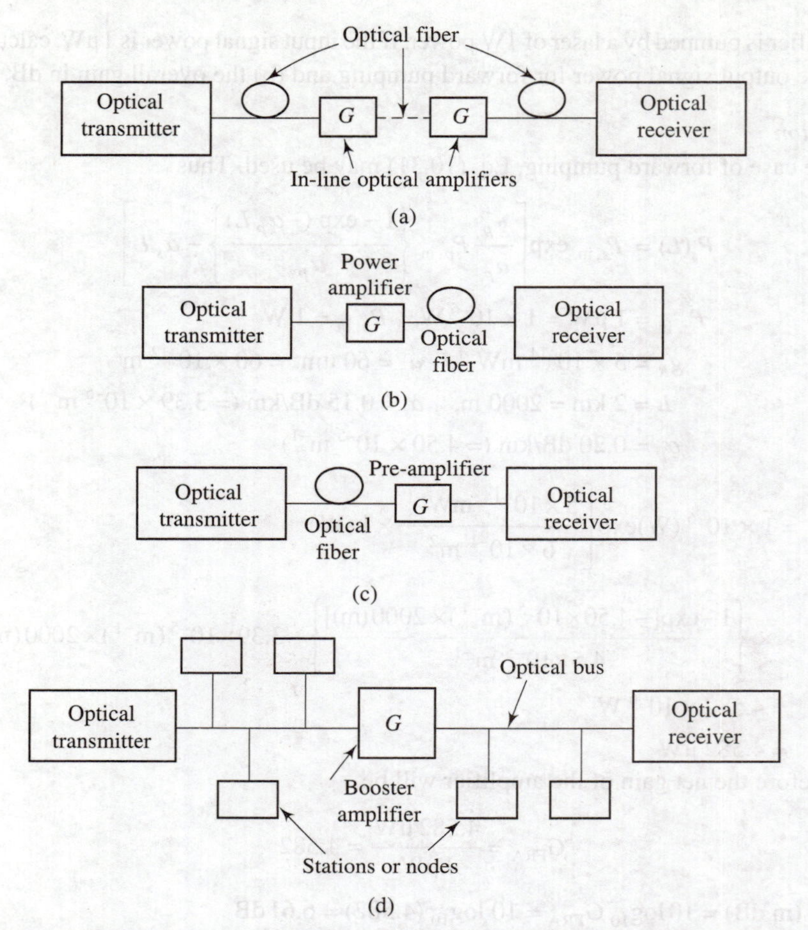

Fig. 10.8 Application of an optical amplifier to (a) increase transmission distance, (b) boost the transmitted power, (c) pre-amplify the received signal, and (d) compensate the distribution losses in a LAN

ranges from 1 to 6 THz. However, the drawback of SOAs is their sensitivity to interchannel crosstalk for channel spacing less than 10 GHz. Raman amplifiers require very high pump powers and long lengths (of the order of kilometres) of optical fibers. In EDFAs, crosstalk does not occur for channel spacing as low as 10 KHz. The required doped fiber length is also quite small (10–30 m). Further, bidirectional propagation is possible in a fiber-optic link carrying multiple wavelengths in both directions through a cascaded chain of EDFAs. Therefore EDFAs have found widespread use in multichannel amplification.

Another way to use an optical amplifier is to increase or boost the transmitted power by placing it just after the transmitter. In this application it may be called a power amplifier or a booster. High pump power will normally be required for this application. It is also possible to use an optical amplifier as a front-end pre-amplifier by putting it just before the receiver to boost the received signal. An optical amplifier may also be used in a local area network (LAN) as a booster amplifier or as a linear-gain block to compensate for coupling insertion loss and power splitting loss.

SUMMARY

- In a fiber-optic communication system, the optical signal gets attenuated as it propagates along the fiber. Therefore, signal strength must be restored at appropriate points along the link. This can be done using either a regenerator or an optical amplifier. A regenerator requires conversion of the optical signal into the electrical domain and reconversion again into the optical domain. Their use is limited to some specific systems. An optical amplifier operates solely in the optical domain.
- There are two main classes of optical amplifiers, namely, (i) semiconductor optical amplifiers (SOAs), which utilize stimulated emission from injected carriers, and (ii) fiber amplifiers, in which gain is provided either by rare-earth dopants or stimulated Raman scattering.
- In an SOA, the light signal passes through the active layer of the forward-biased semiconductor DH and in the process gets amplified. The amplifier gain G is given by

$$G = \frac{P_{\text{out}}}{P_{\text{in}}} = \exp\left[g_0 L - \left(\frac{P_{\text{out}} - P_{\text{in}}}{P_{\text{sat}}} \right) \right]$$

- Optical reflections at the facets of the active region can severely affect the performance of the amplifier, especially when the single-pass gain is high. Several non-linear phenomena in SOAs, lead to interchannel crosstalk in multichannel operation.
- An erbium-doped fiber amplifier (EDFA) operates on the principle of optical pumping of an Er^{3+} ion in silica fiber by either a 0.98-μm or a 1.48 μm source and stimulated emission at 1.55 μm, which is the low-attenuation window of silica-based fibers. The gain of an EDFA depends on several factors, such as doping concentration, fiber length, pump power, etc. These are widely used in multichannel systems.
- A fiber Raman amplifier (FRA) is based on stimulated Raman scattering. Herein, the pump energy at λ_p is transferred to the signal energy at λ_s in a non-resonant process to provide gain at λ_s. For small amplification, the overall gain is given by

$$G_{\text{FRA}} = \exp[(g_0 - \alpha_s)L]$$

where the symbols have their usual meaning.
- Optical amplifiers serve a variety of purposes in the design of fiber-optic communication systems.

MULTIPLE CHOICE QUESTIONS

10.1 What is the difference between a regenerator and an optical amplifier?
(a) A regenerator amplifies as well as restores the signal.
(b) An optical amplifier compensates for transmission loss.
(c) A regenerator converts the optical signal into the electrical domain for amplification and then reconverts it into the optical domain, whereas an optical amplifier operates only in the optical domain.
(d) There is no difference between the two.

10.2 Erbium-doped fiber amplifiers operate at the following windows:
(a) Low-dispersion window (around 1.30 μm)
(b) Low-attenuation window (around 1.55 μm)
(c) Both the windows
(d) None of these

10.3 The structure of a semiconductor optical amplifier differs from a semiconductor laser in the following aspect:
(a) The reflectivity of the end facets of the active region in the SOA is zero.
(b) The reflectivity of the end facets of the active region is 100%.
(c) The SOA is pumped electrically.
(d) There is no difference.

10.4 An SOA differs from an EDFA in the following manner:
(a) An SOA operates in the electrical domain while the EDFA operates in the optical domain.
(b) An SOA is pumped electrically while the EDFA is pumped optically.
(c) An SOA amplifies 1.30 μm while the EDFA amplifies 1.55 μm.
(d) There is no difference.

10.5 Gain in EDFA depends on the following factors
(a) Doping concentration (b) Length of the doped fiber
(c) Pump power (d) All of these

10.6 Which wavelength is most suitable for pumping an EDFA?
(a) 0.85 μm (b) 0.98 μm (c) 1.30 μm (d) 1.55 μm

10.7 In what way does an EDFA differ from a fiber Raman amplifier?
(a) An EDFA requires population inversion while the FRA does not.
(b) An FRA operates on the principle of stimulated Raman scattering.
(c) An EDFA operates on the principle of stimulated emission.
(d) There is no difference.

10.8 In what way are EDFA and FRA similar?
(a) Both of them operate in the all-optical domain.
(b) Both of them can be used around the 1.55-μm window.
(c) Both of them can be employed for multichannel operation.
(d) All of the above.

10.9 Which of the following optical amplifiers is most suited for multichannel bidirectional operation?

(a) SOA (b) EDFA (c) FRA (d) None of these

10.10 Optical amplifiers can be used as

(a) in-line amplifiers to compensate for loss.
(b) power amplifiers to follow the transmitter.
(c) pre-amplifiers to precede the receiver.
(d) all of the above.

Answers

10.1 (c)	10.2 (b)	10.3 (a)	10.4 (b)	10.5 (d)
10.6 (b)	10.7 (a)	10.8 (d)	10.9 (b)	10.10 (d)

REVIEW QUESTIONS

10.1 (a) Explain the basic principle of operation of semiconductor optical amplifiers.
(b) What requirements must be met so that a semiconductor DH functions efficiently as an optical amplifier?

10.2 (a) What is gain ripple in a SOA?
(b) Distinguish between pure TWA and FPA.

10.3 Consider a typical InGaAsP SOA operating at 1.3 μm with the following parameters: active region width = 5 μm, active region thickness = 0.5 μm, active region length = 200 μm, confinement factor $\Gamma = 0.4$, time constant $\tau_c = 1$ ns, $\sigma_g = 3 \times 10^{-20}$ m^2, $N_{tr} = 1.0 \times 10^{24}$ m^{-3}, and bias current $I = 100$ mA. Calculate (a) P_{sat}, (b) the zero-signal gain coefficient, and (c) the zero-signal net gain.
Ans: (a) 31.85 mW (b) 3000 m^{-1} (c) 1.82.

10.4 A SOA has single-pass gain of 10 dB. Calculate the peak-trough ratio of the passband ripple if the facet reflectivities are (i) 0.01% and (ii) 1%.
Ans: (a) 1.0004 (b) 1.4938

10.5 Distinguish between the amplification processes in (a) an erbium-doped fiber amplifier and (b) a fiber Raman amplifier.

10.6 Show that the power conversion efficiency (PCE) of an EDFA, which is defined as

$$\text{PCE} = \frac{P_{s,\text{out}} - P_{s,\text{in}}}{P_{p,\text{in}}}$$

is less than unity; and the maximum value of its quantum conversion efficiency (QCE), which is defined by

$$\text{QCE} = \frac{\lambda_s}{\lambda_p}(\text{PCE})$$

is unity.

10.7 Consider an erbium-doped fiber amplifier being pumped at 0.98 μm with 30 mW pump power. The signal wavelength is 1.55 μm. It is given that the cross-sectional area of the fully doped fiber core is 8.5 μm^2, doping concentration is 5×10^{24} m^{-3}, pump absorption cross section is 2.17×10^{-25} m^2, signal absorption cross section is 2.57×10^{-25} m^2, signal emission cross section is 3.41×10^{-25} m^2, and the input signal power is 200 μW. Assuming that the fiber modes for λ_p and λ_s are fully confined, calculate (a) the rate of absorption per unit volume from the Er^{3+} level E_1 to pump level E_3 due to the pump at λ_p (assuming $N_2 \approx 0$); and (b) the rate of absorption per unit volume from level E_1 to the metastable level E_2 and the rate of stimulated emission per unit volume from level E_2 to level E_1, both due to the signal at λ_s (assuming $N_2 \approx N_1$).

Ans: (a) 1.888×10^{28} m^{-3} s^{-1} (b) 1.78×10^{23} m^{-3} s^{-1}, 1.5640×10^{23} m^{-3} s^{-1}

10.8 Consider that an EDFA being pumped at 0.98 μm is being used as a power amplifier. For an input signal power of 0 dBm at λ_s = 1.55 μm, the output of the amplifier is 20 dBm. Calculate (a) the gain of the amplifier (in dB), and (b) the input pump power required to achieve this gain.

Ans: (a) 20 dB (b) 156.6 mW

10.9 (a) What is the origin of gain saturation in fiber Raman amplifiers? Derive an approximate expression for the saturated amplifier gain.

(b) What are the flexibilities available in FRAs that are not available in SOAs and EDFAs?

Part III: Applications

11 Wavelength-division Multiplexing

After reading this chapter you will be able to understand the following:

- Concepts of wavelength-division multiplexing (WDM) and dense WDM
- Passive components
  - Couplers
  - Multiplexers and demultiplexers
- Active components
  - Tunable lasers
  - Tunable filters

11.1 INTRODUCTION

In the simplest kind of communication system, information is transferred from one point to another in one direction. This requires only one transmitter and one receiver. Such a system is known as a *simplex link*. If the information is to be exchanged, that is, communication to be established, in both the directions, the system needs a transmitter and a receiver at each end. This is called a *duplex link*. Multiplexing is combining two or more signals to be transmitted through the same communication link. Multiplexing of signals may be achieved either electronically or optically. In the electrical domain, digital systems commonly use time-division multiplexing (TDM). In this scheme, signals from different transmitters enter the multiplexer module, which takes a sample of each signal, assigns it a specific time slot, and then combines (multiplexes) all such samples into a single communication link. At the receiver end, a demultiplexer separates these time-slotted samples and directs them to the respective receivers. The latter then restore these to produce original signals. As the multiplexer and demultiplexer modules are electronic circuits, the maximum bit rate at which the system can be operated is limited to about 10 Gbits/s. Analog systems use frequency-division multiplexing (FDM). In this scheme, the channel bandwidth is divided into a number of non-overlapping frequency bands and each signal is assigned to one of these bands. These signals are then combined to generate a signal covering a wider range of frequencies. At the receiver end, individual signals are extracted from the combined FDM signal by appropriate electronic filtering. FDM can also be used in digital system.

The need to send more information through communication links has resulted in the development of better multiplexing schemes. In this hierarchy, wavelength-division multiplexing (WDM) has found major application in fiber-optic communication systems. Therefore, this chapter is devoted to the discussion of WDM and the components needed to implement this scheme.

11.2 THE CONCEPTS OF WDM AND DWDM

In a point-to-point link with a single fiber line, there is an optical source at the transmitting end and an optical detector at the receiving end. With such a scheme, if signals from several optical sources are to be transmitted, each one would require a separate optical fiber. It is important to note at this point that the spectral sources for fiber-optic communication systems, particularly laser diodes, have very narrow spectral bands. Such sources use only a very narrow portion of the available transmision bandwidth of optical fibers as shown in Fig. 11.1. For multichannel operation, therefore, it is possible to operate each source at a different peak emission wavelength (λ_1, λ_2, ..., λ_N) and achieve simultaneous transmission of all the optical signals over the same single fiber. This is the basic concept involved in WDM. This utilizes the transmission capacity of an optical fiber in a better way.

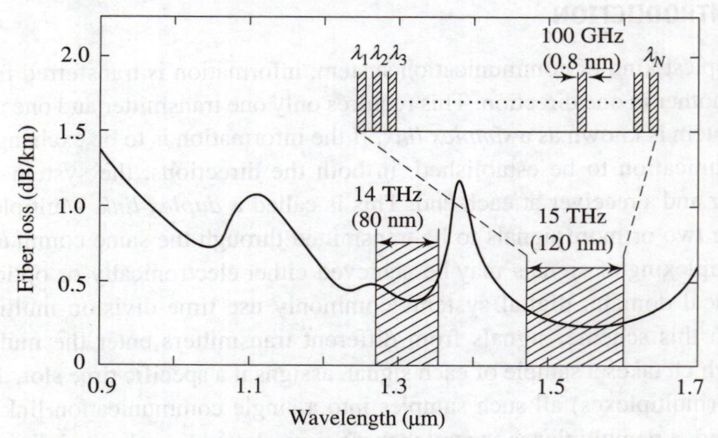

Fig. 11.1 Low-loss second and third transmission windows (around 1.31 and 1.55 μm, respectively) of a typical silica fiber. The inset shows the narrow spectral bands of several sources (typically spaced 0.8 nm apart, which is equivalent to a frequency band of 100 GHz) that can be accommodated in the third window. λ_1, λ_2, etc. denote the peak wavelengths emitted by these sources.

Initially, WDM was used to upgrade the capacity of installed point-to-point fiber-optic links. For example, a 1.3-μm optimized fiber-optic link can be upgraded in capacity by adding another channel at 1.55 μm. Now, with the advent of tunable lasers, which have very narrow spectral line widths, and erbium-doped fiber amplifiers, it has become possible to add many very closely spaced signal bands. For example, for a so-called dry fiber the usable spectral band in the 1.31-μm window is about 80 nm and that in the 1.55-μm window is about 120 nm. This gives a total available bandwidth of about 30 THz in the two windows. Taking into account the availability of lasers with extremely narrow spectral widths, if each source is allotted a frequency band of 100 GHz, one can transmit as many as 300 channels over the same single fiber. This is the concept of dense wavelength-division multiplexing (DWDM). Technically, the ITU-T recommendation G.692 specifies a channel spacing of 100 GHz (0.8 nm at 1552 nm) for DWDM.

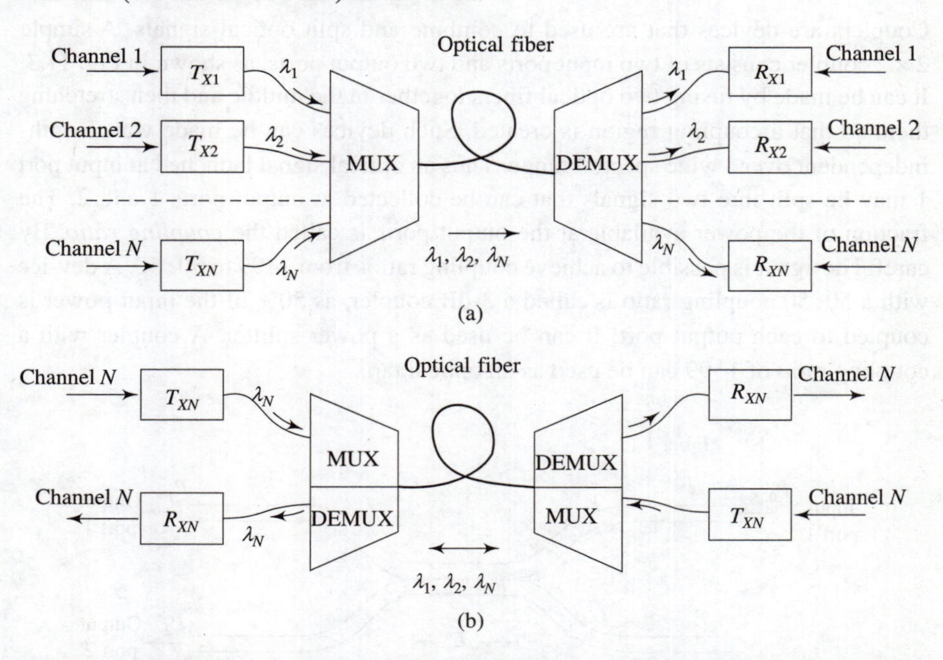

Fig. 11.2 Schematic representation of a (a) unidirectional and (b) bidirectional multichannel fiber-optic communication system using DWDM. Herein T_x and R_x represent the transmitter and receiver, respectively.

The basic configurations of unidirectional and bidirectional multichannel fiber-optic communication systems using DWDM are shown in Fig. 11.2. In a unidirectional link, a wavelength multiplexer (MUX) is used to combine different signal wavelengths (λ_1 to λ_N) at the transmitter end and a wavelength demultiplexer (DEMUX) is used to separate these at the receiver end. A bidirectional scheme, shown in Fig. 11.2(b), involves two-way communication, i.e., sending and receiving information in both the directions. In fact, the implementation of WDM or DWDM requires a variety of passive and active components. These are discussed in the following sections.

11.3 PASSIVE COMPONENTS

The implementation of WDM (or DWDM) requires some optical components that do not need any external control for their operation. These are primarily used to split, combine, or tap off optical signals. The prime components in this category are couplers, multiplexers, and demultiplexers. These are discussed in the following subsections.

11.3.1 Couplers

Couplers are devices that are used to combine and split optical signals. A simple 2×2 coupler consists of two input ports and two output ports, as shown in Fig. 11.3. It can be made by fusing two optical fibers together in the middle and then stretching them so that a coupling region is created. Such devices can be made wavelength-independent over a wide spectral range. Thus an optical signal launched at input port 1 may be split into two signals that can be collected at output ports 1 and 2. The fraction of the power available at the output ports is called the *coupling ratio*. By careful design it is possible to achieve coupling ratios from 1 : 99 to 50 : 50. A dev-ice with a 50 : 50 coupling ratio is called a 3-dB coupler, as 50% of the input power is coupled to each output port. It can be used as a power splitter. A coupler with a coupling ratio of 1 : 99 can be used as an optical tap.

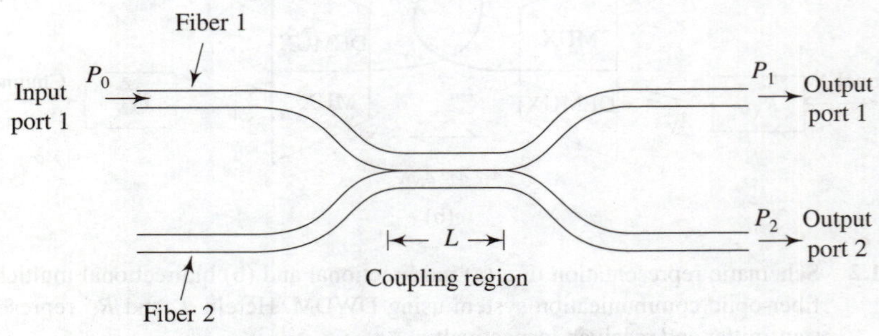

Fig. 11.3 A 2×2 fiber-optic coupler

The coupling mechanism is normally analysed using electromagnetic theory for dielectric waveguides. However, we need not go into that theoretical detail. We can understand this mechanism in a simple manner as follows. We know that the V-parameter of an optical fiber is given by

$$V = \frac{2\pi a}{\lambda} n_1 \sqrt{2\Delta} \tag{11.1}$$

where $2a$ is the core diameter, n_1 is the core index, λ is the wavelength of light propagating through the fiber, and Δ is the relative refractive index difference. In the process of manufacturing a coupler, the fibers are heated, fused together, and stretched. Stretching reduces the core diameter and so also the V-parameter. Thus, the optical power propagating through the core of a single-mode fiber (say) will be less confined. If two identical single-mode fibers are used to make a 2×2 coupler, the power in the single mode propagating through the core of the first fiber will couple to that in the core of the second adjacent fiber in the coupling region (fused portion). By controlling the distance between the fibers it is possible to obtain a desired coupling ratio. Such couplers are called *directional couplers*, because the fibers allow the launched light to pass through them in one direction. If the device allows the light to pass through in two opposite directions, it is called a *bidirectional coupler*. So far, we have considered only wavelength-independent couplers. It is also possible to make the coupling ratio wavelength selective. Such couplers are used to combine or separate two signals of different wavelengths.

Assuming that the above-mentioned 2×2 coupler is loss-less and the two single-mode fibers are identical, the power P_2 coupled from the first fiber into the second fiber over an axial length z is given by (Ghatak & Thyagarajan 1999)

$$P_2 = P_0 \sin^2(\kappa z) \tag{11.2}$$

where P_0 is the power launched at input port 1 (see Fig. 11.3) and κ is the coupling coefficient describing the interaction between the propagating fields in the two fibers.

Assuming that the power is conserved, one can write the following expression for power P_1 delivered to output port 1:

$$P_1 = P_0 - P_2 = P_0\,[1 - \sin^2(\kappa z)] = P_0 \cos^2(\kappa z) \tag{11.3}$$

From Eqs (11.2) and (11.3) one can easily infer that there is a periodic exchange of power between the two fibers. Thus, at $z = m\pi/\kappa$, where $m = 0, 1, 2, \ldots$, $P_1 = P_0$ and $P_2 = 0$, which means that the entire power is in the first fiber; and at $z = (m + 1/2)(\pi/\kappa)$, $m = 0, 1, 2, \ldots$, $P_1 = 0$ and $P_2 = P_0$; that is, the entire power is in the second fiber. The minimum interaction length over which the power is completely transferred from the first fiber to the second fiber is given by

$$z = L_c = \frac{\pi}{\kappa} \tag{11.4}$$

This length L_c is called the coupling length.

Example 11.1 A directional coupler uses two identical single-mode fibers. Determine the interaction length so that the input power P_0 is divided equally at the two output ports.

Solution

$$P_1 = P_2 = P_0/2$$

$\Rightarrow$ $$\sin^2(\kappa L) = \cos^2(\kappa L) = \frac{1}{2}$$

or $$\kappa L = \pi/4$$

This gives the interaction length L to be equal to $\pi/4\kappa$. Note that such a coupler can act as a power divider.

The performance of a directional coupler may be specified in terms of the splitting or coupling ratio, defined as follows (for notations refer to Fig. 11.3):

$$\text{Coupling ratio } (\%) = \left(\frac{P_2}{P_1 + P_2}\right) \times 100 \tag{11.5a}$$

$$\text{Coupling ratio (dB)} = -10\log_{10}\left(\frac{P_2}{P_1 + P_2}\right) \tag{11.5b}$$

So far, we have assumed that the coupler is loss-less. However, in a practical device of this type, some power is always lost when the signal passes through it. There are two basic parameters related to the loss. These are (i) *excess loss*, defined as the ratio of the total output power to the input power, and (ii) *insertion loss* (for a specific port-to-port path), defined as the ratio of power at output port j to power at input port i. Thus, in decibels,

$$\text{Excess loss (dB)} = -10\log_{10}\left(\frac{P_1 + P_2}{P_0}\right) \tag{11.6}$$

and $$\text{Insertion loss (dB)} = -10\log_{10}\left(\frac{P_j}{P_i}\right) \tag{11.7}$$

For a 2 × 2 coupler, for a path from input port 1 to output port 2, using Eqs (11.5b) and (11.6), we can write

$$\text{Insertion loss} = -10\log_{10}\left(\frac{P_2}{P_0}\right)$$

$$= -10\log_{10}\left[\left(\frac{P_2}{P_1 + P_2}\right) \times \left(\frac{P_1 + P_2}{P_0}\right)\right]$$

$$= -10\log_{10}\left[\left(\frac{P_2}{P_1 + P_2}\right) - 10\log_{10}\left(\frac{P_1 + P_2}{P_0}\right)\right]$$

$$= \text{coupling ratio + excess loss} \tag{11.8}$$

Example 11.2 It is required to design a broadband WDM 3-dB coupler so that it splits at λ = 1310 nm and 1550 nm. The two step-index fibers used to make the coupler are identical and single-moded with a core diameter of 8.2 μm, core index n_1 = 1.45, and cladding index n_2 = 1.446. Calculate the position of the output ports with respect to the input port for the two wavelengths.

Solution

From Example 11.1, we know that the interaction length L required to make a 3-dB coupler is $\pi/(4\kappa)$, where κ is the coupling coefficient. A simple empirical relationship given below (Tewari & Thyagarajan 1986) may be used to calculate the value of κ:

$$\kappa = \frac{\pi}{2}\frac{\sqrt{\delta}}{a}\exp[-(A + B\bar{d} + C\bar{d}^2)] \tag{11.9}$$

where

$$A = 5.2789 - 3.663V + 0.3841V^2$$
$$B = -0.7769 + 1.2252V - 0.0152V^2$$
$$C = -0.0175 - 0.0064V - 0.0009V^2$$
$$\delta = \frac{n_1^2 - n_2^2}{n_1^2}, \quad \bar{d} = \frac{d}{a}$$

n_1 is the core refractive index of the fiber, n_2 is the cladding refractive index of the fiber, a is the fiber core radius, and d is the separation between the fiber axis.

Let us take d = 10 μm. With the given parameters, we will have, for λ_1 = 1.31 μm,

$$V_1 = \frac{2\pi a}{\lambda}(n_1^2 - n_2^2)^{1/2} = \frac{\pi \times (8.2\ \mu\text{m})}{(1.31\ \mu\text{m})}[(1.45)^2 - (1.446)^2]^{1/2}$$

or $\quad V_1 = 2.115$

and for λ_2 = 1.55 μm,

$$V_2 = 1.787$$
$$\delta = 5.5096 \times 10^{-3} \text{ and } \bar{d} = 2.439$$

The coupling coefficient for λ_1 will be

$$\kappa_1 = 1.0483\ \text{mm}^{-1}$$

And that for λ_2 will be

$$\kappa_2 = 1.2839\ \text{mm}^{-1}$$

Therefore the interaction lengths L_1 and L_2 for λ_1 = 1.31 μm and λ_2 = 1.55 μm, respectively, will be given by

$$L_1 = \frac{\pi}{4\kappa_1} = \frac{\pi}{4 \times 1.0483} = 0.7488\ \text{mm}$$

and

$$L_2 = \frac{\pi}{4\kappa_2} = \frac{\pi}{4 \times 1.2839} = 0.6114\ \text{mm}$$

Thus the output port positioned at 0.6114 mm with respect to the input port will gather signals at λ_1 = 1310 nm and that positioned at 0.7488 mm will gather signals at λ_2 = 1550 nm. Therefore the coupler can be used as a WDM device.

An obvious generalization of a 3-dB 2×2 coupler is an $N \times N$ star coupler. This device has N input ports and N output ports. In an ideal star coupler the optical power from each input is divided equally among all the output ports. $N \times N$ couplers can be made by fusing together the desired number of fibers as shown in Fig. 11.4 or by suitably interconnecting 3-dB couplers.

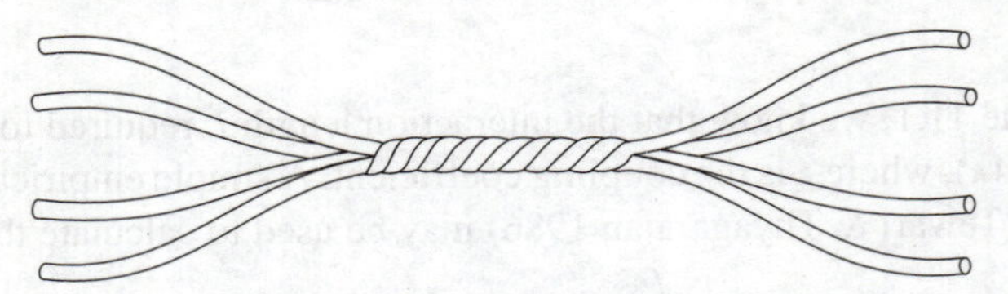

Fig. 11.4 4×4 fused-fiber star coupler

11.3.2 Multiplexers and Demultiplexers

In order to implement a WDM-based system, a multiplexer is required at the transmitting end to combine optical signals from several sources into a single fiber, and a demultiplexer is needed at the receiving end to separate the signals into appropriate channels. As optical sources, e.g., an LED or ILD, do not emit significant amount of optical power outside their designated spectral channel width, interchannel crosstalk is relatively unimportant at the transmitting end. The design problem needing attention here is that the multiplexing device should have low insertion loss. However, there exists a different requirement for demultiplexers, because photodetectors are generally sensitive over a broad range of wavelengths, which may include all the WDM channels. Therefore, a demultiplexer design must be such that it provides good channel isolation of the different wavelengths being used.

In fact, these devices are based on the reversible structure. Hence, any wavelength-division demultiplexer (at least in principle) can also be used as a multiplexer by simply exchanging the input and output directions. Therefore, the following discussion will consider only wavelength-division demultiplexers.

Commonly used wavelength-division demultiplexers (and multiplexers) may be classified into two categories. These are (i) interference filter based devices and (ii) angular dispersion based devices. The two types of devices are discussed, in brief, below.

The basic configuration of a two-wavelength (or two-channel) interference filter demultiplexer is shown in Fig. 11.5.

An interference filter consists of a thin film obtained by depositing several dielectric layers of alternately low and high refractive index. When light propagates through such a structure, it undergoes multiple reflections, giving rise to either constructive or destructive interference depending on the wavelength. Therefore a filter can be designed to produce high transmittance in a given wavelength range and high reflectance outside this range. In Fig. 11.5(a), appropriate conventional microlenses

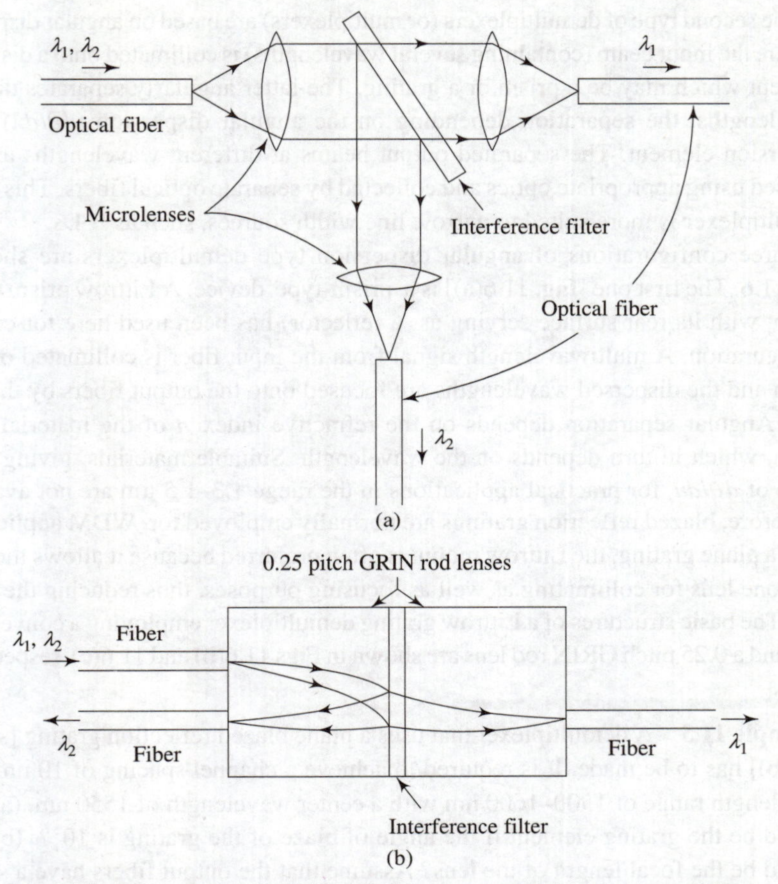

Fig. 11.5 The basic configuration of a two-wavelength (or two-channel) interference filter demultiplexer employing (a) conventional microlenses and (b) GRIN rod lenses.

are used for collimating and focusing light. The incident beam consists of two wavelengths, λ_1 and λ_2. The filter transmits the wavelength λ_1 and reflects the wavelength λ_2, thus demultiplexing the two channels. A compact low-loss two-channel demultiplexer (or multiplexer) may be implemented by employing two 0.25 pitch graded refractive index (GRIN) rod lenses as shown in Fig.11.5(b). These lenses are used for collimating and focusing. The filter is deposited at the interface between these lenses (say, on either of the lens faces). Off-axis entry at the first lens makes it possible to easily separate the reflected beam.

Interference filters can, in principle, be used in series to separate N wavelength channels. However, the complexity involved in cascading the filters and the increase in signal loss that occurs with the addition of each filter generally limit the operation to four or five filters (that is, four or five channels).

The second type of demultiplexers (or multiplexers) are based on angular dispersion. Herein, the input beam (containing several wavelengths) is collimated onto a dispersive element which may be a prism or a grating. The latter angularly separates different wavelengths, the separation depending on the angular dispersion ($d\theta/d\lambda$) of the dispersion element. The separated output beams at different wavelengths are then focused using appropriate optics and collected by separate optical fibers. This type of demultiplexer is more suited to narrow line width sources, such as ILDs.

Three configurations of angular dispersion type demultiplexers are shown in Fig. 11.6. The first one [Fig. 11.6(a)] is a prism-type device. A Littrow prism (a half-prism, with its rear surface serving as a reflector) has been used here for compact configuration. A multiwavelength signal from the input fiber is collimated onto the prism and the dispersed wavelengths are focused onto the output fibers by the same lens. Angular separation depends on the refractive index n of the material of the prism, which in turn depends on the wavelength. Suitable materials, giving a high value of $d\theta/dn$, for practical applications in the range 1.3–1.5 μm are not available. Therefore, blazed reflection gratings are normally employed for WDM applications. With a plane grating, the Littrow mount is often preferred because it allows the use of only one lens for collimating as well as focusing purposes, thus reducing the device size. The basic structures of a Littrow grating demultiplexer employing a conventional lens and a 0.25 pitch GRIN rod lens are shown in Figs 11.6(b) and 11.6(c), respectively.

Example 11.3 A demultiplexer that uses a plane blazed reflection grating [see Fig. 11.6(b)] has to be made. It is required to achieve a channel spacing of 10 nm in the wavelength range of 1500–1600 nm with a center wavelength of 1550 nm. (a) What should be the grating element if the angle of blaze of the grating is 10°? (b) What should be the focal length of the lens? Assume that the output fibers have a spacing of 150 μm.

Solution

A reflection grating has on its surface an array of parallel grooves which are identical in depth and shape, equally spaced, and are provided with a highly reflective coating (see Fig. 11.7). These grooves are inclined at a specific angle (called the angle of blaze, β) with respect to the grating surface. The grating is highly efficient in diffracting wavelengths close to those for which specular reflection occurs. The wavelength for which maximum efficiency is observed is called the blazed wavelength λ_B. When such a grating is used in the Littrow mode (Khare 1993), the angle of incidence (α) and angle of diffraction (θ) are nearly equal to the angle of blaze as shown in Fig. 11.7; that is, $\alpha \approx \theta \approx \beta$. The fundamental grating equation therefore modifies to, taking the blazed wavelength $\lambda = \lambda_B$,

$$m\lambda_B = d[\sin\alpha + \sin\theta] \approx 2d\sin\beta \qquad (11.10)$$

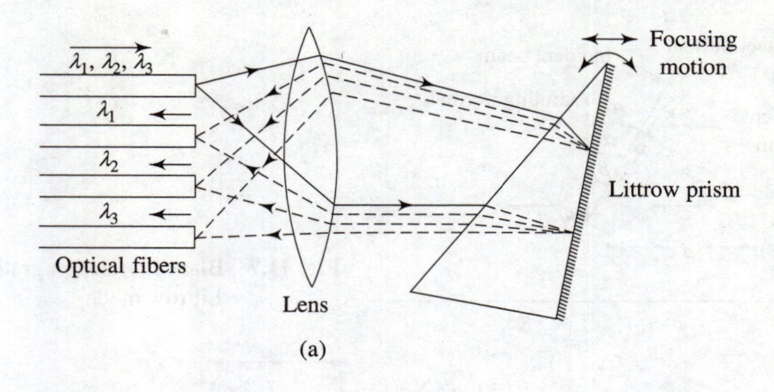

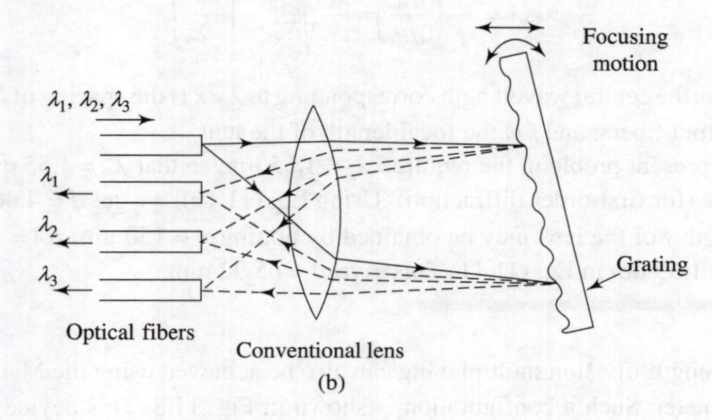

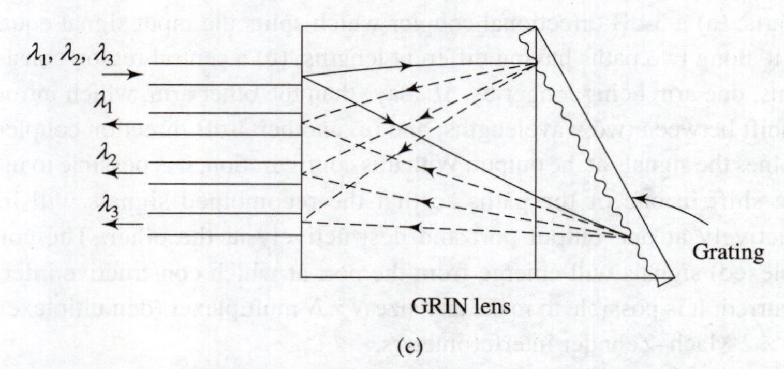

Fig. 11.6 Angular dispersion demultiplexers: (a) Littrow prism type, (b) reflection grating type (with a conventional lens), and (c) reflection grating type (with a GRIN rod lens)

where m is the order of diffraction and d is the grating element (i.e., the distance between two grooves).

With such a demultiplexer using a reflection grating in the Littrow mode, the wavelength channel spacing is given by

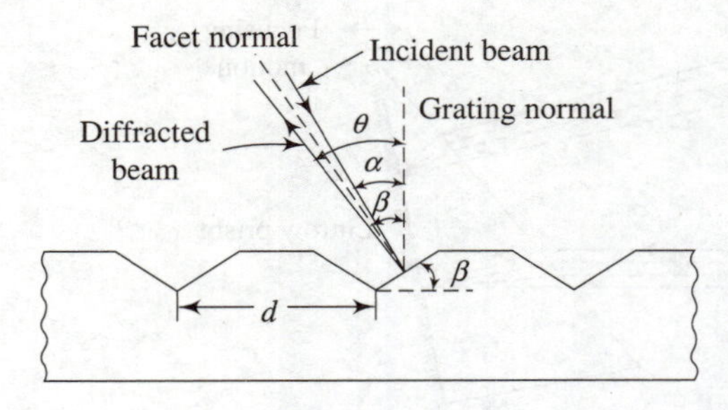

Fig. 11.7 Blazed reflection grating in the Littrow mode

$$\Delta\lambda = \frac{x}{f}\left(\frac{d\lambda}{d\theta}\right) = \frac{x}{f}\sqrt{d^2 - \left(\frac{\lambda_c}{2}\right)^2} \tag{11.11}$$

where λ_c is the central wavelength corresponding to λ_B, x is the spacing of core centres of the output fibers, and f is the focal length of the lens.

In the present problem, the required λ_c = 1.55 μm, so that λ_B = 1.55 μm, β = 10°, and m = 1 (for first-order diffraction). Using Eq. (11.10), we get d = 4.463 μm. The focal length f of the lens may be obtained by putting x = 150 μm, $\Delta\lambda$ = 10 nm, and $\lambda_c = \lambda_B$ = 1.55 μm in Eq. (11.11).This gives f = 65.92 mm.

Wavelength-division multiplexing can also be achieved using the Mach–Zehnder interferometer. Such a configuration is shown in Fig. 11.8. This device consists of three parts: (a) a 3-dB directional coupler which splits the input signal equally and directs it along two paths having different lengths; (b) a central region consisting of two arms, one arm being longer by ΔL (say) than the other arm, which introduces a phase shift between two wavelengths; and (c) another 3-dB direction coupler which recombines the signals at the output. With this configuration, it is possible to introduce a phase shift in one of the paths so that the recombined signals will interfere constructively at one output port and destructively at the other. The combined (multiplexed) signals will emerge from the port at which constructive interference has occurred. It is possible to make any size $N \times N$ multiplexer (demultiplexer) using basic 2 × 2 Mach–Zehnder interferometers.

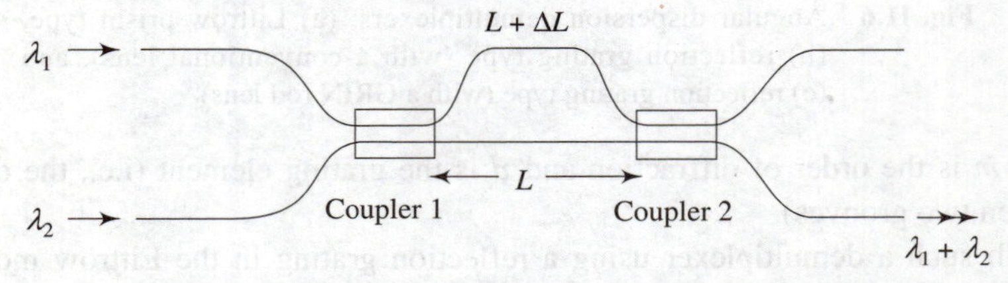

Fig. 11.8 WDM using a 2 × 2 Mach–Zehnder interferometer

A relatively newer and better approach of making integrated demultiplexers is based on a phased array (PHASAR) of optical waveguides, which acts as a grating. Such a structure is shown in Fig. 11.9. It consists of input and output planar waveguide arrays, input and output free propagation regions (FPRs), and planar arrayed waveguides.

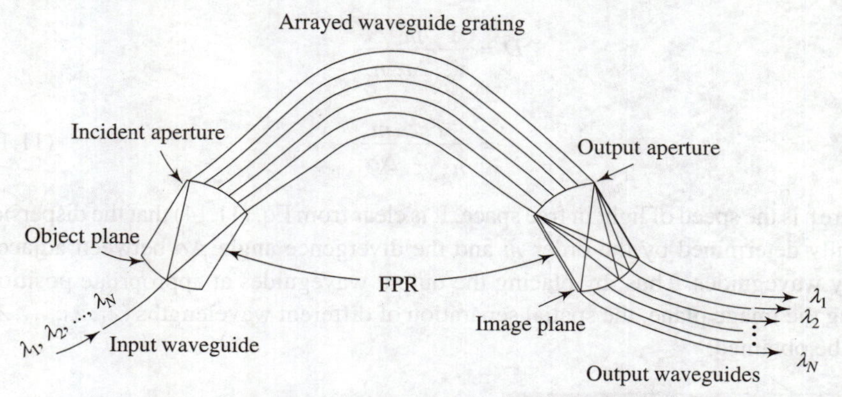

Fig. 11.9 Arrayed waveguide grating demultiplexer

Its operation may be understood as follows. When the optical beam (consisting of wavelengths $\lambda_1, \lambda_2, \ldots, \lambda_N$) propagating through the input waveguide enters the FPR, it is no longer laterally confined but becomes divergent. On arriving at the input aperture, this divergent beam is coupled into the array of waveguides and propagates through them to the output aperture. The length of the individual arrayed waveguides differs from its adjacent waveguides by ΔL, which is chosen such that ΔL is an integral multiple (m) of the central wavelength λ_c of the multiplexer, i.e.,

$$\Delta L = m \frac{\lambda_c}{n_g} \tag{11.12}$$

where the integer m is called the order of the array and n_g is the group index of the guided mode. For this wavelength λ_c, the signals propagating through individual waveguides will arrive at the output aperture with equal phase (apart from an integral multiple of 2π), so that the image of the input field in the object plane will be formed at the centre of the image plane. The dispersion is caused by the length increment ΔL of the adjacent array waveguides, which causes the phase difference $\Delta\phi = \beta\Delta L$ between adjacent array waveguides to vary linearly with signal frequency. Here β is the phase propagation constant of the waveguide mode. As a result, the focal point for different frequencies shifts along the image plane. The spatial shift per unit frequency change (ds/dx) is called the spatial dispersion D of the device. It is given by (Smit & Van Dam 1996)

$$D = \frac{1}{\nu_c}\frac{\Delta L}{\Delta\alpha} \tag{11.13}$$

where ν_c is the central frequency of the PHASAR and $\Delta\alpha$ is the divergence angle between adjacent array waveguides near the input and output apertures. Substituting ΔL from Eq. (11.12) in Eq. (11.13), we get

$$D = \frac{1}{\nu_c}\frac{m}{\Delta\alpha}\frac{\lambda_c}{n_g}$$

$$= \frac{c}{n_g \nu_c^2}\frac{m}{\Delta\alpha} \tag{11.14}$$

where c is the speed of light in free space. It is clear from Eq. (11.14) that the dispersion is fully determined by the order m and the divergence angle $\Delta\alpha$ between adjacent array waveguides. Thus, by placing the output waveguides at appropriate positions along the image plane, the spatial separation of different wavelengths ($\lambda_1, \lambda_2, \ldots, \lambda_N$) can be obtained.

Example 11.4 It is required to make a PHASAR-based demultiplexer for 16 channels with a channel spacing of 100 GHz. The channels are centred around 1.55 μm. Calculate the required order of the arrayed waveguides.

Solution

From our previous discussion, we know that the dispersion of the PHASAR is due to the difference ΔL in the optical path length of adjacent arrayed waveguides, which causes a phase difference $\Delta\phi = \beta\Delta L$ [where $\beta = 2\pi n_g/\lambda_c = (2\pi n_g/c)(\nu_c)$]. Thus $\Delta\phi$ increases with frequency. If the change in frequency is such that $\Delta\phi$ increases by 2π, the transfer will be the same as before. Hence the response of the PHASAR is periodical. This period $\Delta\nu$ in the frequency domain is called the free spectral range (FSR). It can be calculated as follows:

$$\Delta\beta\Delta L = 2\pi$$

or $$\left[\frac{2\pi n_g}{c}(\Delta\nu_c)_{\text{FSR}}\right]\Delta L = 2\pi$$

or $$(\Delta\nu_c)_{\text{FSR}} = \frac{c}{n_g \Delta L} = \frac{\nu_c}{m} \tag{11.15}$$

where we have used Eq. (11.10) for ΔL.

Now, a demultiplexer for 16 channels with a channel spacing of 100 GHz should have an FSR of at least 1600 GHz. Since the centre wavelength is 1.55 μm, the corresponding frequency is

$$\nu_c = \frac{c}{\lambda_c} = \frac{3\times10^8}{1.55\times10^{-6}} = 1.935\times10^{14}\ \text{Hz}$$

Using Eq. (11.13), we get

$$m = \frac{\nu_c}{\Delta\nu_{\text{FSR}}} = \frac{1.935\times10^{14}}{1600\times10^9} \approx 121$$

This means the PHASAR-based demultiplexer would require an array with an order of at least 121.

Fiber Bragg grating shown in Fig. 11.10 can also be used for making an all-fiber demultiplexer. In fact, Ge-doped silica glass exhibits photosensitivity; that is, when this glass is exposed to an intense UV light, the refractive index of the glass is slightly changed. Therefore, by exposing the core of an optical fiber (of Ge-doped silica glass) to the holographic fringe pattern of two interfering UV beams or that generated by a phase plate, it is possible to create a periodic variation in the refractive index of the core, thus creating a phase grating. Figure 11.10 shows the structure of the resulting device and the consequent change in the refractive index of the core. In such a grating, large coupling may occur between the forward and backward propagating modes if the following Bragg condition is satisfied:

$$\lambda_B = 2\Lambda n_{\text{eff}} \tag{11.16}$$

where λ_B is called the Bragg wavelength, Λ is the grating period, and n_{eff} is the effective index of the mode. Thus, proper design can ensure that most of the power is effectively reflected, whereas signals with other wavelengths are transmitted. The advantage of such gratings is that they are fiber-compatible so that the losses generated by connecting them to other fibers are very low.

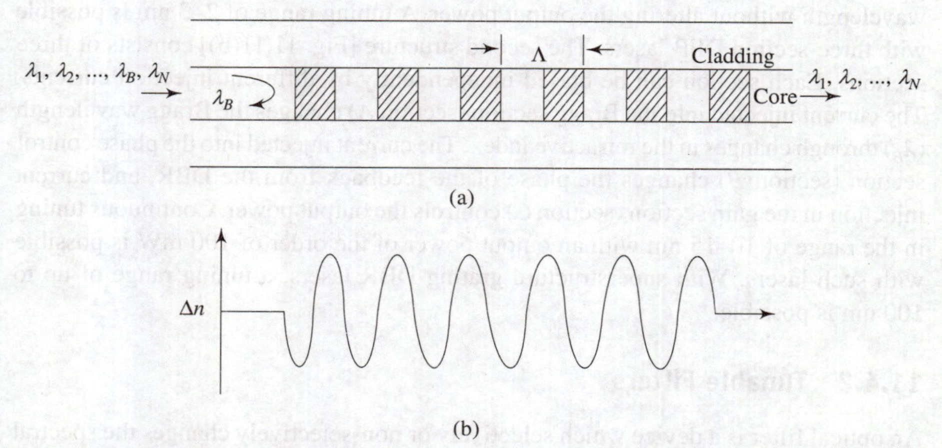

Fig. 11.10 (a) Fiber Bragg grating. (b) Periodic variation of the refractive index of the core.

11.4 ACTIVE COMPONENTS

The performance of active WDM components can be controlled by electronic means. This provides a greater degree of flexibility in the design of optical networks. The prime components in this category are tunable sources, tunable filters, and optical amplifiers. We have already discussed optical amplifiers in Chapter 10. Here we will confine our discussion to tunable sources and filters.

11.4.1 Tunable Sources

In implementing a WDM, it is required to generate several wavelengths (λ_1 to λ_N). One simple option is to use a series of discrete distributed feedback (DFB) or distributed Bragg reflector (DBR) lasers operating at different wavelengths and multiplex their outputs into one fiber using a power combiner or a wavelength multiplexer. But this solution requires a large number of lasers, each of which has to be controlled individually. Further, using a power combiner introduces a loss of at least $10 \log_{10} N$ (dB), where N is the number of wavelength channels. If a multiplexer is used, loss can be reduced, but at the cost of more stringent requirements on the control of emitted wavelengths. Therefore wavelength-tunable lasers are used in modern WDM systems. These devices are based on DFB or DBR structures, discussed in Sec. 7.9.

In fact a controlled variation of emission wavelength is possible by changing the effective refractive index of the cavity or a part of it [refer to Eq. (7.110)]. At least two independent control currents are needed; one in the active region and the other in the tuning region, for the variation of the effective refractive index. Two configurations of lasers using this scheme are shown in Fig. 11.11. The first structure [Fig. 11.11(a)] has a cavity that is subdivided into two to three sections for independent current injection. By controlling the current properly, it is possible to vary the lasing wavelength without altering the output power. A tuning range of 2–3 nm is possible with three-section DFB lasers. The second structure [Fig. 11.11(b)] consists of three sections. Each section can be biased independently by different injection currents. The current injected into the Bragg section (section A) changes the Bragg wavelength (λ_B) through changes in the refractive index. The current injected into the phase control section (section B) changes the phase of the feedback from the DBR, and current injection in the gain section (section C) controls the output power. Continuous tuning in the range of 10–15 nm with an output power of the order of 100 mW is possible with such lasers. With superstructure grating DBR lasers, a tuning range of up to 100 nm is possible.

11.4.2 Tunable Filters

An optical filter is a device which selectively or non-selectively changes the spectral intensity distribution or the state of polarization of the electromagnetic radiation

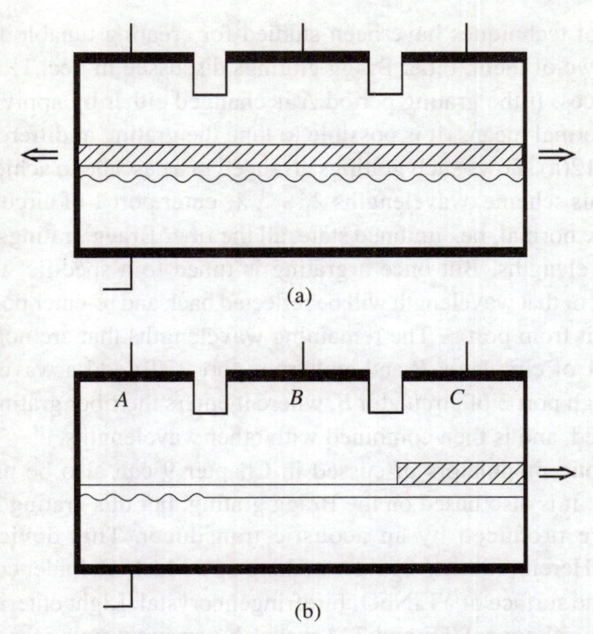

Fig. 11.11 Tunable lasers: (a) multisection DFB laser and (b) multisection DBR laser

incident on it. These filters are normally classified according to the mechanism that is employed for the filtering action. For example, interference filters use the phenomenon of interference of light, polarization filters utilize the phenomenon of polarization, etc. There are two categories of such filters, namely, (i) static filters, whose performance, e.g., transmittance, passband, etc., cannot be changed once the filter is made, and (ii) dynamic or tunable filters, whose properties can be controlled normally by electronic means. In the context of WDM, the role of static or fixed filters is limited. In fact, at the receiver end, it is required to select different wavelengths that have been separated by the demultiplexer. This task is normally performed by tunable filters.

Some important performance parameters of tunable filters are as follows:

1. *Dynamic or tuning range* This is the range of wavelengths over which the filter can be tuned.
2. *Spectral bandwidth* The range of wavelengths transmitted by the filter at the 3-dB level of insertion loss. This is also called the passband of the filter.
3. *Maximum number of resolvable channels* This is the ratio of the total tuning range to the minimum channel spacing required for the transmission of one channel.
4. *Tuning speed* This is the time needed to tune the filter at a specific wavelength.

A variety of techniques have been studied for creating tunable filters. Here we discuss only two of them. Fiber Bragg gratings discussed in Sec. 11.3.3 can be used as active devices. If the grating period Λ is changed either by applying a stretching force or by thermal means, it is possible to tune the grating at different wavelengths λ_B. Figure 11.12(a) shows such gratings arranged in a cascade to achieve an add/drop function. In this scheme, wavelengths $\lambda_1, \ldots, \lambda_N$ enter port 1 of circulator *A* and exit at port 2. In the normal, i.e., untuned state, all the fiber Bragg gratings are transparent to all the wavelengths. But once a grating is tuned to a specific wavelength, say, λ_m $(= \lambda_B)$, light of that wavelength will be reflected back and re-enter port 2 of circulator *A* and then exit from port 3. The remaining wavelengths that are not reflected enter through port 1 of circulator *B* and exit from port 2. To add a wavelength λ_m, it is injected through port 3 of circulator *B*, where it enters the fiber gratings through port 1, gets reflected, and is then combined with other wavelengths.

Acousto-optic modulators discussed in Chapter 9 can also be used for making tunable filters. It is also based on the Bragg grating, but this grating is created by an acoustic wave produced by an acoustic transducer. This device is shown in Fig. 11.12(b). Herein, two waveguides making up a Mach–Zehnder configuration are engraved on the surface of a $LiNbO_3$ birefringent crystal. Light entering the device is separated into orthogonal TE and TM waves by an input polarizer. The transducer produces a surface acoustic wave in the $LiNbO_3$ crystal. This wave sets up a (temporary) dynamic Bragg grating (due to periodic perturbation). The grating period

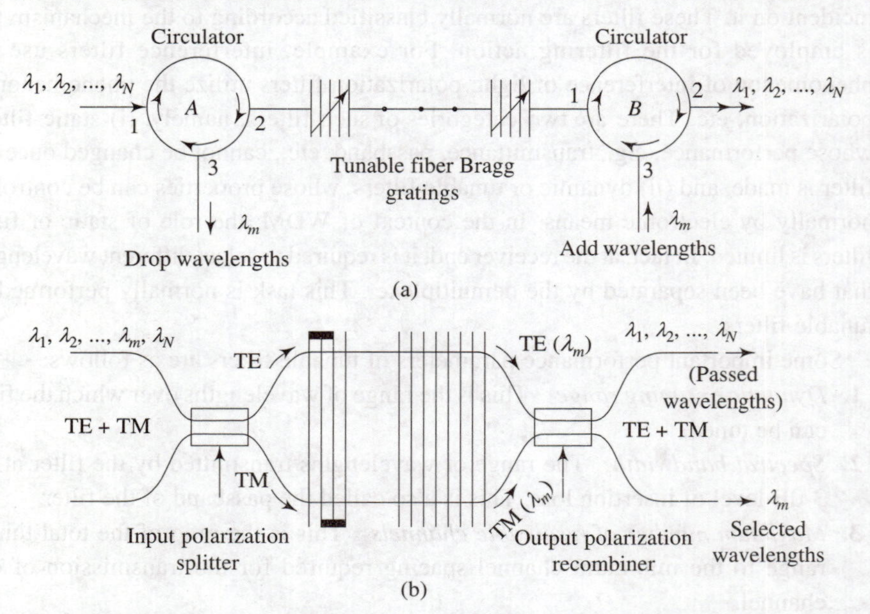

Fig. 11.12 (a) Add/drop multiplexer based on fiber Bragg gratings, (b) acousto-optic tunable filter

is determined by the frequency of the rf signal driving the transducer. An input wavelength, say, λ_m, that matches the Bragg condition of the grating is coupled to the second branch of the filter, whereas other wavelengths pass through the structure without change. For a wavelength λ_m, the Bragg condition is

$$\lambda_m = \Lambda\,(\Delta n) \tag{11.17}$$

where Λ is the grating period and Δn is the difference between the refractive indices of $LiNbO_3$ for the TE and TM modes of that wavelength. An important feature of an acousto-optic tunable filter (AOTF) is that it can select many wavelengths simultaneously. It is possible to induce several gratings on the same interaction length by employing different frequencies.

SUMMARY

- The concept of WDM involves simultaneous transmission of several wavelengths (say, $\lambda_1, \lambda_2, \ldots, \lambda_N$) over the same single optical fiber. In DWDM, the number of wavelength channels is large. Technically, the ITU-T recommendation G.692 specifies a channel spacing of 100 GHz (0.8 nm at 1552 nm) for DWDM.
- The implementation of WDM or DWDM requires a number of passive and active components. The performance of passive components is fixed whereas that of active components can be controlled electronically.
- Directional couplers are used to split or combine two signals at different wavelengths. In a directional coupler made up of two identical single-mode fibers, there is a periodic exchange of power between the two fibers. The minimum interaction length over which the power is completely transferred from the first fiber to the second fiber is called the coupling length.
- The performance of a coupler is specified in terms of the following parameters:

$$\text{Coupling ratio (\%)} = \frac{P_2}{P_1 + P_2} \times 100$$

$$\text{Excess loss (dB)} = -10\log_{10}\left(\frac{P_1 + P_2}{P_0}\right)$$

$$\text{Insertion loss (dB)} = -10\log_{10}\left(\frac{P_j}{P_i}\right)$$

- In order to implement WDM, a multiplexer is required to combine several wavelengths at the transmitting end, and a demultiplexer is needed to isolate the different wavelengths at the receiver end. Several mechanisms of multiplexing (demultiplexing) have been studied. Some of these are as follows:
  - Interference filter based devices
  - Angular dispersion using a Littrow prism or Littrow grating

- Mach–Zehnder interferometer
- Arrayed waveguide grating
- Fiber Bragg grating

- Tunable sources are based on multisection DFB or DBR laser structures. A tuning wavelength range of 10–15 nm is possible with these devices. For wider ranges, an array of tunable lasers has to be used.
- Tunable filters can be made using several different mechanisms. These can be used as add/drop multiplexers in optical networks or demultiplexers in the receiver module.

MULTIPLE CHOICE QUESTIONS

11.1 The function of wavelength-division multiplexer is to
(a) separate signals at different wavelengths and couple them to different detectors.
(b) combine signals at different wavelengths to pass through a single fiber.
(c) tap off part of the energy of the incoming signal.
(d) change the transmission speed of the input signal.

11.2 The scheme of WDM is similar to
(a) FDM for rf transmission. (b) TDM.
(c) SDM. (d) OTDM.

11.3 What is the channel spacing (in nm) specified by the ITU-T recommendation G.692 for DWDM?
(a) 1.6 nm (b) 0.8 nm (c) 0.4 nm (d) 0.2 nm

11.4 What is the channel spacing (in GHz) corresponding to the wavelength of Question 11.3?
(a) 200 GHz (b) 100 GHz (c) 50 GHz (d) 25 GHz

11.5 A 2 × 2 directional coupler has an input power level of 100 μW. The power available at output ports 1 and 2 are, respectively, 45 μW and 45 μW. What is the coupling ratio?
(a) 45% (b) 50% (c) 90% (d) 100%

11.6 For the coupler of Question 11.5, what is the excess loss?
(a) 3 dB (b) 1 dB (c) 0.5 dB (d) 0.46 dB

11.7 For the coupler of Question 11.5, what is the insertion loss for the path from input port 1 to output port 2?
(a) 3.46 dB (b) 5.23 dB (c) 6.92 dB (d) 10 dB

11.8 A 1 × 10 coupler has an input signal 0 dBm. What is the power level at each output port?
(a) 0 dBm (b) –1 dBm (c) –3 dBm (d) –10 dBm

11.9 Which of the following schemes is most suitable for DWDM?
(a) Mach–Zehnder interferometer
(b) Arrayed waveguide grating multiplexer
(c) Fiber Bragg gratings
(d) Blazed reflection gratings

11.10 Which of the following tunable filters is most suitable for DWDM?
(a) Mach–Zehnder interferometer
(b) Fabry–Perot filters
(c) Acousto-optic tunable filters
(d) Fiber Bragg gratings

Answers

11.1 (b)	11.2 (a)	11.3 (b)	11.4 (b)	11.5 (b)
11.6 (d)	11.7 (a)	11.8 (d)	11.9 (b)	11.10 (c)

REVIEW QUESTIONS

11.1 Explain the principles of operation of a 2×2 directional coupler and an $N \times N$ star coupler.

11.2 (a) How can we change the coupling ratio of a 2×2 coupler?
(b) A 2×2 loss-less fiber coupler is using identical single-mode fibers. Calculate the interaction length required to achieve a splitting ratio of 10 : 90.
Ans: (b) $L \approx 1.25\kappa$

11.3 Distinguish between WDM and DWDM. What is the base frequency and channel spacing specified by ITU for DWDM?

11.4 Explain major types of devices for multiplexing/demultiplexing. Compare their merits and demerits.

11.5 (a) Explain the principle of operation of a PHASAR-based demultiplexer.
(b) A PHASAR-based demultiplexer with 32 channels spaced at 50 GHz and a central wavelength of 1.55 μm is to be designed. Calculate the FSR and the order of the array.
Ans: (a) 1600 GHz (b) 121

11.6 (a) Why are tunable sources needed?
(b) Explain the principle of operation of at least two types of tunable lasers.

11.7 (a) What are tunable filters? Where they are used?
(b) Discuss the principle of operation of an AOTF.

11.8 Suggest the design of a tunable filter that is based on the electro-optic effect (discussed in Chapter 9).

11.9 Suggest the possible applications (other than those discussed in this chapter) of fiber Bragg gratings.

12 Fiber-optic Communication Systems

After reading this chapter you will be able to understand the following:

- System design considerations for point-to-point links
  - Digital systems
  - Analog systems
- System architectures
  - Point-to-point links
  - Distribution networks
  - Local area networks
- Non-linear effects and system performance
  - Stimulated Raman scattering
  - Stimulated Brilliouin scattering
  - Self-phase modulation
  - Cross-phase modulation
  - Four-wave mixing
- Dispersion management
- Solitons

12.1 INTRODUCTION

In the preceding chapters, we have discussed the characteristics of individual building blocks/components of fiber-optic communication systems. These include multimode and single-mode optical fibers, cables and connectors, optoelectronic (OE) sources, OE detectors, OE modulators, optical amplifiers, WDM components, etc. Here we will examine how these individual blocks/components may be put together to form a complete system. We will begin with the system design considerations for digital and analog point-to-point links, as such systems are in widespread use within many application areas. Next, we will take up the various types of system architectures.

This is followed by the discussion of non-linear phenomena, their effect on system performance, and the techniques of dispersion management.

12.2 SYSTEM DESIGN CONSIDERATIONS FOR POINT-TO-POINT LINKS

Most of the problems associated with the design of fiber-optic communication systems are due to the unique properties of the optical fiber serving as a transmission medium. However, in common with other communication systems, the major design criteria for a specific application using either digital or analog techniques of transmission are the required transmission distance and the speed of information transfer. These criteria are directly related to two important transmission parameters of optical fibers, namely, (i) fiber attenuation and (ii) fiber dispersion. Keeping these facts in mind, we will discuss system design considerations first for digital communication systems and then for analog systems.

12.2.1 Digital Systems

The simplest kind of fiber-optic link is the simplex (one-directional) point-to-point link having an optical transmitter at one end and an optical receiver at the other (see Fig. 12.1). The key system parameters needed to analyse this link are (i) the desired (or possible) transmission distance (without any repeaters), (ii) the data rate, and (iii) a specified bit-error rate (BER). In order to fulfil these requirements, the system designer has many choices. The major ones are as follows.

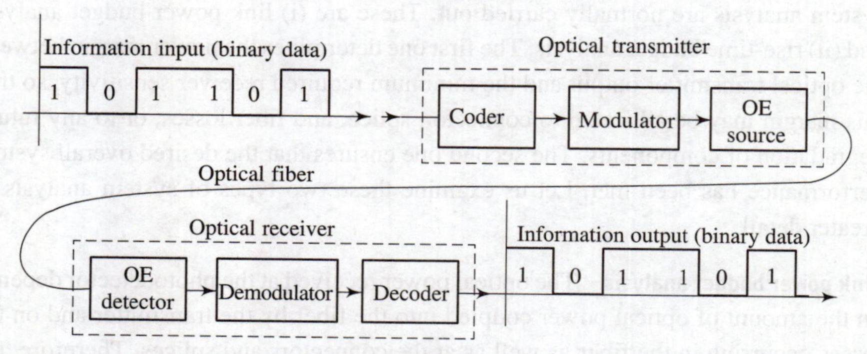

Fig. 12.1 A typical simplex point-to-point digital fiber-optic link. The information input is binary data (shown as a series of 1's and 0's). A coder in the optical transmitter organizes the 1's and 0's and the modulator acts on this data by producing a current that turns the OE source (LED or ILD) on and off. The resulting pulses of light containing the information are transmitted through the optical fiber. At the receiver end, the OE detector converts the pulses of light into pulses of current. A demodulator–decoder combination extracts the information from the electrical pulses.

(i) Optical fiber
   (a) Multimode or single mode
   (b) Core size
   (c) Refractive index profile
   (d) Attenuation
   (e) Dispersion
(ii) Optical transmitter
   (a) Source: LED or ILD
   (b) Operating wavelength (emission wavelength of LED or ILD)
   (c) Output power and emission pattern
   (d) Transmitter configuration
   (e) Modulation and coding
(iii) Optical receiver
   (a) *p-i-n* or avalanche photodiode
   (b) Responsivity at the operating wavelength
   (c) Pre-amplifier design (low impedance, high impedance, or transimpedance front end)
   (d) Demodulation and decoding

The decisions regarding the above requirements are interdependent. The potential choices provide a wide range of economic fiber-optic communication systems. However, it must be kept in mind that the choices are made to optimize the system performance for a particular application.

In order to ensure that the desired system performance can be met, two types of system analysis are normally carried out. These are (i) link power budget analysis and (ii) rise-time budget analysis. The first one determines the power margin between the optical transmitter output and the minimum required receiver sensitivity, so that this margin may be allocated to connector, splice, and fiber losses, or to any future degradation of components. The second one ensures that the desired overall system performance has been met. Let us examine these two types of system analysis in greater detail.

Link power budget analysis The optical power received at the photodetector depends on the amount of optical power coupled into the fiber by the transmitter and on the losses occurring in the fiber as well as at the connectors and splices. Therefore, the power budget is derived from the sequential contributions of all the loss elements in the link. In addition to the link loss contributors, a safety margin of 6–8 dB is normally provided to allow for any future degradation of components and/or future addition of splices, etc. Thus, if we assume that the average power supplied by the transmitter is P_{tx}, the sensitivity of the receiver is P_{rx}, the total link loss or channel loss (including the fiber splice and connector losses) is C_L, and the system's safety margin is M_S, then the following relationship must be satisfied:

$$P_{tx} = P_{rx} + C_L + M_S \tag{12.1}$$

The channel loss C_L may be expressed by the following equation:

$$C_L = \alpha_f L + \alpha_{\text{con}} + \alpha_{\text{splice}} \tag{12.2}$$

where α_f is the fiber loss (in dB/km), L is the link length, α_{con} is the sum of the losses at all the connectors in the link, and α_{splice} is the sum of the losses at all the splices in the link. In Eqs (12.1) and (12.2), P_{tx} and P_{rx} are expressed in dBm, and C_L, M_S, α_{con}, and α_{splice} are expressed in dB. These two equations can be used to estimate the maximum transmission distance for a given choice of components.

Rise-time budget analysis A rise-time budget analysis is useful for determining the dispersion limitation of a fiber-optic link. This is particularly important in the case of digital systems, where it is to be ensured that the system will be able to operate satisfactorily at the desired bit rate. The rise time t_r of a linear system is defined as the time during which the system's response increases from 10% to 90% of the maximum output value when its input changes abruptly (a step function). Let us consider a simple *RC* circuit as an example of a linear system (see Fig. 12.2). When the input voltage (V_{in}) across this circuit changes abruptly (i.e., in a step) from 0 to V_0, the output voltage (V_{out}) changes with time as

$$V_{\text{out}}(t) = V_0[1 - \exp(-t/RC)] \tag{12.3}$$

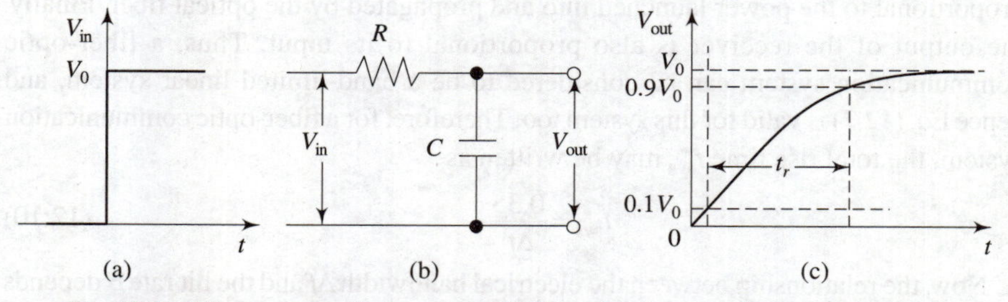

Fig. 12.2 The response of a low-pass *RC* filter circuit to a voltage step input V_0

where R is the resistance and C the capacitance of the circuit. The rise time t_r is given by

$$t_r = 2.2RC \tag{12.4}$$

The transfer function $H(f)$ for this circuit may be obtained by taking the Fourier transform of Eq. (12.3) and is given by

$$H(f) = \frac{1}{(1 + 2\pi jfRC)} \tag{12.5}$$

The 3-dB electrical bandwidth Δf for the circuit corresponds to the frequency at which $|H(f)|^2 = 1/2$ and is given by the expression

$$\Delta f = \frac{1}{2\pi RC} \tag{12.6}$$

Therefore, using Eqs (12.4) and (12.6), we can relate t_r to Δf by the expression

$$t_r = \frac{2.2}{2\pi\Delta f} = \frac{0.35}{\Delta f} \tag{12.7}$$

In a fiber-optic communication system, there are three building blocks, as shown in Fig. 12.1, and each block has its own rise time associated with it. Therefore, the total rise time of the system, t_{sys}, is obtained by taking the root sum square of the rise times of each block. If we assume that the rise times associated with the transmitter, fiber, and receiver are t_{tx}, t_f, and t_{rx}, then

$$t_{sys} = [t_{tx}^2 + t_f^2 + t_{rx}^2]^{1/2} \tag{12.8}$$

The rise time of the optical fiber should include the contribution of intermodal dispersion ($t_{intermodal}$) and intramodal dispersion ($t_{intramodal}$) through the relation

$$t_f = [t_{intermodal}^2 + t_{intramodal}^2]^{1/2} \tag{12.9}$$

In the absence of mode coupling, $t_{intermodal}$ and $t_{intramodal}$ are normally approximated by the time delays (ΔT) caused by intermodal and intramodal dispersion, respectively.

As we know, the optical power generated by the optical transmitter is generally proportional to its input current, and the optical power received by the receiver is proportional to the power launched into and propagated by the optical fiber. Finally, the output of the receiver is also proportional to its input. Thus, a fiber-optic communication system can be considered to be a band-limited linear system, and hence Eq. (12.7) is valid for this system too. Therefore, for a fiber-optic communication system, the total rise time t_{sys} may be written as

$$t_{sys} = \frac{0.35}{\Delta f} \tag{12.10}$$

Now, the relationship between the electrical bandwidth Δf and the bit rate B depends on the digital pulse format. For the return-to-zero (RZ) format, $\Delta f = B$ and for the non-return-to-zero (NRZ) format $\Delta f = B/2$. Therefore, for digital systems, t_{sys} should be below its maximum value given by

$$t_{sys} \leq \begin{cases} \dfrac{0.35}{B} & \text{for the RZ format} \\[2ex] \dfrac{0.70}{B} & \text{for the NRZ format} \end{cases} \tag{12.11}$$

(It should be mentioned here that the RZ and NRZ formats, to be discussed in the next subsection, are used for signal encoding.)

Example 12.1 Consider the design of a typical digital fiber-optic link which has to transmit at a data rate of 20 Mbits/s with a BER of 10^{-9} using the NRZ code. The transmitter uses a GaAlAs LED emitting at 850 nm, which can couple on an average 100 μw (–10 dBm) of optical power into a fiber of core size 50 μm. The fiber cable consists of a graded-index fiber with the manufacture's specification as follows: $\alpha_f = 2.5$ dB/km, $(\Delta T)_{\text{mat}} = 3$ ns/km, $(\Delta T)_{\text{modal}} \doteq 1$ ns/km. A silicon *p-i-n* photodiode has been chosen, for detecting 850-nm optical signals, for the front end of the receiver. The detector has a sensitivity of –42 dBm in order to give the desired BER. The source along with its drive circuit has a rise time of 12 ns and the receiver has a rise time of 11 ns. The cable requires splicing every 1 km, with a loss of 0.5 dB/splice. Two connectors, one at the transmitter end and the other at the receiver end, are also required. The loss at each connector is 1 dB. It is predicted that a safety margin of 6 dB will be required. Estimate the maximum possible link length without repeaters and the total rise time of the system for assessing the feasibility of the desired system.

Solution

Using Eq. (12.2), the total channel loss C_L may be calculated as follows:

$$C_L = \alpha_f L + (\text{splice loss per km}) \times L + (\text{loss per connector}) \times \text{no. of connectors}$$
$$= (2.5 \text{ dB/km}) \times L \text{ (km)} + (0.5 \text{ dB/splice}) \times (1 \text{ splice/km}) \times L \text{ (km)} + (1 \text{ dB}) \times 2$$
$$= (3L + 2) \text{ dB}$$

Here, $P_{tx} = -10$ dBm, $P_{rx} = -42$ dBm, and $M_S = 6$ dB. Substituting the values of P_{tx}, P_{rx}, C_L, and M_S in Eq. (12.1), we get

$$-10 = -42 + (3L + 2) + 6$$

or
$$L = 8 \text{ km}$$

Therefore, a maximum transmission path of 8 km is possible without repeaters.

Let us now calculate the total rise time t_{sys} using Eqs (12.8) and (12.9). It is given that $t_{tx} = 12$ ns, $t_{rx} = 11$ ns. In the case of multimode fibers, intramodal dispersion, is primarily due to material dispersion, and hence $t_{\text{intramodal}} \approx t_{\text{mat}}$.

$$t_{\text{mat}} = (3 \text{ ns/km}) \times L = (3 \text{ ns/km}) \times (8 \text{ km}) = 24 \text{ ns}$$

$$t_{\text{intermodal}} = (1 \text{ ns/km}) \times L = (1 \text{ ns/km}) \times (8 \text{ km}) = 8 \text{ ns}$$

Therefore,
$$t_{\text{sys}} = [(12)^2 + (24)^2 + (8)^2 + (11)^2]^{1/2} = 30 \text{ ns}$$

The maximum allowable rise time t_{sys} for our 20-Mbits/s NRZ data stream [from Eq. (12.11)], is

$$t_{\text{sys}} \le \frac{0.70}{B} = \frac{0.70}{20 \times 10^6} \text{ s} = 35 \text{ ns}$$

Since t_{sys} (= 30 ns) for the proposed link is less than the maximum allowable limit, the choice of components is adequate to meet the system design criteria.

Line coding An important criterion in the design of a digital fiber-optic link is that the decision circuit in the receiver must be able to extract precise timing information from the incoming optical signal. The precise timings are required in order to (i) allow the signal to be sampled by the receiver at a time when the signal-to-noise ratio is maximum, (ii) maintain proper pulse spacing, and (iii) indicate the start and end of each timing interval. Further, channel noise and the distortion mechanism may cause errors in the signal detection process. Therefore, it is desirable for the transmitted optical signal to have an inherent error-detecting capability. It is possible to incorporate these features into the data stream by restructuring or encoding the signal. The main function of time coding is to introduce redundancy into the data stream for the sake of minimizing errors, particularly those resulting from channel interference effects.

In order to understand line codes, let us first be familiar with some commonly used terms. A *digital signal* comprises a series of discrete voltage pulses. An individual pulse in the total signal is called a *signal element*. In fact binary data are transmitted by encoding each data bit into signal elements. The element may be a positive or a negative voltage pulse. If the signal consists of both positive and negative voltage pulses, it is said to be *bipolar*. If only one polarity of the voltage pulse is present, the signal is said to be *unipolar*. The *data rate* R is the transmission rate of data in bits per second. The *bit duration* T_b $(=1/R)$ is the time taken by the transmitter to transmit one bit. A variety of wave shapes (signal elements) may be used to represent binary data. The mapping of binary data bits to signal elements is called *encoding*. Three popular encoding schemes (shown in Fig. 12.3) are used in fiber-optic systems. These are (i) NRZ, (ii) RZ, and (iii) biphase (Manchester).

The simplest NRZ code is the NRZ level (or NRZ-L), shown in Fig. 12.3(a), in which 1's and 0's of a serial data stream are represented by voltage levels that are constant during the bit period T_b, with a 1 represented by high voltage level and a 0 represented by low voltage level. That is, for a 1, there will be a light pulse filling the entire bit period, and for a 0, no light pulse will be transmitted. These codes are easier to generate and decode but do not posses an inherent error-monitoring capability. However, they make efficient use of bandwidth. The RZ code differs from NRZ codes in that only half the bit period is used for data, while the voltage is zero in the second half of the bit period. Thus, in a unipolar RZ data format, shown in Fig. 12.3(b), a 1 is represented by a half-period optical pulse that occurs in the first half of the bit period and a 0 is represented by no signal during the bit period. The disadvantages of the unipolar RZ format are that it requires double the bandwidth of the NRZ-L format, and a long string of 0's can cause loss of timing information. A data format which possesses the virtues of easy time synchronization, no dc component, and some inherent facility for error detection is biphase-L (or the optical Manchester code), shown in Fig. 12.3(c). In this code, there is a transition in the

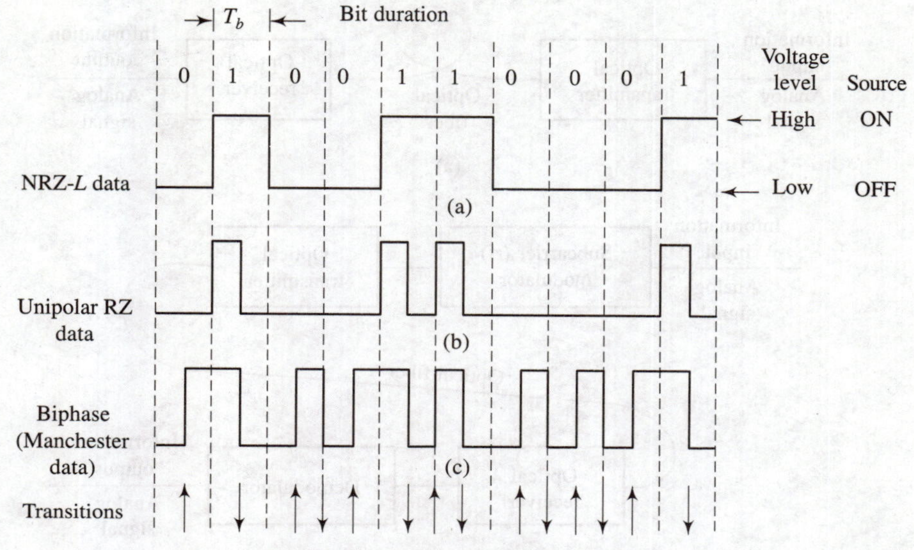

Fig. 12.3 Popular encoding schemes

middle of the bit interval. During one half of the bit interval, the voltage level is high for a 1 and low for a 0. In this scheme, therefore, a transition from high to low in the middle of the bit interval represents a 1 and a transition from low to high represents a 0. These codes are widely used in fiber-optic systems.

12.2.2 Analog Systems

For long-haul communication links, digital systems with single-mode fibers are usually considered superior, even with their expensive terminal equipment for coding, multiplexing, timing, etc. The reason for this is that an analog fiber-optic system requires 20–30 dB higher signal-to-noise ratio as compared with that required for a similar digital fiber-optic system. However, for short-haul and medium-haul links, analog fiber-optic systems can be very attractive, especially for the transmission of video signals, because of their simplicity and cost effectiveness. There are many other application of analog systems. It has been observed that for most applications, analog transmitters use laser diodes, and hence we will concentrate on this source here. In the design implementation of an analog system, the main parameters that need to be considered are carrier-to-noise ratio, bandwidth, and the signal distortion resulting from non-linearities in the transmission system. In such systems, carrier-to-noise ratio analysis is used instead of signal-to-noise ratio, because the information signal is normally superposed on an rf carrier.

Figure 12.4 shows the basic elements of two types of analog fiber-optic links. In the first system [Fig. 12.4(a)] the optical transmitter contains either an LED or ILD as

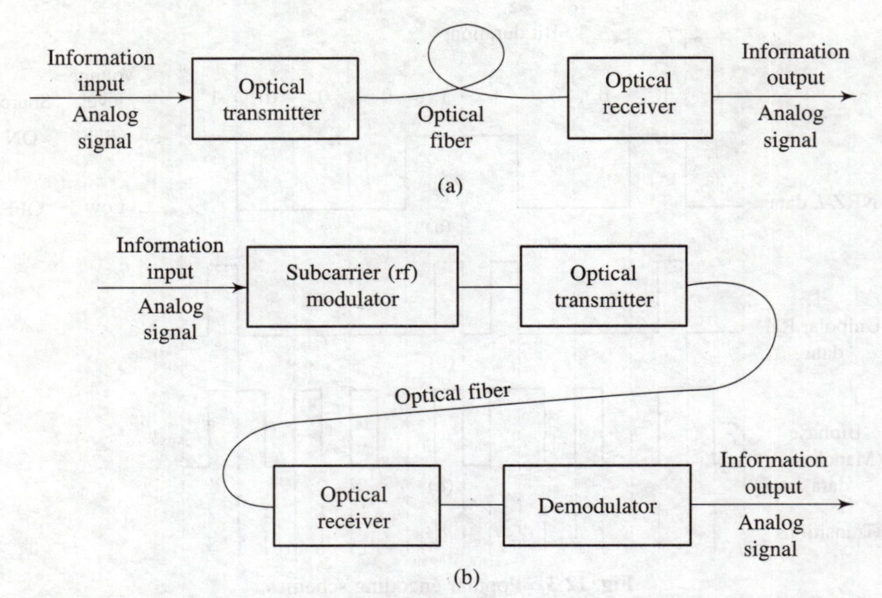

Fig. 12.4 An analog fiber-optic communication system: (a) type I—optical intensity is directly modulated by the analog signal, (b) type II—A subcarrier is modulated by the analog signal

the optical source. The output intensity of the source is directly changed or modulated by the information-carrying analog signal. It is necessary to first set a bias point on the source approximately in the middle of the linear output region. The analog signal can then be transmitted using one of the several modulation techniques, the simplest one being direct intensity modulation. In this scheme, the optical output from the source is modulated by varying the current around the bias point in proportion to the level of the information signal. Thus the signal is directly transmitted in the baseband. The modulated signal travels down the optical fiber and is demodulated at the receiver.

There also exists a more efficient way of modulation in which the baseband signal is first translated into an electrical subcarrier prior to intensity modulation as shown in Fig. 12.4(b). This is accomplished using standard techniques of amplitude modulation (AM), frequency modulation (FM), or phase modulation (PM). These modulation techniques are employed when there is a need to send multiple analog signals over the same fiber, as in the case of broadband common antenna television (CATV) supertrunks.

The performance of analog systems is normally analysed by calculating the carrier-to-noise ratio (CNR) of the system. It is defined as the ratio of the rms carrier power to the rms noise power (resulting from the source, detector, amplifier, and inter-modulation) at the output of the receiver; thus

$$\text{CNR} = \frac{\text{rms carrier power}}{\text{rms noise power}} \tag{12.12}$$

For single-channel transmission the noise contributions of the source, detector, and amplifier are considered, whereas for the transmission of multiple information channels through the same fiber, the intermodulation factor is also considered. Here we are examining a simple single-channel amplitude-modulated signal sent at baseband frequencies. In this analysis we have closely followed Keiser (2000).

In order to determine carrier power, let us consider a laser transmitter. The optical signal variation by the source is caused by the drive current (through the source), which is a sum of the fixed bias and an analog input signal (a time-varying sinusoid), as shown in Fig. 12.5. An ILD acts as a square-law device, so that the envelope of the output optical power $P(t)$ has the same waveform as the input drive current. If we assume that the time-varying analog drive signal is $s(t)$, then we may write

$$P(t) = P_B[1 + ms(t)] \tag{12.13}$$

where P_B is the output optical power at the bias current level (I_B) and m is the modulation index defined in terms of the peak optical power P_{peak} as follows:

$$m = \frac{P_{\text{peak}}}{P_B} \tag{12.14}$$

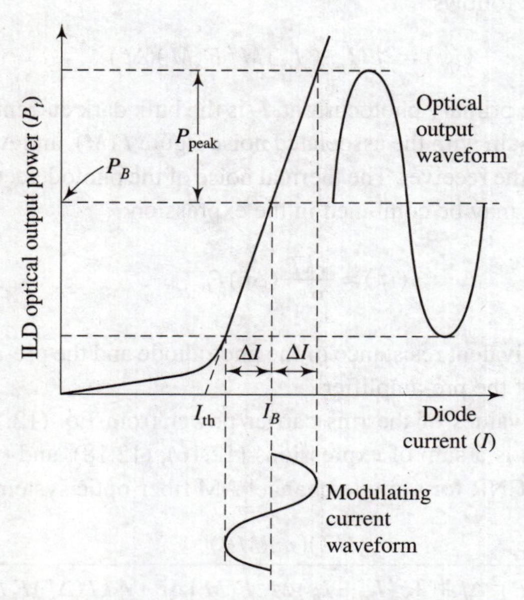

Fig. 12.5 Schematic representation of the biasing conditions of an ILD and its response to analog signal modulation

For a time-varying sinusoidally received signal, the rms carrier power C at the output of the receiver (in units of A^2) is given by

$$C = \frac{1}{2}(mRM\bar{P})^2 \tag{12.15}$$

where R is the unity gain responsivity of the photodetector, M is the gain of the APD ($M = 1$ for *p-n* and *p-i-n* photodiodes), and $\bar{P}$ is the average received optical power.

The rms noise power in Eq. (12.12) is the sum of the noise powers arising due to the source, photodetector, and pre-amplifier. The source noise, in the present case, will be given by

$$\langle i_{\text{source}}^2 \rangle = \text{RIN}(R\bar{P})^2 \Delta f \tag{12.16}$$

where RIN is the laser relative intensity noise measured in dB/Hz and is defined by

$$\text{RIN} = \frac{\langle (\Delta P_L)^2 \rangle}{\bar{P}_L^{\,2}} \tag{12.17}$$

where $\bar{P}_L$ is the average laser light intensity and $\langle (\Delta P_L)^2 \rangle$ is the mean square intensity fluctuation of the laser output. Δf is the effective noise bandwidth. This type of noise decreases as the injection current level increases.

The photodiode noise arises mainly due to shot noise and bulk dark current noise, given by Eqs (8.20) and (8.21). Thus, combining the two we get an expression for photodiode noise as follows:

$$\langle i_N^2 \rangle = 2e(I_p + I_d)M^2 F(M)(\Delta f) \tag{12.18}$$

Here, $I_p = R\bar{P}$ is the primary photocurrent, I_d is the bulk dark current of the detector, M is the detector's gain with the associated noise figure $F(M)$, and Δf is the effective noise bandwidth of the receiver. The thermal noise of the photodetector and the noise of the pre-amplifier may be combined in the expression

$$\langle i_T^2 \rangle = \frac{4kT}{R_{\text{eq}}}(\Delta f)F_t \tag{12.19}$$

where R_{eq} is the equivalent resistance of the photodiode and the pre-amplifier, and F_t is the noise factor of the pre-amplifier.

Substituting the values of the rms carrier power from Eq. (12.15) and the rms noise power [which is a sum of expressions (12.16), (12.18), and (12.19)] into Eq. (12.12), we get the CNR for a single-channel AM fiber-optic system:

$$\text{CNR} = \frac{(1/2)(mRM\bar{P})^2}{\text{RIN}(R\bar{P})^2 \Delta f + 2e(I_p + I_d)M^2 F(M)\Delta f + 4kT(\Delta f)F_t / R_{\text{eq}}} \tag{12.20}$$

Most of the general design considerations for digital fiber-optic systems outlined in Sec. 12.2.1 may be applied to analog systems as well. However, one must take extra care to ensure that the optical source and the photodetector have linear input–output characteristics, in order to avoid optical signal distortion. A careful link power budget analysis is a must because analog systems require a higher SNR at the receiver than their digital counterparts. The temporal response of analog systems may be determined by rise-time calculations similar to those done for digital systems. In this case too, the maximum attainable bandwidth Δf is related to t_{sys} by Eq. (12.10).

Example 12.2 A type-I intensity-modulated analog fiber-optic link employs a laser transmitter which couples a mean optical power of 0 dBm into a multimode optical fiber cable. The cable exhibits an attenuation of 3.0 dB/km with splice losses estimated at 0.5 dB/km. A connector at the receiver end shows a loss of another 1.5 dB. The *p-i-n* photodiode receiver has a sensitivity of –25 dBm for a CNR of –50 dB with a modulation index of 0.5. A safety margin of 7 dB is required. The rise times of the ILD and *p-i-n* diode are 1 ns and 5 ns, respectively, and the intermodal and intramodal rise times of the fiber cable are 9 ns/km and 2 ns/km, respectively. (a) What is the maximum possible link length without repeaters? (b) What is the maximum permitted 3-dB bandwidth of the system?

Solution

(a) *Link power budget*

The mean optical power coupled into the fiber cable by the laser transmitter (P_{tx}) = 0 dBm, the mean optical power required at the *p-i-n* receiver (P_{rx}) = –25 dBm, and the total system margin ($P_{tx} - P_{rx}$) = 25 dB.

Assume that the repeaterless link length is L. Then, using Eq. (12.2), the total channel loss C_L may be calculated as follows:

$$C_L = (\text{attenuation/km}) \times L + (\text{splice loss/km}) \times L + \text{connector loss}$$
$$= (3 \text{ dB/km}) \times L + (0.5 \text{ dB/km}) \times L + 1.5 \text{ dB}$$
$$= (3.5L + 1.5) \text{ dB}$$

Therefore, from Eq. (12.1), we have

$$P_{tx} - P_{rx} = C_L + M_S$$

$$\Rightarrow \quad 25 \text{ dB} = [(3.5L + 1.5) + 7] \text{ dB}$$

Thus $$L = \frac{16.5}{3.5} \approx 4.7 \text{ km}$$

(b) *Rise-time budget*

$$t_f^2 = [(9 \text{ ns/km} \times 4.7 \text{ km})^2 + (2 \text{ ns/km} \times 4.7 \text{ km})^2] = 1877.65 \text{ ns}^2$$

$$t_{sys} = (t_{tx}^2 + t_f^2 + t_{rx}^2)^{1/2}$$
$$= [(1 \text{ ns})^2 + 1877.65 \text{ ns}^2 + (5 \text{ ns})^2]^{1/2}$$
$$= 43.63 \text{ ns}$$

Therefore, the system bandwidth

$$\Delta f = \frac{0.35}{t_{sys}} = \frac{0.35}{43.6 \times 10^{-9}} \text{ Hz}$$
$$= 8 \times 10^6 \text{ Hz} = 8 \text{ MHz}$$

Thus the proposed link length without repeaters is 4.7 km with a 3-dB bandwidth of 8 MHz.

12.3 SYSTEM ARCHITECTURES

Fiber-optic communication systems may be classified into three broad categories. These are (i) point-to-point links, (ii) distribution networks, and (iii) local area networks.

12.3.1 Point-to-point Links

Our previous discussion of digital and analog fiber-optic communication systems has been based on essentially point-to-point links. Their role is to transport information from one point to another as accurately as possible. For short-haul applications (say, less than 10 km), the attenuation, dispersion, and bandwidth of optical fibers are not of major concern. In such cases, optical fibers are used primarily because of their immunity to electromagnetic interference and radio-frequency interference. However, for long-haul applications, e.g., transoceanic light wave systems, the low loss, low dispersion, and large bandwidth of optical fibers are important factors. Therefore, whenever the link length exceeds a certain value, it becomes essential to compensate for the fiber loss and/or dispersion. Such a compensation is normally carried out by optical amplifiers, dispersion-compensating fibers, or other means, which are discussed in Sec. 12.5.

12.3.2 Distribution Networks

There are many applications of fiber-optic systems which require the information not only to be transmitted but also distributed to a group of subscribers. Such applications include the local distribution of telecommunication services and local broadcast of multiple video channels over cable television (CATV). Two commonly used topologies for distribution networks are shown in Fig. 12.6.

In the case of hub topology, channel distribution is done at the central locations or the hubs, where an automatic cross-connect facility switches channels in the electrical domain. The optical fiber is used mainly to connect different hubs. Telephone networks within a city normally employ hub topology for the distribution of audio channels.

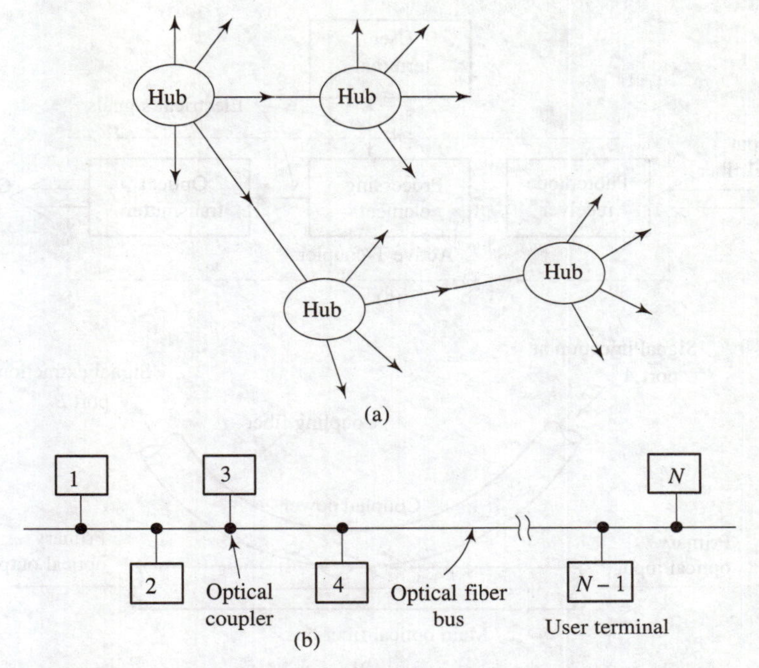

Fig. 12.6 Configuration of distribution networks: (a) hub topology and (b) bus topology

Since the bandwidth of optical fibers is large, several hubs can share the same single fiber needed for the main hub. The drawback of such a topology is that the outage of a single-fiber cable can affect services to a large portion of the network.

The bus topology usually employs a single-fiber cable that carries multichannel optical signals throughout the area of service. Distribution is done using optical taps, which divert a small fraction of optical power to each station or subscriber. As compared with the coaxial cable bus, an optical fiber based bus network is more difficult to implement. The main problem is the ready availability of bidirectional optical taps which can efficiently couple optical signals into and out of the main optical fiber trunk. Access to an optical data bus is normally achieved by means of either an active or a passive coupler, as shown in Fig. 12.7.

In the case of an active coupler, a front-end photodiode receiver converts the optical signal from the bus into an electrical signal. The processing element removes or copies a part of this signal for transmission to the user terminal and sends the remainder to the optical transmitter. The latter, in turn, converts the electrical signal back into the optical bit stream, which gets coupled into the output fiber that is connected to the next terminal. The advantage of such a linear fiber bus network is that every accessing terminal acts as a repeater. Therefore, at least in principle, an active bus can accommodate an unlimited number of terminals. However, the reliability of each

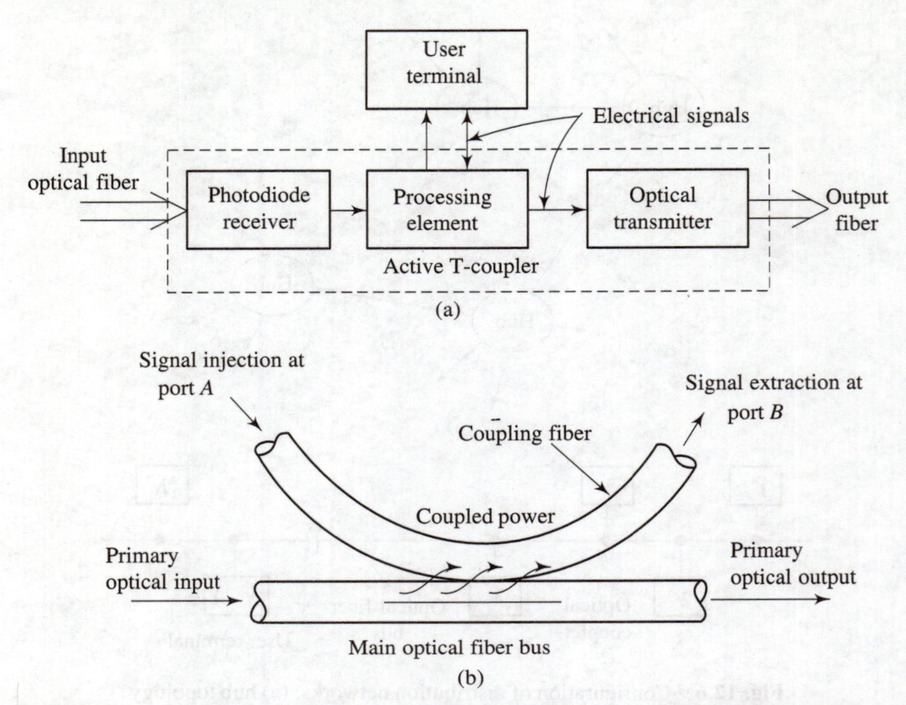

Fig. 12.7 (a) An active T-coupler and (b) a passive T-coupler. The tilted arrows show that some optical power has been tapped off the main fiber bus into the receiver at port *B*.

repeater is critical to the operation of a single-fiber bus network. The failure of any one repeater will stop all the traffic. This problem may be overcome by using some bypass scheme, so that if one repeater fails, the bypass ensures optical continuity from the preceding transmitter to the next terminal.

In the case of a passive coupler no repeaters are used. At each terminal node a passive coupler is used to remove a fraction of the optical signal from the main fiber bus trunk line or to inject additional optical signals into the trunk. A major problem with this type of coupler is that the optical signal is not regenerated at each terminal node. Therefore, optical losses at each tap coupled with the fiber losses between the taps limit the size of the network to a small number of terminals.

12.3.3 Local Area Networks

The large bandwidth offered by optical fibers has motivated system designers to use this technology in networks which have a large number of users within a local area (e.g., a university campus, factory office, etc.) interconnected in such a way that any user can access the network randomly to transmit data to any other user. Such networks are referred to as local area networks (LANs). The prime difference between a distribution network and a LAN is related to the random access offered to multiple users by LAN. There are three basic LAN topologies, namely, the bus, ring, and star configuration. The bus topology is similar to that shown in Fig. 12.6(b). The ring and star topologies are shown in Fig. 12.8. In the ring topology, the nodes are connected by point-to-point links to form a single closed ring as shown in Fig. 12.8(a). Information data packets (including address bits) are transmitted from node to node around the ring consisting of an optical fiber cable. The interface at each node consists of an active device which can recognize its own address in a data packet in order to accept messages. The interface also serves as a repeater for retransmitting messages that are addressed to other nodes. A popular interface for ring topology in a fiber-optic LAN, known as a fiber distributed data interface (FDDI), operates at 100 Mbits/s by using 1.3-μm multimode fiber and LED-based transmitters.

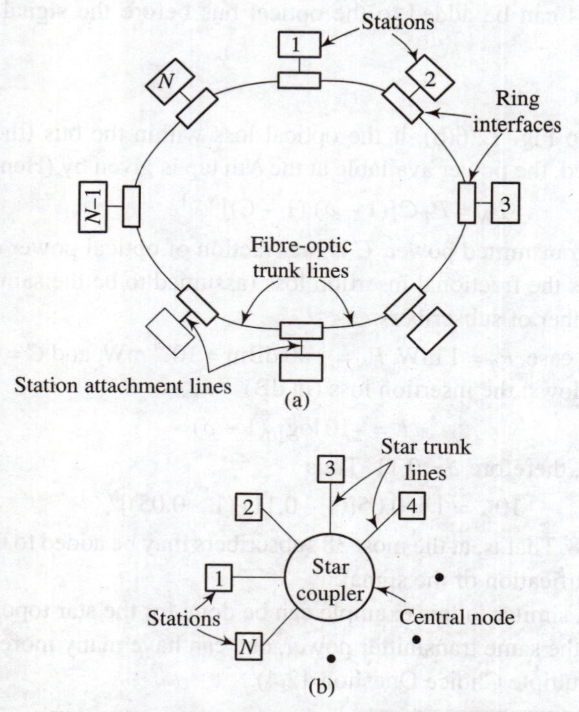

Fig. 12.8 (a) Ring and (b) star LAN topologies

In the star topology, all nodes are joined through point-to-point links to a single point called the central node or hub. The central node may be an active or a passive device. In an active central node, all incoming optical signals are converted into electrical signals through photodiode receivers. The electrical signal is then distributed to drive different node transmitters. The switching operation can also be performed at this node. In a star configuration with a passive central node the distribution takes place in the optical domain, through devices such as directional couplers. In this case, the input from one node is distributed to many output nodes, and hence the power transmitted to each node depends on the number of users.

The network topologies discussed above are also used by metropolitan area networks (MANs), which connect users within a city or in a metropolitan area around the city, and wide area networks (WANs), which provide user interconnection over a large geographical area.

Example 12.3 A university campus CATV system uses an optical bus to distribute video signals to the subscribers. The transmitter couples 0 dBm (1 mW) of optical power into the bus. Each receiver has a sensitivity of −40 dBm. Each optical tap couples 5% of optical power to the subscriber and has a 0.5 dB insertion loss. How many subscribers can be added to the optical bus before the signal needs in-line amplification?

Solution

With reference to Fig. 12.6(b), if the optical loss within the bus (the optical fiber itself) is neglected, the power available at the Nth tap is given by (Henry et al. 1988)

$$P_N = P_T C[(1 - \delta)(1 - C)]^{N-1} \tag{12.21}$$

where P_T is the transmitted power, C is the fraction of optical power coupled out at each tap, and δ is the fractional insertion loss (assumed to be the same) at each tap, and N is the number of subscribers.

In the present case, $P_T = 1$ mW, $P_N = -40$ dBm $= 10^{-4}$ mW, and $C = 0.05$. δ may be calculated as follows: the insertion loss (in dB)

$$L = -10\log_{10}(1 - \delta)$$

Here $L = 0.5$ dB, therefore $\delta = 0.11$. Thus

$$10^{-4} = 1 \times 0.05[(1 - 0.11)(1 - 0.05)]^{N-1}$$

This gives $N = 38$. That is, at the most 38 subscribers may be added to the bus without any in-line amplification of the signal.

A calculation similar to this example can be done for the star topology. It will be found that with the same transmitter power, one can have many more subscribers in this case (see Multiple Choice Question 12.4).

12.4 NON-LINEAR EFFECTS AND SYSTEM PERFORMANCE

We have so far assumed fiber-optic systems to be band-limited linear systems. This assumption is valid when these systems are operated at moderate power levels (say, few milliwatts) and bit rates (say, up to about 2.5 Gbits/s). However, if the transmitted power level is high or the bit rates exceed 10 Gbits/s, non-linear effects become important and must be considered while designing a fiber-optic communication system for a specific application. In this section, we will discuss two types of non-linear effects that place limitations on system performance. These are (i) non-linear inelastic processes, e.g., stimulated Raman scattering and stimulated Brillioun scattering, and (ii) non-linear effects arising from intensity-dependent variation in the refractive index of the fiber, e.g., self-phase modulation, cross-phase modulation, and four-wave mixing (FWM).

Non-linear processes are difficult to model because they depend on various factors, such as the transmission length, cross-sectional area of the fiber, transmitted power level, etc. The problem can be understood as follows. If the signal power is assumed to be constant, the effect of a non-linear process increases with distance. However, we know that due to attenuation within the fiber, the signal power does not remain constant but decreases with distance. As a consequence, the non-linear process diminishes in magnitude. In practice, therefore, it is fairly reasonable to assume that the power is constant over a certain effective fiber length, which is less than the actual length of the fiber, and also take into account the exponential decay in power due to absorption. This effective length L_{eff} is given by (Ramaswami & Sivarajan 1998)

$$L_{\text{eff}} = \frac{1}{P_0}\int_0^L P(z)dz = \frac{1}{P_0}\int_0^L P_0 e^{-\alpha z} dz = \frac{1-e^{-\alpha L}}{\alpha} \tag{12.22}$$

where L is the actual length of the fiber, α is the attenuation coefficient, P_0 is the power at the input end of the fiber, and $P(z)$ is the power at a distance z. For large values of L, $L_{\text{eff}} \rightarrow 1/\alpha$. Thus if we take α to be typically around 0.22 dB km^{-1} (which is equivalent to a coefficient of 0.0507 km^{-1}) at 1.55 μm, $L_{\text{eff}} \approx 20$ km. If the link incorporates optical amplifiers and the total amplified link length is L_A and the total span length of the fiber between amplifiers is L, the effective length is given approximately by the following relation (Ramaswami & Sivarajan 1998):

$$L_{\text{eff}} = \frac{1-e^{-\alpha L}}{\alpha}\frac{L}{L_A} \tag{12.23}$$

Thus the total effective length decreases as the amplifier span increases.

It has been said above that the effect of non-linear processes increases with increase in light intensity transmitted through the fiber. This intensity, however, is inversely proportional to the cross-sectional area of the fiber core. In Chapter 5 we have seen that for a single-mode fiber, this power is not distributed uniformly over the entire cross-sectional area of the core. In practice, therefore, it is convenient to use an effective cross-sectional area A_{eff}. If the mode field radius is w, $A_{\text{eff}} \approx \pi w^2$. Typical effective areas for a conventional single-mode fiber (CSF), dispersion-shifted fiber (DSF), and dispersion-compensating fiber (DCF) are 80 μm^2, 50 μm^2, and 20 μm^2, respectively.

12.4.1 Stimulated Raman Scattering

Stimulated Raman scattering (SRS) is the result of the inelastic scattering of a light wave (propagating through a silica-based optical fiber) by silica molecules. When a photon of energy $E_1 = h\nu_1$ (where the symbols have their usual meaning) interacts with a silica molecule, some of its energy, depending on the vibrational frequency of the molecule, is absorbed by the latter and the photon is scattered. As the original photon has lost some energy, the energy of the scattered photon becomes less, say, $E_2 = h\nu_2$. This change in the frequency of the interacting photon from ν_1 to ν_2 ($\nu_1 > \nu_2$) is called *Stokes' shift*. Since the light wave propagating through the fiber is a source of interacting photons, it is normally called a *pump wave*. This process generates scattered light at a wavelength longer than the pump wave ($\lambda_2 > \lambda_1$ as $\nu_2 < \nu_1$). If two or more signals at different wavelengths are simultaneously injected into the fiber, SRS can cause power to be transferred from the lower wavelength signals to the higher wavelength signals. This effect is shown in Fig. 12.9. As a consequence, SRS can severely affect the performance of a multichannel fiber-optic communication system by transferring energy from shorter wavelength channels to other longer wavelength channels. This effect occurs in both the directions.

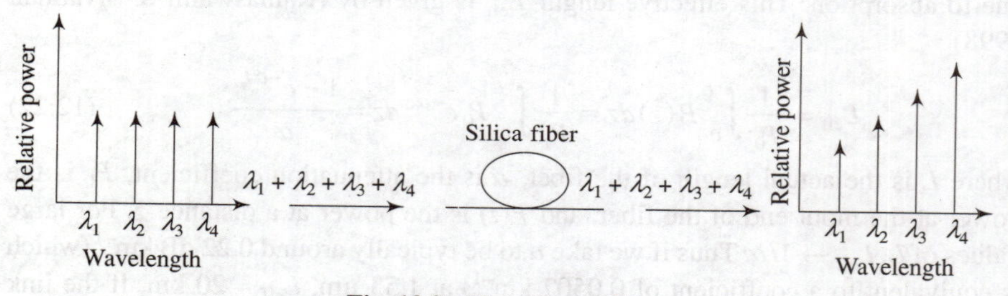

Fig. 12.9 The effect of SRS

The effect of SRS can be estimated following Buck (1995) as follows. Consider a WDM system with N equally spaced channels, 0, 1, 2, ..., $(N-1)$, with a channel spacing of $\Delta\lambda_s$. With the assumptions that the same power is transmitted in all the channels, the Raman gain increases linearly, and that there is no interaction between other channels, the fraction of power coupled from channel 0 to channel i is given approximately by

$$F(i) = g_R \frac{i\Delta\lambda_s}{\Delta\lambda_c} \frac{P_0 L_{\text{eff}}}{2 A_{\text{eff}}}$$

where g_R is the peak Raman gain coefficient, $\Delta\lambda_c$ is the total channel spacing, and the other symbols have their usual meaning. Therefore, the fraction of power coupled from channel 0 to all the other channels will be given by

$$F = \sum_{i=1}^{N-1} F(i) = \frac{g_R \Delta\lambda_s}{\Delta\lambda_c} \frac{P_0 L_{\text{eff}}}{2 A_{\text{eff}}} \frac{N(N-1)}{2} \tag{12.24}$$

The power penalty for channel 0 is then $-10\log_{10}(1-F)$. Thus, in order to keep this penalty below 0.5 dB, the fraction F should be less than 0.1.

SRS is not a serious problem in systems with small number of channels. However, it can create severe problems in WDM systems with large numbers of wavelength channels. In order to alleviate the effect of SRS, (i) the channels should be spaced as closely as possible and (ii) the transmitted power level in each channel should be kept below the threshold (in other words, the distance between the optical amplifiers in the link should be reduced).

Example 12.4 Calculate the total power that should be transmitted over 32 channels of a WDM system that are spaced 0.8 nm apart at 1.55 μm for a repeaterless distance of 80 km. Take $L_{\text{eff}} = 20$ km, $\Delta\lambda_c = 125$ nm, $g_R = 6\times10^{-14}$ m/W, and $A_{\text{eff}} = 55\ \mu\text{m}^2$.

Solution

If we assume that the power penalty is to be kept below 0.5 dB, we must have $F \le 0.1$. Therefore, from Eq. (12.24), we can calculate the total transmitted power P_{tot} as follows:

$$P_{\text{tot}} = NP_0 = \frac{4F\Delta\lambda_c A_{\text{eff}}}{g_R \Delta\lambda_s L_{\text{eff}}(N-1)}$$

or

$$P_{\text{tot}} = \frac{0.1\times(125\text{ nm})\times4\times(55\times10^{-12}\text{ m}^2)\times(10^{-3}\text{ km/m})}{(6\times10^{-14}\text{ m/W})\times(0.8\text{ km})\times(20\text{ nm})\times(32-1)}$$

$$= 0.0924\text{ W} = 92.4\text{ mW}$$

The transmitted power per channel, P_0, therefore, should be less than 2.887 mW. In this calculation it has been assumed that there is no dispersion in the system.

12.4.2 Stimulated Brillioun Scattering

Stimulated Brillioun scattering (SBS) may be viewed as the scattering of a pump wave by an acoustic wave (generated by the oscillating electric field of the pump wave). This process creates a Stokes' wave of lower frequency, which travels in the backward direction. The Stokes' wave experiences gain at the expense of the depletion of the signal power of the forward propagating signal (i.e., the pump wave). The frequency shift due to SBS is called the Brillioun shift and is given by

$$\nu_B = 2\bar{n}\,V_A/\lambda_p \tag{12.25}$$

where $\bar{n}$ is the mode index of the fiber, V_A is the velocity of the acoustic wave, and λ_p is the wavelength of the pump wave. If we take typical values of $\bar{n} = 1.46$ for the silica fiber, $V_A = 5960\text{ m s}^{-1}$, and $\lambda_p = 1.55\ \mu\text{m}$, then $\nu_B = 11.22$ GHz. This interaction occurs over a very narrow line width of $\Delta\nu_B = 20$ MHz at $\lambda_p = 1.55$ μm. There are two important features of SBS: (i) it does not cause any interaction between different wavelengths, as long as the wavelength spacing is much greater than 20 MHz, and (ii) its effect can create significant distortion within a single channel, especially when the amplitude of the scattered wave is comparable with the signal power.

A simple criterion for determining the impact of SBS is to consider the SBS threshold power P_{th}, which is defined as the signal power at which the backscattered power equals the fiber input power. It is given by the following approximate expression:

$$P_{th} \approx \frac{21bA_{eff}}{g_B L_{eff}}\left[1 + \frac{\Delta\nu_{source}}{\Delta\nu_B}\right] \tag{12.26}$$

where $\Delta\nu_{source}$ is the line width of the source, g_B is the Brillioun gain, and the value of b lies between 1 and 2, depending on the relative polarizations of the pump and Stokes' waves. Thus P_{th} increases with increase in the source line width.

Example 12.5 Calculate the SBS threshold power for the worst and best possible cases if the line width of the source as 100 MHz, $g_B = 4 \times 10^{-11}$ m/W, $A_{eff} = 55 \times 10^{-12}$ m^2, $L_{eff} = 20$ km, and $\lambda_p = 1.55$ μm.

Solution

As discussed earlier, at $\lambda_p = 1.55$ μm, $\Delta\nu_B = 20$ MHz. For the worst case, $b = 1$. Therefore, using Eq. (12.26), we get

$$P_{th} = \frac{21 \times 1 \times 55 \times 10^{-12}}{6 \times 10^{-11} \times 20 \times 10^3}\left[1 + \frac{100 \times 10^6}{20 \times 10^6}\right] = 8.66 \times 10^{-3} \text{ W}$$

or $\quad P_{th} = 8.66$ mW

For the best possible case, $b = 2$ and we have

$$P_{th} = 17.32 \text{ mW}$$

12.4.3 Four-wave Mixing

In a WDM system, if three waves with angular frequencies ω_i, ω_j, and ω_k co-propagate inside a silica fiber simultaneously, then the non-linear susceptibility of the silica fiber generates new waves at angular frequencies $\omega_i \pm \omega_j \pm \omega_k$. This phenomenon is known as *four-wave mixing* because three waves at frequencies ω_i, ω_j, and ω_k combine to produce a fourth wave at a frequency $\omega_i \pm \omega_j \pm \omega_k$. In principle, several frequencies corresponding to the combinations of plus and minus signs are possible. However, most of them do not build up due to the lack of a phase-matching condition. But frequency combinations of the form $\omega_i + \omega_j - \omega_k$ (with $i, j \neq k$) are often troublesome for WDM systems, as they can become phase-matched when the wavelength channels are closely spaced or are spaced near the dispersion zero of the fiber. Such frequency combinations are defined as

$$\omega_{ijk} = \omega_i + \omega_j - \omega_k \quad (i, j \neq k) \tag{12.27}$$

For N wavelength channels co-propagating through the fiber, the number of generated frequencies is

$$M = \frac{N^2}{2}(N-1) \tag{12.28}$$

If the wavelength channels are equally spaced, the new waves overlap the original injected frequencies. This causes severe crosstalk and the depletion of the original signal waves, thus degrading the system performance.

In general, the penalty due to four-wave mixing can be reduced by (i) making the channel spacing unequal, (ii) increasing the channel spacing, and (iii) using a non-zero dispersion-shifted fiber instead of a dispersion-shifted fiber.

12.4.4 Self- and Cross-phase Modulation

Self-phase modulation (SPM) arises because the refractive index n of the fiber depends on the intensity I (which is equivalent to the power per unit effective area of the fiber). The relation is as follows:

$$n = n_0 + n_2 I = n_0 + n_2\left(\frac{P}{A_{\text{eff}}}\right) \tag{12.29}$$

where n_0 is the ordinary refractive index of the fiber core and n_2 is the non-linear index coefficient, P is the optical power, and A_{eff} is the effective area of the fiber. Depending on the dopant, the value of n_2 for a silica fiber varies from 2.2 to 3.4 × 10^{-8} $\mu m^2/W$.

To understand the effect of SPM, let us consider a Gaussian pulse propagating through a fiber with a non-linear index of refraction given by Eq. (12.29). The pulse shape is shown in Fig. 12.10. The time axis is normalized to the time parameter t_0, the pulse half-width measured at $1/e$ intensity point. As is evident from the figure, the intensity of the pulse first rises from zero to a maximum and then falls to zero again. Because the refractive index of the fiber is dependent on intensity, it will also vary with time. This variation of n with t will give rise to a temporally varying phase change in exactly the same fashion. Thus, different parts of the pulse undergo different phase shifts, which gives rise to what is known as *frequency chirping*; that is, the rising edge of the pulse shifts towards higher frequencies (red shift) and the trailing edge shifts towards lower frequencies (blue shift). This pulse chirping, in turn, enhances the group velocity dispersion (GVD) induced pulse broadening. Moreover, this effect

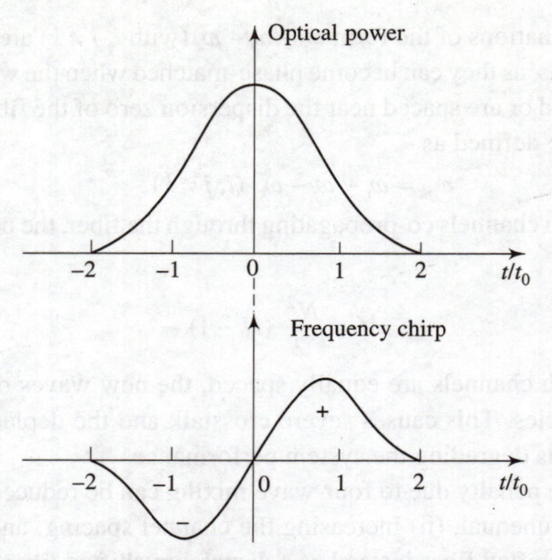

Fig. 12.10 SPM-induced frequency chirp for a Gaussian pulse

is proportional to the transmitted signal power, and hence SPM effects are more pronounced in long-haul systems where the transmitted powers are high.

In WDM systems, the intensity-dependent refractive index of the fiber gives rise to another kind of non-linear effect, called cross-phase modulation (CPM). In this process, the power fluctuation in one channel produces phase fluctuations in other co-propagating channels. The effect can be significant if the system is using dispersion-shifted fibers and is operating above 10 Gbits/s. The effect may be reduced by employing non-zero dispersion single-mode fibers.

12.5 DISPERSION MANAGEMENT

As we have seen in Chapter 10, erbium-doped fiber amplifiers have opened a new era of optical transmission technologies, allowing us to use WDM or DWDM with compact and economical approaches. However, the price that has to be paid for this success is the combat with the accumulated impact of the dispersive and non-linear effects of the transmission fiber, which grows with the transmission distance. Dispersion management techniques have been developed to solve this problem. The basic idea is to use two types of fibers with opposite signs of dispersion to produce a sawtooth pattern of the dispersion map as shown in Fig. 12.11. The condition for perfect dispersion compensation is

$$D_1L_1 + D_2L_2 = 0 \tag{12.30}$$

where D_1 and D_2 are the dispersion parameters of the two types of fibers and L_1 and L_2 are their respective lengths. A special type of fiber called the dispersion-

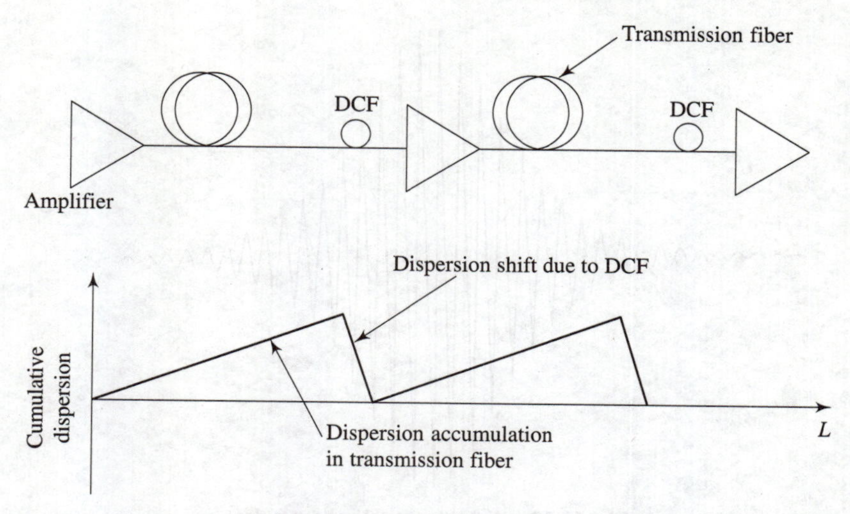

Fig. 12.11 Dispersion management using DCF

compensating fiber (DCF) has been developed for this purpose. Typically, a small length of the DCF may be placed just before the optical amplifier. If the transmission fiber has a low positive dispersion, the DCF should have a large negative dispersion. With this approach, the total cumulative dispersion is made zero (or small) so that the dispersion-induced penalties are negligible, but the dispersion is non-zero everywhere along the link so that the penalties due to non-linear effects are also reduced.

12.6 SOLITONS

Solitons are very narrow optical pulses with high peak powers that retain their shapes as they propagate along the fiber. We have seen in Chapter 5 that owing to GVD, the pulse propagating through the fiber gets broadened. In the anomalous regime (where chromatic dispersion is positive, say, above 1.32 μm in silica-based optical fibers) SPM causes the pulse to narrow, thereby partly compensating the chromatic dispersion. If the relative effects of SPM and GVD for an appropriate pulse shape are controlled properly, the compression of the pulse resulting from SPM can exactly balance the broadening of the pulse due to GVD. Therefore, the pulse shape either does not change or changes periodically as the pulse propagates down the fiber. The family of pulses that do not undergo any change in shape are called fundamental solitons and the pulses that undergo periodic changes are known as higher order solitons.

The shape of fundamental solitons that are being experimented upon along with optical amplifiers for communication is shown in Fig. 12.12. In fact the solitons overcome the detrimental effect of chromatic dispersion, and optical amplifiers negate the attenuation. Hence using the two together offers the promise of very high bit rate transmission with large repeaterless distances.

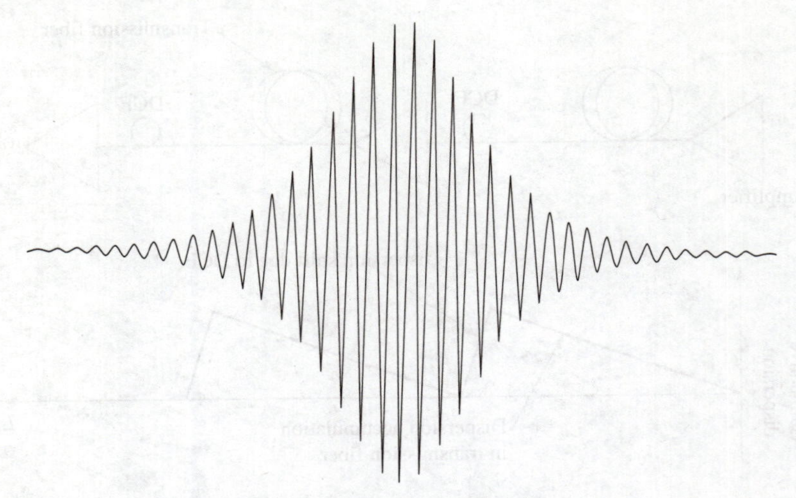

Fig. 12.12 A fundamental soliton pulse

SUMMARY

- System design considerations for point-to-point links are related to two important transmission parameters of optical fibers. These are (i) fiber attenuation and (ii) fiber dispersion.
- In order to ensure that the designed system performance has been met, two types of system analysis are carried out. These are (i) link power budget analysis and (ii) rise-time budget analysis.

  To estimate maximum repeaterless transmission distance, the following relation is used:

$$P_{tx} = P_{rx} + C_L + M_s$$

  For the desired bit rate, the total rise time of the digital system should be below its maximum value given by

$$t_{sys} \le \begin{cases} \dfrac{0.35}{B} & \text{for RZ format} \\[2ex] \dfrac{0.70}{B} & \text{for NRZ format} \end{cases}$$

  The performance of analog systems is normally analyzed by the calculating the carrier-to-noise ratio (CNR) of the system.
- Fiber-optic communication systems may be classified into three broad categories. These are (i) the point-to-point link, (ii) distribution networks, and (iii) local area networks.

- There are two types of non-linear effects that place limitations on system performance particularly at high transmitted power levels or at high bit rates exceeding 10 Gbits/s. These are (i) non-linear inelastic processes, e.g., SRS and SBS, and (ii) non-linear effects arising from the intensity-dependent variation in the refractive index of the fiber core, e.g., SPM, CPM, and FWM.
- Dispersion in long-haul systems can be compensated using two types of fibers with opposite signs of dispersion.
- Solitons are very narrow optical pulses with high peak powers that retain their shape as they propagate along the fibers. Fundamental solitons along with optical amplifiers offer the promise of very high bit rate transmission over large repeaterless distances.

MULTIPLE CHOICE QUESTIONS

12.1 A fiber-optic link of length 50 km has two splices each exhibiting a loss of 1 dB. The fiber itself has a rated 0.2 dB/km loss. If the minimum power required to run a photodetector is 20 nW, what power must be supplied by the optical source?
(a) 20 nW (b) 0.317 μW (c) 0.317 mW (d) 0.50 mW

12.2 A fiber used in a typical link has $(\Delta T)_{mat}$ = 3 ns/km and $(\Delta T)_{modal}$ = 1 ns/km. The link length is 8 km. The rise times of the transmitter and receiver are, respectively, 12 ns and 11 ns, respectively. The total rise time of the system is
(a) 30 ns (b) 35 ns (c) 23 ns (d) 27 ns

12.3 What is the bit duration of a 2.5-Gbits/s signal?
(a) 2.5 ns (b) 1 ns (c) 0.4 ns (d) 0.1 ns

12.4 For a passive star network, the total optical power supplied by the central node is 1 mW and that received at the terminal nodes is 0.1 μW. If the fractional insertion loss at each coupler is 0.05, what is the number of subscribers (nodes)?
(a) 50 (b) 100 (c) 250 (d) 500

12.5 Which of the following is a non-linear inelastic process?
(a) SRS (b) SPM (c) CPM (d) FWM

12.6 Which of the following non-linear effects arise from the intensity-dependent variation of the refractive index of a fiber?
(a) SPM (b) CPM (c) FWM (d) All of these

12.7 Assuming the fiber link span to be large, what will be its effective length if the attenuation per unit length is taken to be 0.20 dB/km?
(a) 95.6 km (b) 50.2 km (c) 22.2 km (d) 10 km

12.8 A typical standard dispersion-shifted fiber has a dispersion parameter of 16 ps nm^{-1} km^{-1}. The repeaterless fiber length is 50 km. What should be the

dispersion parameter of a DCF of length 2 km in order to compensate for the accumulated dispersion of the DSF for a repeaterless distance?
(a) 80 ps nm^{-1} km^{-1} (b) –80 ps nm^{-1} km^{-1}
(c) –200 ps nm^{-1} km^{-1} (d) –400 ps nm^{-1} km^{-1}

12.9 If the number of wavelength channels in a WDM system is 10, what will the number of frequencies created by four-wave mixing be?
(a) 30 (b) 100 (c) 450 (d) 900

12.10 Which of the following process is used to compensate for the GVD-induced dispersion in a soliton?
(a) FWM (b) SPM (c) CPM (d) All of these

Answers

12.1 (b)	12.2 (a)	12.3 (c)	12.4 (d)	12.5 (a)
12.6 (d)	12.7 (c)	12.8 (d)	12.9 (d)	12.10 (c)

REVIEW QUESTIONS

12.1 What are link power budget and rise-time budget analyses? Perform these analyses for a fiber-optic link that uses a conventional single-mode fiber that has been upgraded by DCF loops and optical amplifiers.

12.2 It is desired to design a 1.55-μm single-mode fiber-optic digital link to be operated at 500 Mbits/s. A single-mode InGaAsP laser can couple on an average an optical power of –13 dBm into the fiber. The fiber exhibits an attenuation of 0.5 dB/km and is available in pieces of 5 km each. An average splice loss of 0.1 dB is expected. The receiver is an InGaAs avalanche photodetector with a sensitivity of –40 dBm. The coupling loss at the receiver end is expected to be 0.5 dB. Set up a link power budget keeping a safety margin of 7 dB and calculate the transmission distance without repeaters.
Ans: 37.5 km

12.3 A 1.3-μm single-mode fiber-optic communication system is designed to operate at 1.5 Gbits/s. It uses an InGaAsP laser capable of coupling 0 dBm (1 mW) of optical power into the fiber. The fiber cable loss (including the splicing losses) is rated as 0.5 dB/km. The connectors at each end of the link exhibit a loss of 1 dB. The receiver is a InGaAs *p-i-n* diode with a sensitivity of –35 dBm. Set up a link power budget keeping a safety margin of 6 dB and calculate the repeaterless distance.
Ans: 54 km

12.4 A typical 12-km 0.85-μm digital fiber-optic link is designed to operate at 100 Mbits/s. The link uses a GaAs LED transmitter and an Si-pin receiver, which have rise times of 10 ns and 12 ns, respectively. The spectral half-width of the LED is 30 nm. The cable uses a step-index multimode fiber which has a core index n_1 of 1.46, a relative refractive index difference Δ of 0.01, and a material dispersion parameter of 80 ps km^{-1} nm^{-1}. Prepare a rise-time budget of the link. Is it possible to operate the system with an NRZ format?

Ans: $\tau_{sys} = 0.58$ μs, which is much greater than 0.70/B; hence the system cannot operate with an NRZ format.

12.5 Discuss different types of system architectures. Suggest the application of each of these topologies.

12.6 A typical CATV system uses an optical bus to distribute video signals to subscribers. The transmitter is coupling 0 dBm of optical power into the bus. Each receiver has a sensitivity of –42 dBm. Each optical tap couples 5% of optical power to the subscriber and has an insertion loss of 0.3 dB. On an average, the optical loss within the optical fiber bus itself is 0.01 dB/m and the minimum distance between two subscribers is 50 m. How many subscribers can be added to the bus before the signal needs an in-line amplification? How many subscribers can be added if the passive couplers are replaced by active couplers?

Ans: 29

12.7 A typical fiber-optic communication link is 100 km long (see Fig. 12.13). The fibers between stations *A* and *B* and those between *C* and *D* are GI fibers with an NA of 0.20, α of 2.0, and exhibiting a total attenuation of 0.1 dB/km; the fiber between stations *B* and *C* is also a GI fiber but with an NA of 0.17, α of 1.9, and exhibiting a total attenuation of 0.12 dB/km. There is perfect splicing of optical fibers at *B* and *C*; the link has two couplers (one each at stations *A* and *D*), each giving a loss of 0.5 dB. A safety margin of 7 dB is desired. Fresnel reflection losses at *B* and *C* may be assumed negligible.

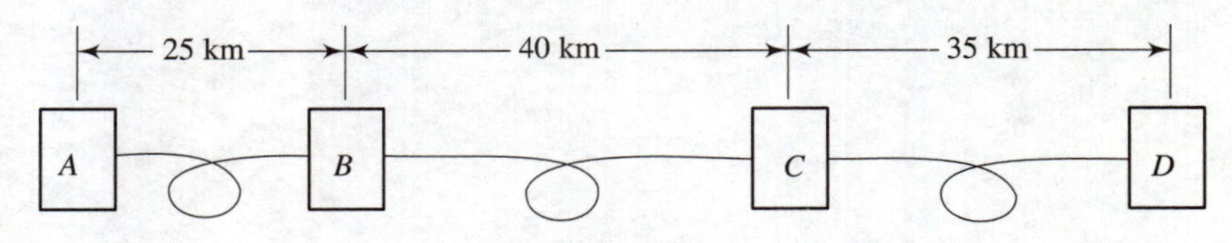

Fig. 12.13

(a) Calculate the minimum optical power that must be launched by the transmitters on the two sides for a duplex link (i.e., for two-way communication) if the receivers on either side require a minimum of 0.5 μW for signal recovery.

(b) Assume that the receivers and transmitters have rise times of 10 ns and 12 ns. The multipath time dispersion in all the fibers is almost negligible. The sources used in the transmitters have a spectral half-width of 20 nm.

The material dispersion parameter for the fibers between *A* and *B* and those between *C* and *D* is 70 ps km^{-1} nm^{-1}, and that for the fiber between *B* and *C* is 80 ps km^{-1} nm^{-1}. Calculate the maximum bit rate at which the system can be operated using the NRZ format.

Ans: (a) 54 μW (b) 4.7 Mbits/s

12.8 Why are non-linear effects observed in optical fibers? Why do they become pronounced at high power levels?

12.9 What is FWM? What are the negative effects of FWM in WDM systems? Can it be used in a beneficial way? How?

12.10 (a) Define SPM and CPM. How can SPM be used to produce fundamental solitons?

(b) What are the unique properties of solitons?

13 Fiber-optic Sensors

After reading this chapter you will be able to understand the following:

- Fiber-optic sensors
- Classification of fiber-optic sensors
- Intensity-modulated sensors
- Phase-modulated sensors
- Spectrally modulated sensors
- Distributed sensors
- Fiber-optic smart structures
- Industrial applications

13.1 INTRODUCTION

In recent years, fiber optics has found major application in sensor technology due to the inherent advantages of optical fibers, e.g., immunity to electromagnetic interference or radio-frequency interference, electrical isolation, chemical passivity, small size, low weight, and their ability to interface with a wide variety of measurands. Developments in fiber optics, which have been driven largely by the communication industry, along with those in the field of optoelectronics have enabled fiber-optic sensor technology to reach an ideal potential for many industrial applications.

This chapter discusses (i) fiber-optic sensors, (ii) their classification, (iii) their configurations, and (iv) their applications. Thus, starting with the basic classification based on common fiber-optic modulation techniques, we proceed to elaborate various mechanisms of point sensing under these categories. Distributed sensors are taken up next. Fiber-optic smart structures along with applications of point and distributed FOS are described in the end.

13.2 WHAT IS A FIBER-OPTIC SENSOR?

A fiber-optic sensor (FOS) is a device which uses light guided within an optical fiber to detect any external physical, chemical, biomedical, or any other parameter.

A generalized configuration of a fiber-optic sensor is shown as a block diagram in Fig. 13.1. It consists of an optoelectronic source, optical fiber(s), a modulating element, an optoelectronic detector, a signal processor, and finally a read-out device.

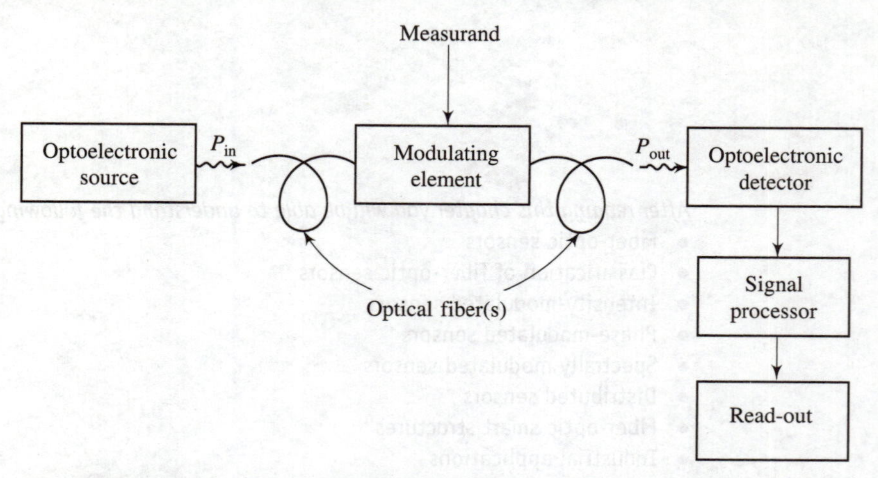

Fig. 13.1 Generalized configuration of a fiber-optic sensor (P_{in} and P_{out} are input and output optical powers, respectively.)

Let us consider a simple FOS in which the measurand (which may be displacement, force, pressure, temperature, etc.) modulates the intensity of light propagating through the optical fiber and the modulating element combination. The modulated light changes the detector output, which can be further processed and calibrated to give the value of the measurand. A variety of schemes have been suggested for this kind of modulation, some of which will be discussed in Sec. 13.4. A careful examination of Fig. 13.1 reveals that the signal S developed by the detector, to a good approximation, may be given by the relation

$$S \approx P(\lambda)\eta T(\lambda, l)\, M(I, \phi \text{ or } \lambda)R(\lambda) \tag{13.1}$$

where $P(\lambda)$ is the power furnished by the optoelectronic source as a function of the wavelength λ; η is the coupling efficiency of the input/output fiber(s) with the modulating element; $T(\lambda, l)$ is the transmission efficiency of the optical fiber(s), which will depend on the wavelength λ and length l of the fiber; $M(I, \phi \text{ or } \lambda)$ is the response of the modulating element, which may modulate the intensity (I), phase (ϕ), or spectral distribution (λ) (in the present case, we are considering only intensity modulation); and $R(\lambda)$ is the responsivity of the photodetector. The main assumptions here are that (i) the system is using a source that provides fairly monochromatic light, (ii) the optical fiber is a single-mode fiber so that the LP_{01} (linearly polarized) mode is propagating through the fiber (if a multimode fiber is used, all the modes are excited uniformly), and (iii) the response of the modulating element and that of the detector are almost linear.

This simplified expression [Eq. (13.1)] provides a basis for exploring possible methods for optimizing the design of FOSs. The choice of components is largely governed by the selection of the modulation scheme rather than the measurand, because the same measurand can be sensed using various modulating mechanisms. Therefore, prior to discussing various types of sensors, let us look at the common schemes of modulation which classify FOSs into different categories.

13.3 CLASSIFICATION OF FIBER-OPTIC SENSORS

There are two ways in which fiber-optic sensors can be classified. The first is to group them into the following two categories:

(i) *Extrinsic sensors* The light from an optical source is launched into the fiber (see Fig. 13.1) and is guided to a point where the measurement is to be performed. At this point the light is allowed to exit the fiber and get modulated by the measurand in a separate zone before being relaunched into the same or a different fiber. The devices based on this principle are called extrinsic sensors.

(ii) *Intrinsic sensors* The light launched into the fiber gets modulated in response to the measurand whilst still being guided in the fiber. Such devices are called intrinsic sensors.

The second and more logical way is to classify them according to the modulation scheme that is used in making the sensor. Accordingly we can group them into the following three major categories:

(i) *Intensity-modulated sensors* in which the intensity of light launched into the fiber is changed either intrinsically or extrinsically by the measurand.

(ii) *Phase-modulated sensors* in which the phase of monochromatic light propagating through the fiber is changed (normally intrinsically) by the measurand.

(iii) *Spectrally modulated sensors* in which the wavelength of light is changed (normally extrinsically) by the measurand.

Although several mechanisms of fiber-optic sensing using these modulation schemes have been suggested, we will discuss only a few representative ones to get an insight into the favourable and unfavourable features of these types of devices.

13.4 INTENSITY-MODULATED SENSORS

A variety of schemes have been suggested for intensity modulation. One of the important schemes among these involves the displacement of one fiber relative to the other by the measurand. Light is launched into one fiber, which is kept fixed, and a second fiber is made to undergo either longitudinal (or axial) displacement, or lateral (or transverse) displacement, or an angular displacement with respect to the first one, by the measurand, as shown in Fig. 13.2.

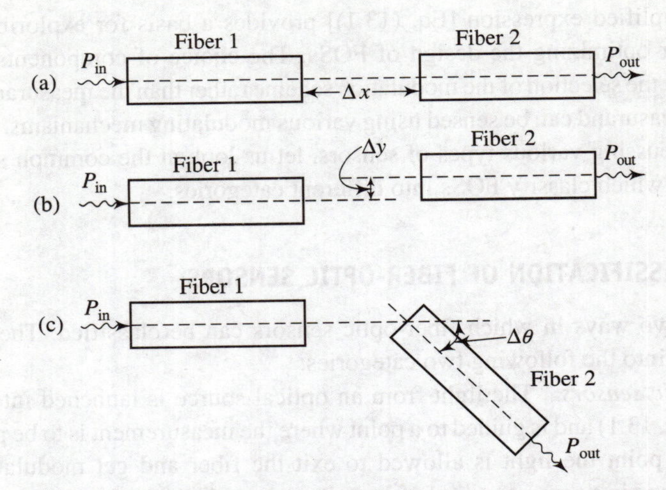

Fig. 13.2 (a) Longitudinal (or axial), (b) lateral (or transverse), and (c) angular displacement of fiber 2 with respect to fiber 1 caused by the measurand

The coupling efficiency η of Eq. (13.1) in this case will depend on many factors. In fact, for the three types of displacements shown in Figs 13.2(a), 13.2(b), and 13.2(c), the coupling efficiencies for two compatible multimode fibers are given by Eqs (13.2), (13.3), and (13.4), respectively (you may recall the expressions for misalignment losses discussed in Chapter 6). Thus the coupling efficiency η_{long} for longitudinal displacement Δx between two fibers is given by

$$\eta_{\text{long}} \approx \frac{16k^2}{(1+k)^4}\left[1 - \frac{\Delta x\,(\text{NA})}{4\,an}\right] \tag{13.2}$$

Here $k = n_1/n$, n_1 is the refractive index of the core of the fiber, n is the refractive index of the medium surrounding the fiber (e.g., air), a is the core radius, and NA is the numerical aperture of the fibers. The coupling efficiency η_{lat} for a lateral offset Δy between the axes of the two fibers is given by

$$\eta_{\text{lat}} \approx \frac{16k^2}{(1+k)^4}\,\frac{2}{\pi}\left[\cos^{-1}\left(\frac{\Delta y}{2a}\right) - \left(\frac{\Delta y}{2a}\right)\left\{1 - \left(\frac{\Delta y}{2a}\right)^2\right\}^{1/2}\right] \tag{13.3}$$

Finally, the coupling efficiency η_{ang} for an angular displacement $\Delta\theta$ between the axes of the two fibers is given by

$$\eta_{\text{lat}} \approx \frac{16k^2}{(1+k)^4}\left[1 - \frac{n\Delta\theta}{\pi\,(\text{NA})}\right] \tag{13.4}$$

With a specific configuration of the sensing device (i.e., with an LED or ILD of specific wavelength, fibers of fixed length, diameter, etc., and a photodiode combination), the terms $P(\lambda)$, $T(\lambda, l)$, and $R(\lambda)$ in Eq. (13.1) may be taken to be constant. With a specific medium (e.g., dry air or an index-matched fluid) between the two fiber ends, $k = n_1/n$ may also be considered constant. Thus, employing Eqs (13.1) and (13.2), the signal developed by the detector (normalized to that corresponding to $\Delta x = 0$, $\Delta y = 0$, and $\Delta\theta = 0$) for longitudinal displacement Δx, keeping $\Delta\theta = 0$, $\Delta y = 0$, may be written as (Khare et al. 1995)

$$S_{\text{long}}\,(\text{normalized}) = 1 - \frac{\Delta x(\text{NA})}{4an} \tag{13.5}$$

Similarly, the signals developed by the detector for lateral displacement Δy (normalized to that for $\Delta y = 0$) keeping $\Delta x = 0$ and $\Delta\theta = 0$ and for angular displacement $\Delta\theta$ (normalized to that for $\Delta\theta = 0$), keeping $\Delta x = 0$ and $\Delta y = 0$ are given, respectively, by the following expressions:

$$S_{\text{lat}}\,(\text{normalized}) = \frac{2}{\pi}\left[\cos^{-1}\left(\frac{\Delta y}{2a}\right) - \left(\frac{\Delta y}{2a}\right)\left\{1 - \left(\frac{\Delta y}{2a}\right)^2\right\}^{1/2}\right] \tag{13.6}$$

and

$$S_{\text{ang}}(\text{normalized}) = 1 - \frac{n\Delta\theta}{\pi\,(\text{NA})} \tag{13.7}$$

Here, $\cos^{-1}(\Delta y/2a)$ in Eq. (13.6) and $\Delta\theta$ in Eq. (13.7) are expressed in radians.

Thus any parameter that can be transformed into any one of these three types of displacements (e.g., longitudinal, lateral, or angular) can be measured with the sensor.

Example 13.1 Assume that the fibers used in Fig. 13.2 are compatible multimode fibers with the following specifications: core diameter = 100.2 μm and NA = 0.30. The medium between the two fibers is air. Plot the theoretical normalized signals S_{long}, S_{lat}, and S_{ang} developed by the detector as a function of displacements Δx, Δy, and $\Delta\theta$, respectively.

Solution

The plots are given in Figs 13.3–13.5.

Notice the linearity of response of the sensor in Figs 13.3 and 13.5. The response of mechanism (b) of Fig. 13.2 can be made linear by the differential arrangement discussed below.

Figure 13.6(a) shows the design of a differential fiber-optic sensor. Herein, light from an optical source is launched into the transmitting fiber. Two separate compatible

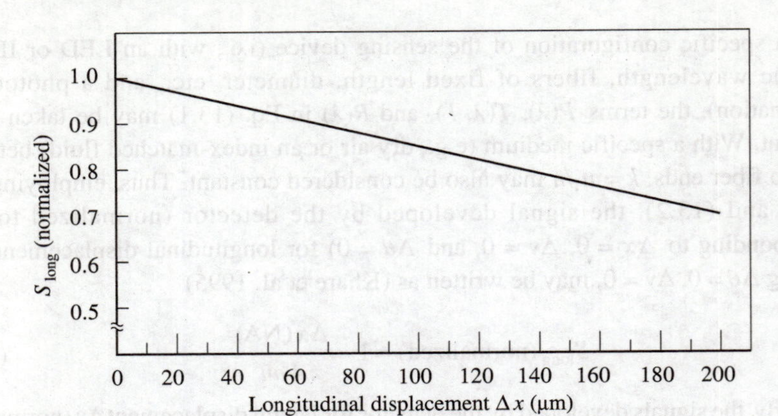

Fig. 13.3 Variation of S_{long} as a function of Δx (with $\Delta y = 0$ and $\Delta\theta = 0$)

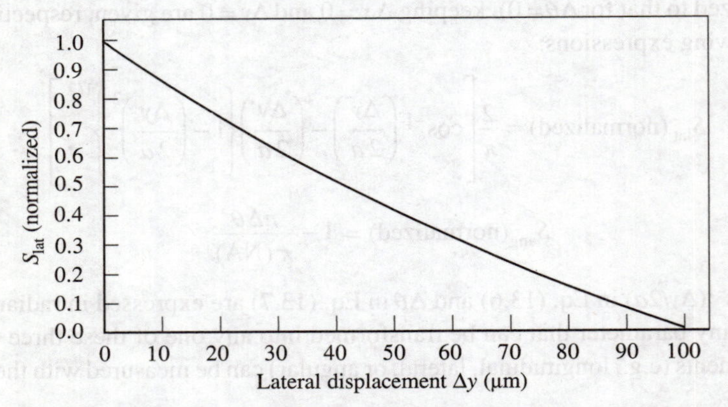

Fig. 13.4 Variation of S_{lat} as a function Δy (with $\Delta x = 0$ and $\Delta\theta = 0$)

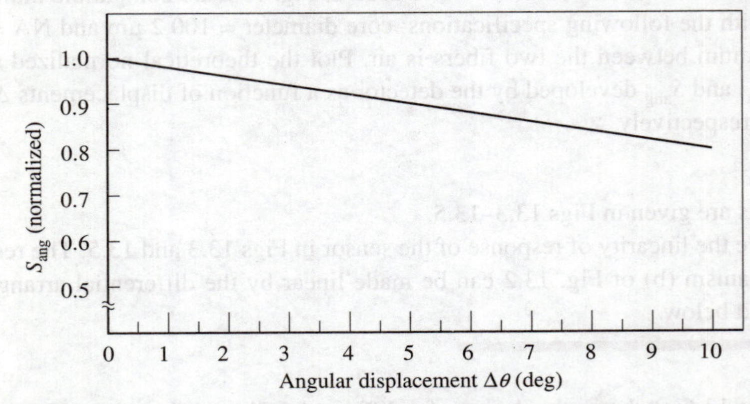

Fig. 13.5 Variation of S_{ang} as a function of $\Delta\theta$ (with $\Delta x = 0$ and $\Delta y = 0$)

fibers of equal length are placed with respect to the transmitting fiber such that both these fibers receive equal amount of power in the equilibrium position. The output of these two fibers is detected by two identical detectors. When the transmitting fiber is displaced by the measurand through a distance d, the area of overlap of receiving fiber 1 increases and that of receiving fiber 2 decreases. In other words, the lateral offset (Δy_1) between the axes of the transmitting fiber and receiving fiber 1 decreases, whereas the offset (Δy_2) between the axes of the transmitting fiber and receiving fiber 2 increases, as shown in Figs 13.6(a) and 13.6(b). In fact, $\Delta y_1 = a - d$ and $\Delta y_2 = a + d$, where a is the radius of each fiber. Consequently, the coupling efficiencies η_1 and η_2 of receiving fibers 1 and 2 with respect to the transmitting fiber increase and decrease, respectively. This, in turn, causes the outputs V_1 and V_2 of the two detectors to vary accordingly. If we take $V_1 - V_2 = V_c$, it can be shown that the normalized value of V_c may be written as

$$(V_c)_{\text{normalized}} = \frac{V_c}{(V_c)_{\max}} = \frac{\eta_1 - \eta_2}{(\eta_1 - \eta_2)_{\max}} \tag{13.8}$$

The theoretical and experimental variation of $(V_c)_{\text{normalized}}$ [Fig. 13.6(c)] is almost linear in the range of $d = \pm a$.

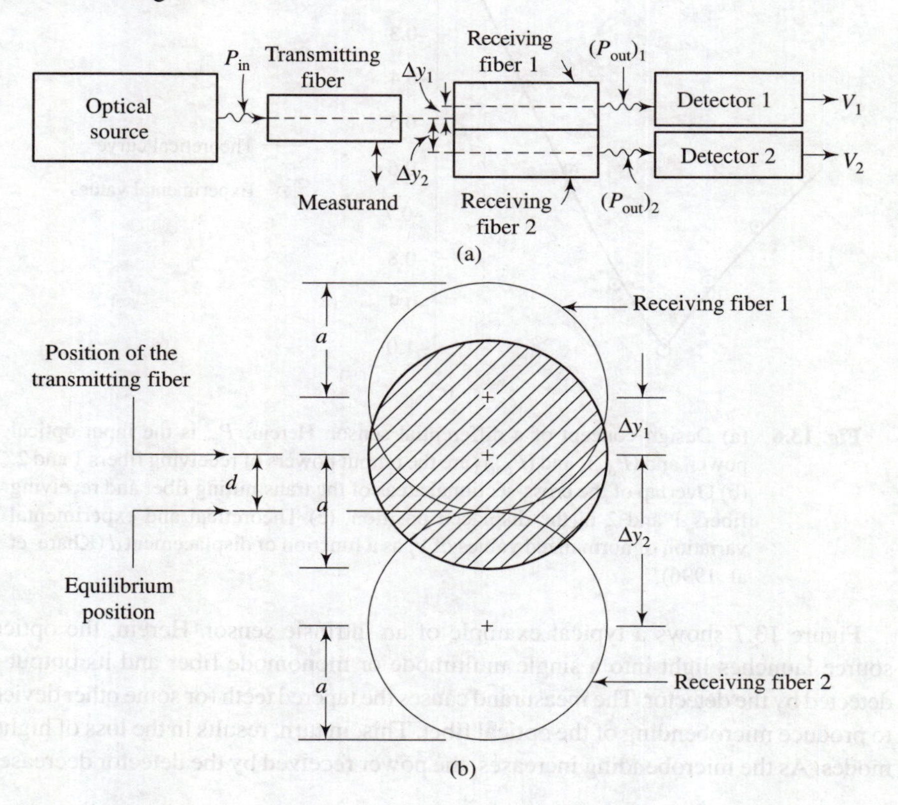

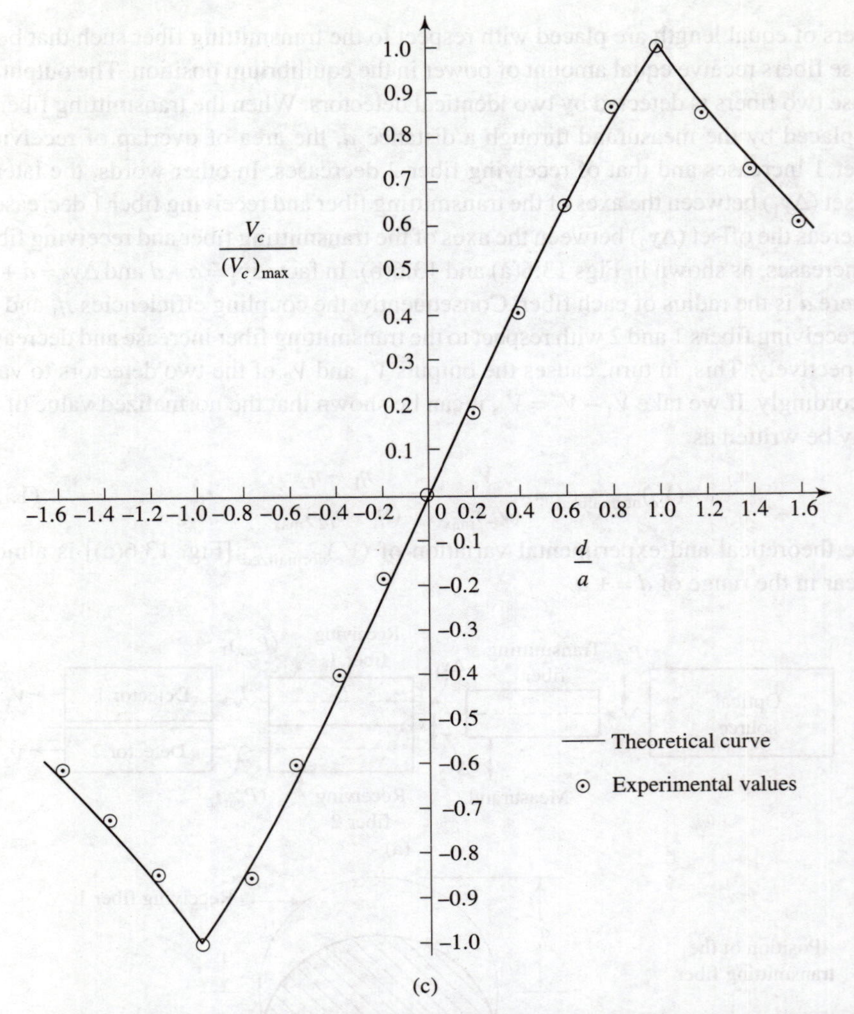

Fig. 13.6 (a) Design concept of a differential sensor. Herein, P_{in} is the input optical power, and $(P_{out})_1$ and $(P_{out})_2$ are the output powers of receiving fibers 1 and 2. (b) Overlap of the cross-sectional areas of the transmitting fiber and receiving fibers 1 and 2 in the displaced position. (c) Theoretical and experimental variation of normalized values of V_c as a function of displacement d (Khare et al. 1996).

Figure 13.7 shows a typical example of an intrinsic sensor. Herein, the optical source launches light into a single multimode or monomode fiber and its output is detected by the detector. The measurand causes the tapered teeth (or some other device) to produce microbending of the optical fiber. This, in turn, results in the loss of higher modes. As the microbending increases, the power received by the detector decreases.

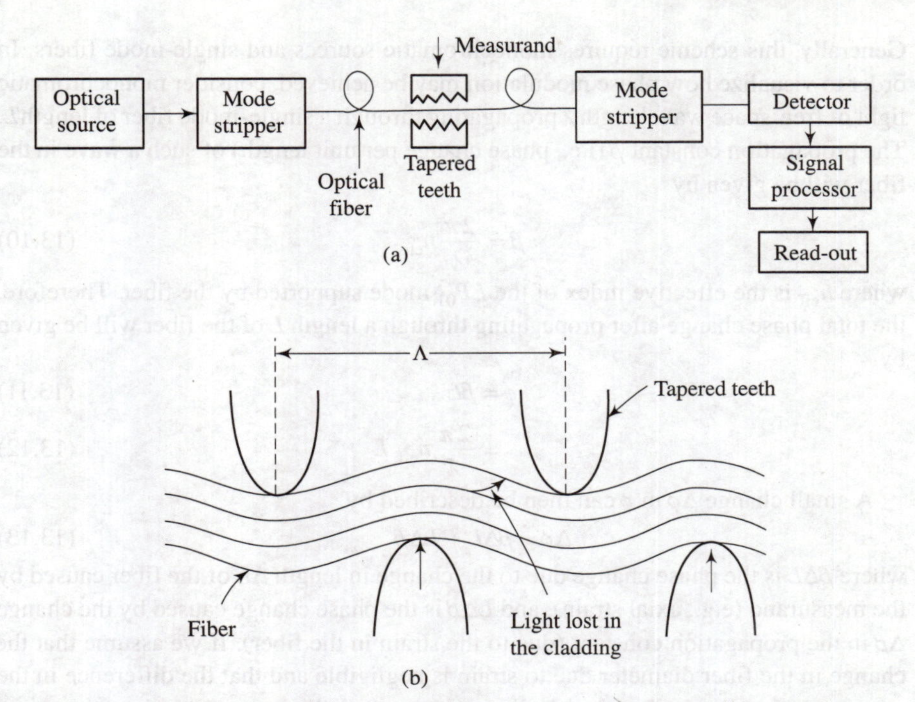

Fig. 13.7 (a) An intrinsic-type microbend sensor. (b) Details of the loss mechanism.

It can be shown that the loss is maximum for periodic microbending, with the bend pitch Λ given by

$$\Lambda = \frac{\alpha \pi a n}{\mathrm{NA}} \tag{13.9}$$

where α is the profile parameter of the core refractive index of the fiber, a is the core radius, n is the refractive index of the fiber core, and NA is the numerical aperture of the fiber.

Intensity-modulated sensors offer the virtues of simplicity of construction, reliability, low cost, and compatibility to multimode fiber technology. However, the signal developed by the detector depends on a large number of factors, some of which may not be under full control. Therefore, absolute measurements with these sensors may not be possible. For good performance they need some form of referencing. In fact such sensors are most suitable for switching applications, and applications in which digital (on–off) or modulation frequency encoders are used.

13.5 PHASE-MODULATED SENSORS

In this category of sensors, the measurand (e.g., temperature, pressure, strain, magnetic field, etc.) causes the phase modulation of an optical wave guided by the fiber.

Generally, this scheme requires monochromatic sources and single-mode fibers. In order to visualize how phase modulation may be achieved, consider monochromatic light of free-space wavelength λ propagating through a single-mode fiber of length L. The propagation constant β (i.e., phase change per unit length) of such a wave in the fiber will be given by

$$\beta = \frac{2\pi}{\lambda} n_{\text{eff}} \tag{13.10}$$

where n_{eff} is the effective index of the LP_{01} mode supported by the fiber. Therefore, the total phase change after propagating through a length L of the fiber will be given by

$$\phi = \beta L \tag{13.11}$$

$$= \frac{2\pi}{\lambda} n_{\text{eff}}\, L \tag{13.12}$$

A small change $\Delta\phi$ in ϕ can then be described by

$$\Delta\phi = \beta\Delta L + L\Delta\beta \tag{13.13}$$

where $\beta\Delta L$ is the phase change due to the change in length ΔL of the fiber caused by the measurand (e.g., axial strain), and $L\Delta\beta$ is the phase change caused by the change $\Delta\beta$ in the propagation constant (due to the strain in the fiber). If we assume that the change in the fiber diameter due to strain is negligible and that the difference in the core and cladding refractive indices is very small, we may write to a good approximation $n_{\text{eff}} \approx n$, where n is the core index of the fiber. Hence $\beta \approx (2\pi/\lambda)n$ and $\Delta\beta \approx (2\pi/\lambda)\Delta n$.

Substituting the values of β and $\Delta\beta$ in Eq. (13.13), we get

$$\Delta\phi = \frac{2\pi}{\lambda} n\Delta L + \frac{2\pi}{\lambda} L\Delta n$$

$$= \frac{2\pi}{\lambda}(n\Delta L + L\Delta n) \tag{13.14}$$

and

$$\phi \approx \frac{2\pi}{\lambda} nL$$

Therefore

$$\frac{\Delta\phi}{\phi} = \frac{\Delta L}{L} + \frac{\Delta n}{n} \tag{13.15}$$

Example 13.2 A light of wavelength $\lambda = 0.633$ μm is propagating through a single-mode silica-based optical fiber. Assume that the measurand is temperature, which changes the refractive index of silica at the rate of $10^{-5}\,°C^{-1}$. The nominal refractive index of the core is $n = 1.45$ and the fractional change in the length of the fiber per degree change in temperature is $5.1 \times 10^{-7}\,°C^{-1}$. Calculate the phase change per unit length per degree rise in the temperature of the fiber.

Solution

Using Eq. (13.14), the phase modulation due to the temperature T may be expressed by the relation

$$\frac{\Delta\phi}{\Delta T} = \frac{2\pi}{\lambda}\left(n\frac{\Delta L}{\Delta T} + L\frac{\Delta n}{\Delta T}\right)$$

The phase change per unit length per degree rise in temperature can then be given by

$$\frac{1}{L}\frac{\Delta\phi}{\Delta T} = \frac{2\pi}{\lambda}\left[\frac{n}{L}\frac{\Delta L}{\Delta T} + \frac{\Delta n}{\Delta T}\right] \tag{13.16}$$

Substituting the values of

$$n = 1.45, \quad \frac{\Delta n}{\Delta T} = 10^{-5}\,°\text{C}^{-1}, \text{ and } \quad \frac{1}{L}\frac{\Delta L}{\Delta T} = 5.1\times 10^{-7}\,°\text{C}^{-1}$$

we get

$$\frac{1}{L}\frac{\Delta\phi}{\Delta T} = \frac{2\pi}{0.633\times 10^{-6}}[1.45\times 5.1\times 10^{-7} + 10^{-5}]$$
$$= 106.5 \text{ rad m}^{-1}\,°\text{C}^{-1}$$

This is indeed a large magnitude.

Similar calculations for phase modulation due to pressure and strain can also be performed (see Review Questions 13.5 and 13.6). In fact, sensors based on this technique are very sensitive. In the following subsections we discuss a few representative sensors of this type.

13.5.1 Fiber-optic Mach–Zehnder Interferometric Sensor

The basic configuration of an all-fiber Mach–Zehnder interferometric sensor is shown in Fig. 13.8. Herein, the light from a laser source is split by a 3-dB coupler and sent equally to the sensing fiber and reference fiber arms (both of which are single-mode fibers). The outputs of the two fibers are recombined at the second 3-dB coupler. The sensing arm is in direct contact with the measurand, while the reference arm is shielded

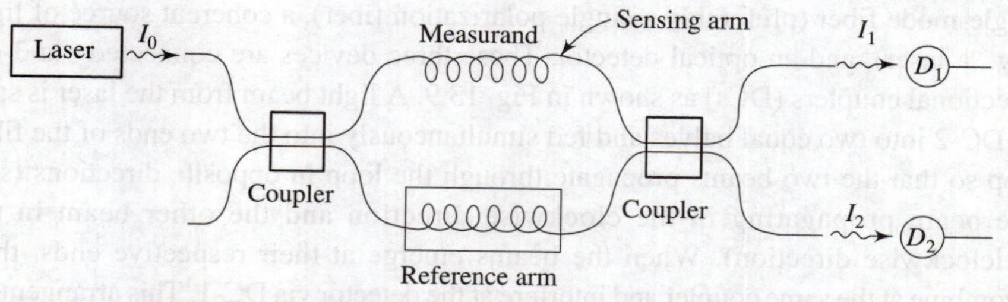

Fig. 13.8 Fiber-optic Mach–Zehnder interferometric sensor

from external perturbation. The measurand acts on the sensing arm and changes the phase of the light wave either by changing its length, its refractive index, or both. As the reference arm is not affected by the measurand, there appears a phase difference $\Delta\phi$ between the two waves arriving at the second coupler. The intensity I_1 and I_2 of light arriving at the two output ports of the second coupler will depend on this phase difference between the two waves emerging from the sensing and reference arms.

If we assume that the coupling coefficients of the two couplers are same, i.e., $k_1 = k_2 = k = 0.5$, and the fibers exhibit negligible loss, it can be shown that

$$I_1 = (I_0/2)(1 + \cos\Delta\phi) = I_0\cos^2(\Delta\phi/2) \quad (13.17)$$

and

$$I_2 = (I_0/2)(1 - \cos\Delta\phi) = I_0\sin^2(\Delta\phi/2) \quad (13.18)$$

where $I_0 = I_1 + I_2$ is the input intensity.

If the sensing and reference arms introduce identical phase shifts, $\Delta\phi$ will be zero and $I_1 = I_0$ and $I_2 = 0$; that is, the entire launched power appears at output port 1. On the other hand, if $\Delta\phi = \pi$, $I_1 = 0$, and $I_2 = I_0$, the entire launched power appears at output port 2. For other values of $\Delta\phi$, the power will be divided into the two ports according to Eqs (13.17) and (13.18).

In practice, the sensor is operated at the quadrature point corresponding to $\Delta\phi = \pi/2$. Thus a fixed bias of $\pi/2$ is induced and the phase change θ caused by the measurand is related to the output intensity. We may, therefore, write in this case

$$\Delta\phi = \pi/2 + \theta$$

If θ is very small, substituting this value of $\Delta\phi$ in Eq. (13.18), we get

$$\begin{aligned} I_2 &= (I_0/2)[1 - \cos(\pi/2 + \theta)] \\ &= (I_0/2)[1 + \sin\theta] \\ &\approx (I_0/2)(1 + \theta) \end{aligned} \quad (13.19)$$

Similarly,

$$I_1 \approx (I_0/2)(1 - \theta) \quad (13.20)$$

Thus, for small values of θ, I_1 and I_2 vary almost linearly with the phase change θ caused by the measurand.

13.5.2 Fiber-optic Gyroscope

A fiber-optic gyroscope is essentially a rotation sensor. It consists of a loop of a single-mode fiber (preferably a single polarization fiber), a coherent source of light (e.g., a laser), and an optical detector. These three devices are connected via 3-dB directional couplers (DCs) as shown in Fig. 13.9. A light beam from the laser is split by DC-2 into two equal halves and fed simultaneously into the two ends of the fiber loop so that the two beams propagate through the loop in opposite directions (say, one beam propagating in the clockwise direction and the other beam in the anticlockwise direction). When the beams emerge at their respective ends, they recombine at the same coupler and interfere at the detector via DC-1. This arrangement

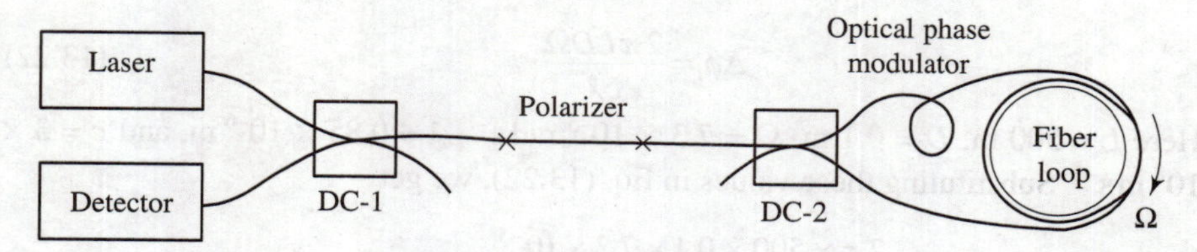

Fig. 13.9 A fiber-optic gyroscope

may be thought of as a special form of the Mach–Zehnder interferometer, in which the two arms lie within the same single fiber, but the two beams travel in opposite directions. If we assume that the couplers are lossless and the fiber loop is stationary, the counterpropagating beams will take the same time to emerge, resulting in all the power returning towards DC-1.

Now suppose the entire arrangement is rotating clockwise at an angular velocity Ω. Then the clockwise travelling beam will view the end of fiber receding from it as it travels, and hence it will have to travel farther to emerge. Conversely, the anticlockwise travelling beam will see its corresponding end approaching, and hence it will have to travel a smaller path to emerge. Consequently, there is a relative phase shift between the two beams when they emerge at the respective ends. This causes a corresponding shift in the interference pattern formed at the detector. This phenomenon of phase shift is known as the *Sagnac effect*. This effect can, therefore, be used to measure the rotational speed Ω.

It can be shown that the phase difference between the two counterpropagating beams, when the fiber loop is rotating, is given by

$$\Delta\phi = \frac{8\pi NA\Omega}{c\lambda} \tag{13.21}$$

where N is the number of turns in the loop, A is the area of a single turn, c is the speed of light in free space, and λ is the free-space wavelength of light.

The arrangement shown in Fig. 13.9 is called the *minimum configuration design*. The function of the polarizer and the phase modulator is to ensure that the detection bias is maintained at the point of maximum sensitivity. Such a configuration of a fiber-optic gyroscope has several applications ranging from ballistic missiles, through vehicle navigation systems, to industrial machinery.

Example 13.3 A fiber-optic gyroscope has a circular coil of diameter 0.1 m and the total length of optical fiber used in the coil is 500 m. If it is operating at $\lambda = 0.85$ μm, what is the phase shift corresponding to the earth's rotational speed ($\Omega = 7.3 \times 10^{-5}$ m rad s^{-1})?

Solution

For the total length L of the fiber and diameter D of the coil, Eq. (13.21) can be written as

$$\Delta\phi = \frac{2\pi L D \Omega}{c\lambda} \tag{13.22}$$

Here $L = 500$ m, $D = 0.1$ m, $\Omega = 7.3 \times 10^{-5}$ rad s^{-1}, $\lambda = 0.85 \times 10^{-6}$ m, and $c = 3 \times 10^{8}$ m s^{-1}. Substituting these values in Eq. (13.22), we get

$$\Delta\phi = \frac{2\pi \times 500 \times 0.1 \times 7.3 \times 10^{-5}}{3 \times 10^{8} \times 0.85 \times 10^{-6}} = 8.99 \times 10^{-5} \text{ rad}$$

This value is quite small indeed but can be easily measured.

13.6 SPECTRALLY MODULATED SENSORS

In this category of sensors, the wavelength of light is modulated by the measurand. Several schemes of wavelength modulation have been suggested. Here we describe only two important techniques.

13.6.1 Fiber-optic Fluorescence Temperature Sensors

Fluorescence is the phenomenon of emission of light (other than thermal radiation) by a material (normally called a 'phosphor') upon absorption of suitable electromagnetic radiation (from the ultraviolet, visible, or infrared range). In this process, a photon of higher energy (say, $h\nu_a$) is absorbed and a photon of lower energy (say, $h\nu_e$) is emitted. Therefore the emitted wavelength λ_e is greater than the absorbed wavelength λ_a. This is called Stokes' fluorescence. The intensity of fluorescence, in most phosphors, varies with temperature. Therefore, by selecting an appropriate fluorescent material, a temperature sensor based on this technique may be designed.

Figure 13.10 shows the design of such a sensor developed by ASEA (a Swedish company). This device utilizes a small crystal of GaAs sandwiched between GaAlAs layers in the sensor head. This is made to fluoresce by absorption of light emitted by a GaAs LED. As the temperature at the sensor head is increased, the emission band broadens and shifts towards the longer wavelength side. Two narrow bandpass filters are used to select two portions of the emission spectrum, and the intensity at each band is measured by respective detectors (photodiodes), their ratio obtained and correlated with temperature.

Another commercial device based on fluorescence is shown in Fig. 13.11. It uses a UV source for excitation and a rare-earth phosphor in the sensor head. The change in the ratio of the emission at two different wavelengths (λ_1 and λ_2) emitted by the sensor is calibrated as a function of temperature.

In sensors of this type, another channel is required for referencing, because there are several factors other than temperature that may cause the fluorescence intensity

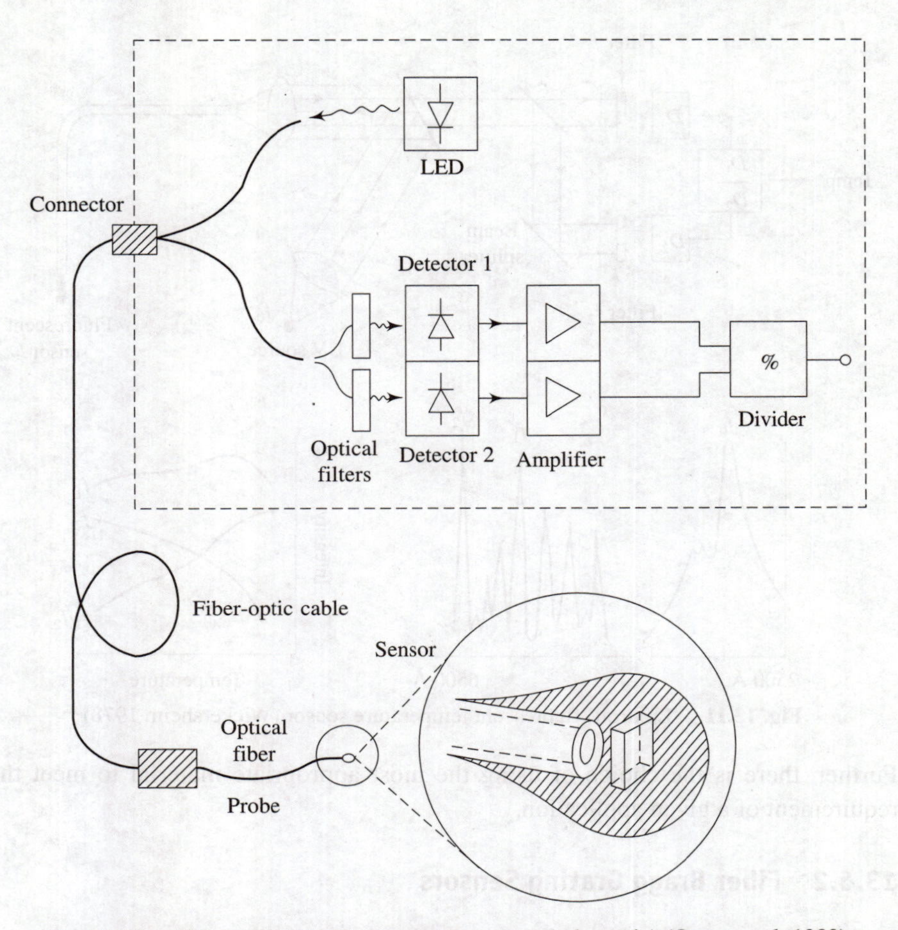

Fig. 13.10 ASEA model 1010 temperature sensor (schematic) (Ovren et al. 1983)

to vary. Consequently, a sensing technique based on the measurement of the lifetime of fluorescence has been developed, which has found wider application. Figure 13.12 illustrates the fundamental principle of this technique. The fluorescent material is excited by a pulse of suitable wavelength. The intensity rises exponentially with time. After the cessation of excitation, the intensity I falls again exponentially with time t and may be expressed by the following relation:

$$I(t) = I_0 \exp(-t/\tau) \tag{13.23}$$

where I_0 is the maximum intensity at time $t = 0$ and τ is the time constant, i.e., the time required for the intensity to decay to I_0/e. This constant is also called the lifetime of fluorescence. The lifetime is an intrinsic parameter of the fluorescent material, and hence is not affected by the change in fluorescence intensity. With such a system, any configuration of the probe can be used depending on the design requirement.

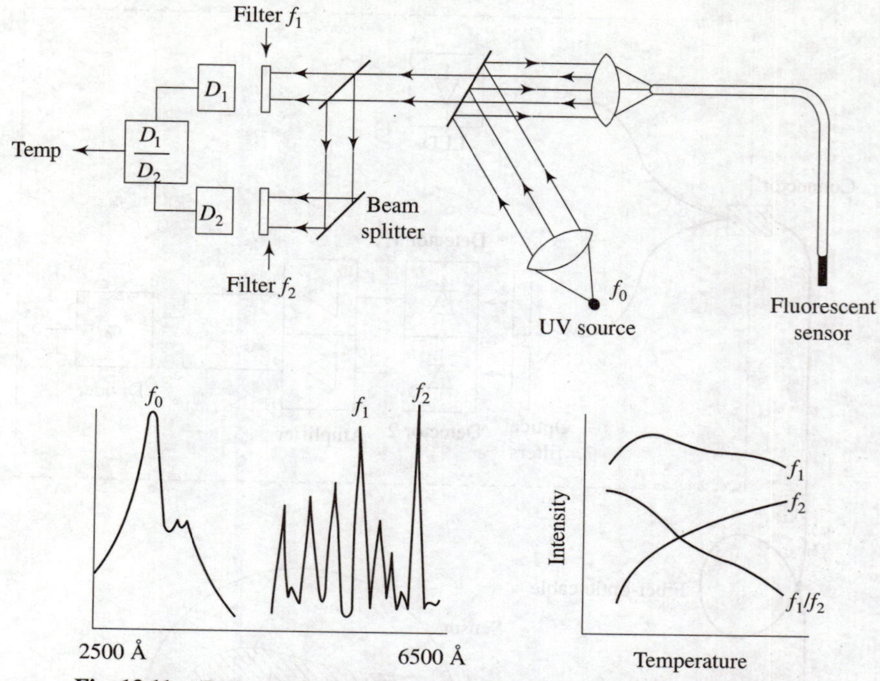

Fig. 13.11 'Luxtron' fluorescent temperature sensor (Wickersheim 1978)

Further, there is the choice of using the most appropriate material to meet the requirement of a given application.

13.6.2 Fiber Bragg Grating Sensors

In Chapter 11, we have discussed fiber Bragg gratings (FBGs) in connection with their application in WDM systems. To recall, an FBG is written into a segment of a Ge-doped single-mode fiber in which the refractive index of the core is made to vary periodically by exposure to a spatial pattern of UV light. When the FBG is illuminated by an optical source of broad spectral width, a Bragg wavelength λ_B is reflected by it. This wavelength should satisfy the following condition:

$$\lambda_B = 2n_{\text{eff}}\Lambda \tag{13.24}$$

where n_{eff} is the effective mode index and Λ is the spatial period of index modulation. Such gratings can be used for sensing strain, temperature, pressure, etc. Indeed, the FBG central wavelength λ_B varies with changes in these parameters and the corresponding wavelength shifts are given as follows (Rao 1998):

(a) For a longitudinal strain $\Delta\varepsilon$ applied to an FBG, the shift in λ_B is given by

$$(\Delta\lambda_B)_{\text{strain}} = \lambda_B(1 - \rho_\alpha)\Delta\varepsilon \tag{13.25}$$

where ρ_α is the photoelastic coefficient of the fiber.

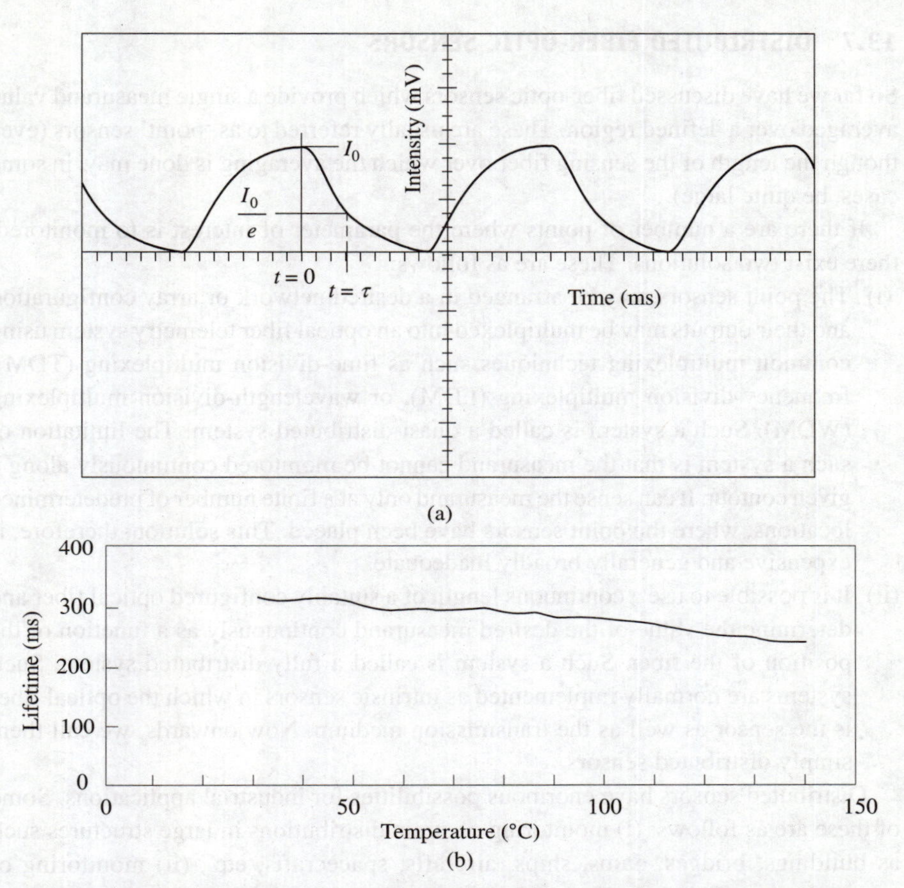

Fig. 13.12 (a) Rise and decay of fluorescence for SrS:Cu phosphor at room temperature. (b) Variation of τ with temperature for the same phosphor (Courtesy of the Fiber Optic Sensor Group, BITS, Pilani).

(b) For a temperature change of ΔT, the corresponding shift in λ_B is given by

$$(\Delta\lambda_B)_{\text{temp}} = \lambda_B(1 + \xi)\Delta T \tag{13.26}$$

where ξ is the thermo-optic coefficient of the fiber.

(c) For a pressure change of ΔP, the corresponding shift in λ_B is given by

$$(\Delta\lambda_B)_{\text{pressure}} = \lambda_B \left[\frac{1}{\Lambda}\frac{\partial \Lambda}{\partial P} + \frac{1}{n_{\text{eff}}}\frac{\partial n_{\text{eff}}}{\partial P} \right] \Delta P \tag{13.27}$$

Apart from these parameters, FBG sensors have found an important application in 'fiber-optic smart structures' for monitoring strain distributions.

13.7 DISTRIBUTED FIBER-OPTIC SENSORS

So far we have discussed fiber-optic sensors which provide a single measurand value averaged over a defined region. These are usually referred to as 'point' sensors (even though the length of the sensing fiber over which the averaging is done may, in some cases, be quite large).

If there are a number of points where the parameter of interest is to monitored, there exist two solutions. These are as follows:

(i) The point sensors may be arranged in a desired network or array configuration and their outputs may be multiplexed into an optical fiber telemetry system using common multiplexing techniques such as time-division multiplexing (TDM), frequency-division multiplexing (FDM), or wavelength-division multiplexing (WDM). Such a system is called a quasi-distributed system. The limitation of such a system is that the measurand cannot be monitored continuously along a given contour. It can sense the measurand only at a finite number of predetermined locations, where the point sensors have been placed. This solution, therefore, is expensive and generally broadly inadequate.

(ii) It is possible to use a continuous length of a suitably configured optical fiber and determine the value of the desired measurand continuously as a function of the position of the fiber. Such a system is called a fully distributed system. Such systems are normally implemented as intrinsic sensors in which the optical fiber is the sensor as well as the transmission medium. Now onwards, we call them simply distributed sensors.

Distributed sensors have enormous possibilities for industrial applications. Some of these are as follows: (i) monitoring of strain distributions in large structures such as buildings, bridges, dams, ships, aircrafts, spacecrafts, etc., (ii) monitoring of temperature profiles in boilers, power transformers, generators, furnaces, etc., and (iii) mapping of electric- and magnetic-field distributions and the intrusion alarm system for homes and industrial machines.

In distributed sensing, the monitoring of the measurand along the contour of the optical fiber requires some means of identifying the signal originating from a given section of the fiber. There are several methods by which this can be done. We discuss below a commonly used technique of optical time domain reflectometry (OTDR). Herein, a pulsed signal is transmitted into one end of the fiber and the backscattered signals from different parts of the fiber are recovered at the same fiber end. We have already discussed OTDR in connection with the field measurements in Chapter 6. Here we will see how this method may be used for sensing.

A distributed sensor based on monitoring backscattered power with OTDR is shown in Fig. 13.13. An optical pulse of high intensity is launched into the fiber through a DC. As it propagates, it is backscattered due to Rayleigh scattering. The backscattered signal is coupled to the detector again through the DC, processed, and finally read out

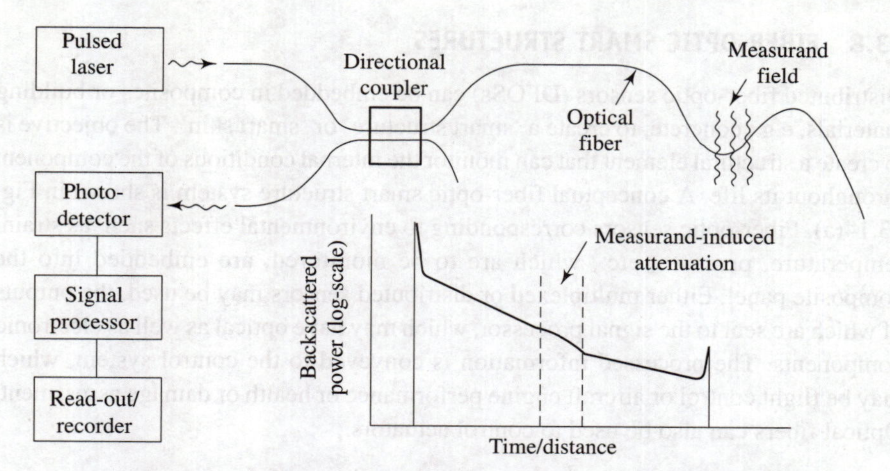

Fig. 13.13 A distributed sensor based on monitoring attenuation with OTDR

or recorded. A plot of backscattered power as a function of time/distance is shown in Fig. 13.13. If the fiber is homogenous and subject to a uniform environment, the backscattered power is given by the following relation (Barnoski & Jenson 1976):

$$P(l) = \frac{1}{2} P_0 WS\alpha_s(l) V_g \exp\left\{-\int_0^l 2\alpha(z)\,dz\right\} \tag{13.28}$$

where P_0 is the power launched into the fiber, W is the pulse width, $P(l)$ is the backscattered power coupled to the detector as a function of length l of the fiber, where $l = ct/2n$ is the location of the forward-travelling pulse at the time of generation of the detected backscattered signal.

Thus, OTDR can sense the parameter that can change the total attenuation coefficient α, keeping the scattering coefficient α_s and the capture fraction S constant. Obviously, then Eq. (13.28) will be modified to

$$P(l) = A\exp\left[-\int_0^l \alpha(z)\,dz\right] \tag{13.29}$$

where A is a constant. Alternatively, it is also possible to sense parameters that can change the scattering coefficient α_s if α and S are kept constant. Then Eq. (13.28) will take the form

$$P(l) = A'\alpha_s(l)\exp(-B'l) \tag{13.30}$$

where A' and B' are constants.

Indeed, there are many parameters, e.g., pressure, strain, temperature, etc., that can cause α or α_s to vary, and hence they can easily be measured or monitored by this technique.

13.8 FIBER-OPTIC SMART STRUCTURES

Distributed fiber-optic sensors (DFOSs) can be embedded in composites or building materials, e.g., concrete, to create a 'smart structure' or 'smart skin'. The objective is to create a structural element that can monitor the internal conditions of the component throughout its life. A conceptual fiber-optic smart structure system is shown in Fig. 13.14(a). Fiber-optic sensors corresponding to environmental effects such as strain, temperature, pressure, etc., which are to be monitored, are embedded into the composite panel. Either multiplexed or distributed sensors may be used, the outputs of which are sent to the signal processor, which may have optical as well as electronic components. The processed information is conveyed to the control system, which may be flight control or aircraft engine performance or health or damage assessment. Optical fibers can also be used to control actuators.

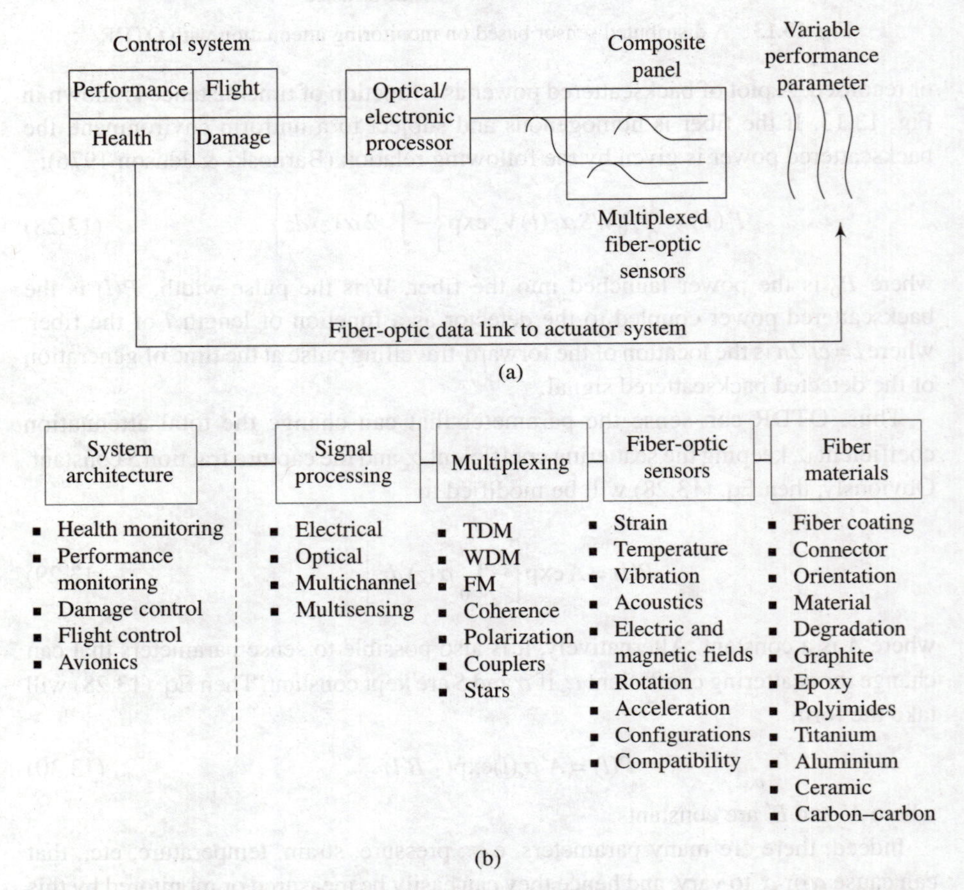

Fig. 13.14 (a) A fiber-optic smart structure. (b) Technologies associated with 'smart structures' and 'smart skins' (Udd 1991).

The relevant technologies associated with fiber-optic smart structures and skins are listed in Fig. 13.14(b). The prime issues needing consideration are as follows:

(i) Embedding fibers into composite materials, which means the selection of appropriate coatings compatible with the composites, orientation of the fiber in the material, and the means to access the ends of optical fibers so that they can be connected to other parts or to fiber-optic links outside the structure.

(ii) Selecting the technique of fiber-optic sensing from the available ones. (Normally interferometric sensors are employed, as they are highly sensitive, accurate, and compatible with multiplexing techniques. FBG sensors are also strong candidates for smart skins. Distributed sensors can serve the purpose in a better way.)

(iii) Designing the signal processing unit, which may be used to process the outputs of embedded sensors and also monitor the performance of the finished structure throughout its life.

Some typical applications of fiber-optic smart structures are shown in Fig. 13.15. Figure 13.15(a) shows a space-based habitat having arrays of fiber-optic sensors arranged to find the location and extent of impact damage by monitoring strain distribution, acoustics, or fiber breakage. Several other sensors can also be used to measure, say, leakage, radiation dose, etc. Figure 13.15(b) illustrates the use of DFOSs in monitoring the performance of pressurized tanks.

Indeed, fiber-optic smart structures have the potential to revolutionize the field of future composite materials and intelligent structures.

13.9 INDUSTRIAL APPLICATIONS OF FIBER-OPTIC SENSORS

From the point of view of industrial applications, fiber-optic sensors possess the virtues of excellent sensitivity, dynamic range, low cost, high reliability, immunity to electromagnetic interference, electromagnetic pulses, and radio-frequency interference, small size, and low weight. Almost all the parameters needed for industrial process control, e.g., temperature, pressure, strain, fluid level, flow rate, displacement/position, vibration, *p*H, electric and magnetic fields, voltage and current, etc., can be measured or monitored by FOSs. A number of commercial devices are now available, a detailed description of which may be found in the literature by Culshaw and Dakin (1989), Grattan and Meggitt (1998), and Udd (1995). New DFOS designs have also been taken to the stage of commercial products. For example, Herga pressure mats (by Herga Ltd) based on distributed microbend losses are being used to protect personnel working close to robots or to detect collisions of remotely controlled vehicles. The distributed cryogenic leak detection system, fully distributed temperature sensor system, DTS-II (by York Ltd), distributed cable strain monitor (by NTT), fiber-optic hydrophone array (by Plessey Ltd), etc. are already available in the market.

The performance of these sensors has been demonstrated to be more than sufficient for industrial applications. The future is heading towards all-fiber smart structures.

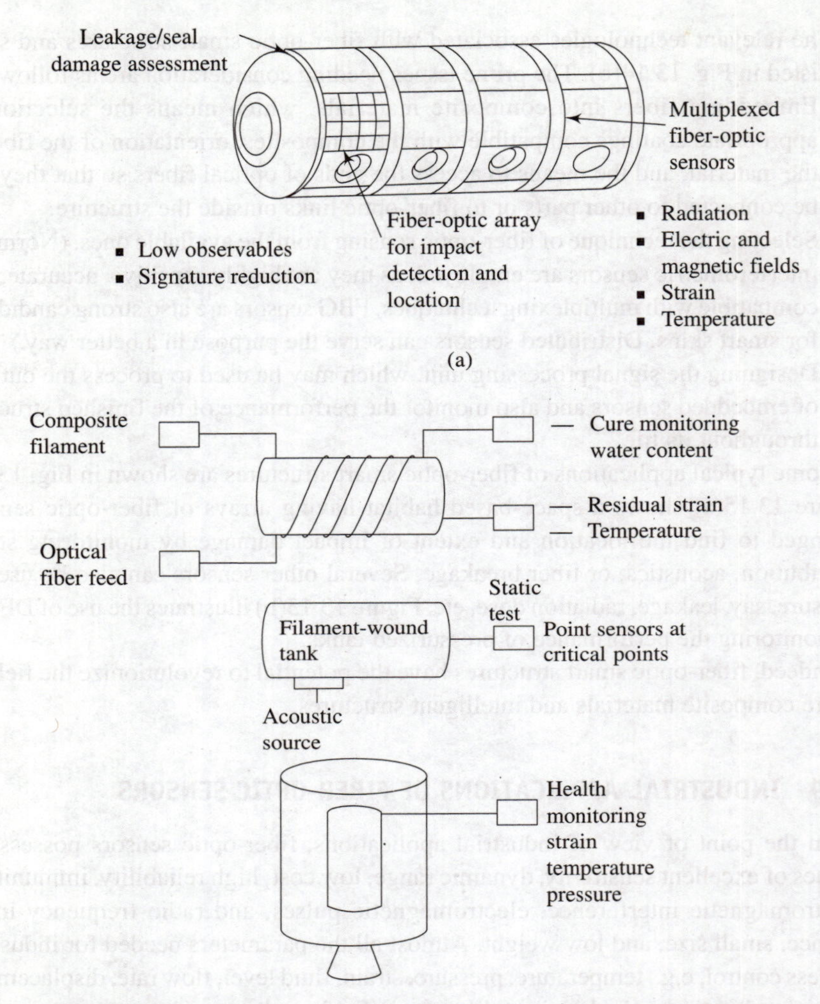

Fig. 13.15 Fiber-optic smart structures technology for (a) space-based habitat and (b) pressurized tanks (Udd 1991)

SUMMARY

- A fiber-optic sensor (FOS) is a device that uses light guided within an optical fiber to detect any external physical, chemical, or other parameter.
- FOSs may be classified into two groups, namely, (i) extrinsic sensors and (ii) intrinsic sensors.
- In extrinsic sensors, light from an optical source is launched into the fiber and guided to a point where the measurement is to be performed. Here the light gets modulated by the measurand and is relaunched into the same (or other) fiber.

- In intrinsic sensors, the light launched into the fiber gets modulated in response to the measurand whilst still being guided in the fiber.
- Another way of classifying FOSs is based on the scheme of modulation used for making the sensor. Thus the important categories of FOSs are
  (i) intensity-modulated sensors,
  (ii) phase-modulated sensors, and
  (iii) spectrally modulated sensors
- Intensity modulation can be achieved in a variety of ways. Again, there are extrinsic and intrinsic type of mechanisms for this kind of modulation.
- Phase modulation is more difficult to implement but gives more accurate results. Mach–Zehnder type of sensors can be used for the measurement of pressure and strain. Fiber-optic gyroscopes can be used for rotation sensing as well as some other parameters.
- Spectral modulation can also be achieved in several ways but this scheme has been prominently used for making fiber-optic fluorescent thermometers.
- Apart from point sensors, distributed sensing is required in many application areas. OTDR is the main technique used in implementing DFOSs. Temperature, pressure, strain, etc. can be sensed along a given contour continuously using these sensors.
- Point or distributed FOSs can be embedded in composite materials or structures to make smart skins and smart structures.
- Almost any industrial parameter, e.g., temperature, pressure, liquid level, displacement/position, pH, etc., can be measured or monitored using FOSs or DFOSs.

MULTIPLE CHOICE QUESTIONS

13.1 Fiber-optic sensors use the following as a prime means of measurement.
(a) Electric field (b) Magnetic field
(c) Optical field (d) Gravitational field

13.2 Which of the following measurands cannot be measured by a microbend sensor?
(a) Displacement (b) Temperature
(c) Pressure (d) Electric current

13.3 Which of the following measurands can change the refractive index of a silica-based fiber?
(a) Pressure (b) Temperature (c) Acoustic wave (d) All of these

13.4 What kind of change can be measured by an all-fiber interferometer?
(a) Intensity (b) Phase (c) Wavelength shift (d) None of these

13.5 Which of the following is an example of an intensity-modulated sensor?
(a) A sensor based on the relative displacement of two fibers
(b) A fiber-optic gyroscope

(c) A Mach–Zehnder interferometer
(d) All of the above

13.6 Which of the following is an example of a wavelength-modulated sensor?
(a) A microbend sensor
(b) A fiber-optic gyroscope
(c) A fluorescence temperature sensor
(d) None of these

13.7 DFOSs can, in principle, be based on the following property (properties) of a silica fiber.
(a) Attenuation of light (b) Scattering of light
(c) Absorption of light (d) All of these

13.8 A fiber-optic smart structure can be used for the following task.
(a) Monitoring internal strain(s)
(b) Monitoring the structural integrity of the completed component
(c) Mapping of thermal profiles
(d) All of the above

13.9 Fiber Bragg gratings cannot be used for the following measurement.
(a) Pressure (b) Liquid level (c) Strain (d) Temperature

13.10 A strain profile of a dam (to be constructed) is to be mapped. Which of the following arrangements would give best results?
(a) Using point strain sensors at different points and multiplexing their outputs
(b) Using an all-fiber DFOS after the dam is constructed
(c) Using embedded fiber-optic smart structures during the construction process
(d) All of the above

Answers

13.1 (c)	13.2 (d)	13.3 (d)	13.4 (b)	13.5 (a)
13.6 (c)	13.7 (d)	13.8 (d)	13.9 (b)	13.10 (c)

REVIEW QUESTIONS

13.1 (a) How are fiber-optic sensors classified?
(b) Suggest a criterion for designing an intensity-modulated fiber-optic sensor. On what factors does the signal developed by the detector depend in this case?

13.2 Suggest design(s) of displacement sensors based on (a) misalignment losses between two fibers and (b) microbending losses.

13.3 The angular displacement sensor of Fig. 13.2(c) is employing identical fibers with a core index of 1.46 and a cladding index of 1.45. If the range of angular

deviation to be detected varies from 0° to 10°, calculate (a) the range of S_{ang} (normalized), (b) the range of loss (in dB), and (c) the minimum power that must be launched by the source if the photodetector has a sensitivity of –30 dBm. (Assume negligible loss in the fibers.)
Ans: (a) 1–0.673 (b) 0 to 2.027 dB (c) 1.6 μW

13.4 Suggest the possible applications of the configuration shown in Fig. 13.9, other than rotation sensing.

13.5 In the case of a fiber-optic Mach–Zehnder interferometric (MZI) pressure sensor, show that the phase change $\Delta\phi$ per unit length in the sensing arm due to change in pressure ΔP is given approximately by

$$\frac{\Delta\phi}{L\Delta P} = \frac{2\pi}{\lambda}\left[n\frac{\Delta L}{L\Delta P} + \frac{\Delta n}{\Delta P}\right]$$

where the symbols have their usual meaning.

13.6 Derive an expression for an all-fiber MZI strain sensor for the phase change per unit length in the sensing arm due to change in strain.

13.7 A fiber-optic gyroscope has a circular coil of diameter 12 cm. The total length of the fiber used in the coil is 400 m. If it is operating at $\lambda = 0.633$ μm, what is the phase shift corresponding to the angular speed of 5×10^{-4} rad s^{-1}?
Ans: 0.066 rad

13.8 Suggest the design of the apparatus that may be required to calibrate a fluorescent temperature sensor.

13.9 For an FBG sensor, show that the shift $\Delta\lambda_B$ in the Bragg wavelength λ_B due to pressure change ΔP may be given by

$$\frac{\Delta\lambda_B}{\Delta P} = \lambda_B\left[\frac{1}{\Lambda}\frac{\partial\Lambda}{\partial P} + \frac{1}{n_{eff}}\frac{\partial n_{eff}}{\partial P}\right]$$

where Λ is the spatial period of index modulation and n_{eff} is the effective mode index.

13.10 What are fiber-optic smart structures? What could be the possible applications of such structures?

14 Laser-based Systems

After reading this chapter you will be able to understand the following:

- Classification of lasers
- Solid-state lasers
  - Ruby laser
  - Nd:YAG laser
- Gas lasers
  - He–Ne laser
  - CO_2 laser
- Dye lasers
- *Q*-switching and mode-locking
- Laser-based systems for different applications

14.1 INTRODUCTION

During the past two decades, there has been considerable progress in the design, development, and application of various types of laser-based systems. In Chapter 7, we have discussed some simplified aspects of laser action, and semiconductor lasers that are primarily used in fiber-optic communication systems. This chapter introduces some other types of lasers and their applications in various fields.

Lasers may be classified in a variety of ways. One way is to classify them according to their mode of operation, e.g., pulsed lasers, *Q*-switched lasers, or continuous-wave (cw) lasers. The second way is to categorize them according to the mechanism by which population inversion and gain are achieved. Thus we have three-level lasers or four-level lasers. A more logical way is to classify them according to the state of the active medium used. Thus, broadly, there are four classes of lasers; namely, (i) solid-state lasers, (ii) gas lasers, (iii) dye (liquid) lasers, and (iv) semiconductor lasers. Here we will discuss only the first three types of lasers, as the last one has already been described in detail in Chapter 7.

14.2 SOLID-STATE LASERS

A solid-state laser employing optical pumping is shown in Figs 14.1(a) and 14.1(b). A linear pumping source and the laser rod are placed parallel to each other inside an elliptical reflector. The pumping lamp is placed along one principal axis (focus) and the laser rod along the other. This configuration achieves a higher concentration of light flux from the pumping lamp onto the laser rod. The lamps employed for optical pumping are basically discharge tubes, whose configuration may be linear, π-shaped, or helical as shown in Fig. 14.2. Linear and π-shaped discharge tubes are suitable for systems shown in Fig. 14.1. If a helical lamp is employed, the laser rod is placed along the axis of the helix, and the entire system is kept inside a cylindrical reflector.

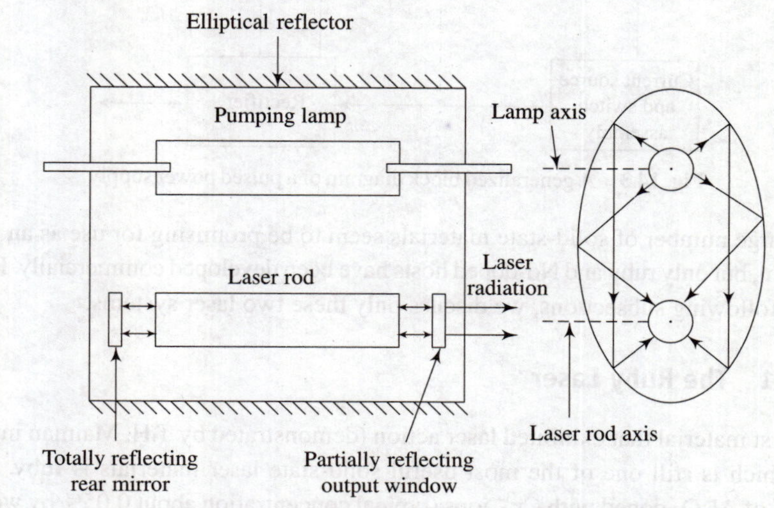

Fig. 14.1 (a) Solid-state laser with elliptical reflector. (b) Focusing of pumping light onto the laser rod.

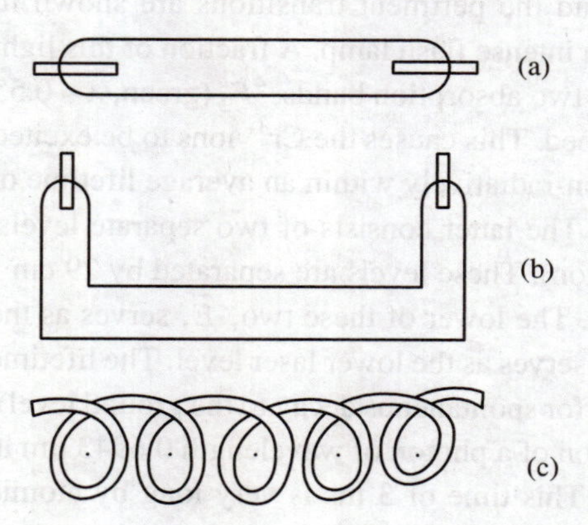

Fig. 14.2 Lamps for optical pumping: (a) linear, (b) π-shaped, and (c) helical

Solid-state lasers normally operate in the pulsed mode, and hence use pulsed power sources to supply power to the flash tubes. A generalized block diagram of a pulsed power supply is shown in Fig. 14.3. It consists of a current source, a rectifier, a control circuit, and a flash tube trigger circuit. The actual circuit configuration may depend on the objectives sought.

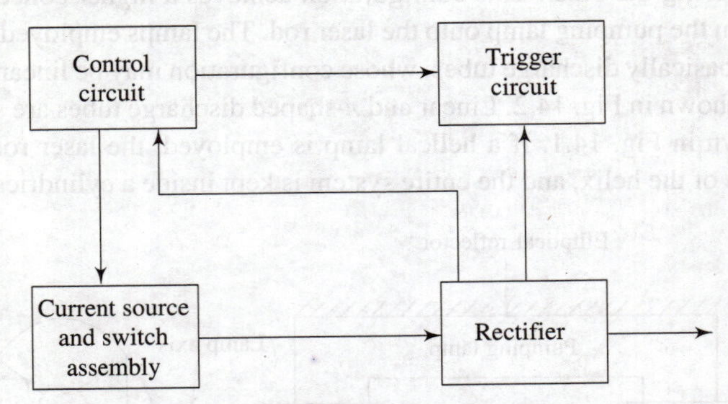

Fig. 14.3 A generalized block diagram of a pulsed power supply

A large number of solid-state materials seem to be promising for use as an active medium, but only ruby and Nd-doped hosts have been developed commercially. Hence, in the following subsections, we discuss only these two laser systems.

14.2.1 The Ruby Laser

The first material that exhibited laser action (demonstrated by T.H. Maiman in 1960) and which is still one of the most useful solid-state laser materials is ruby. It is a crystal of Al_2O_3 doped with Cr^{3+} ions (typical concentration about 0.05% by weight). The latter serve as active centres in the active medium Al_2O_3. A simplified energy-level diagram of the Cr^{3+} ion in ruby and the pertinent transitions are shown in Fig. 14.4. Ruby is pumped optically by an intense flash lamp. A fraction of this light that corresponds to the frequencies of the two absorption bands, $^4F_2$ (green, $\lambda \approx 0.55$ μm) and $^4F_1$ (blue, $\lambda \approx 0.40$ μm), is absorbed. This causes the Cr^{3+} ions to be excited to these levels. The excited ions decay non-radiatively within an average lifetime of about 50 ns to the upper lasing level ^2E. The latter consists of two separate levels, labelled $2\bar{A}$ and $\bar{E}$ (a conventional notation). These levels are separated by 29 cm^{-1} (or 8.7×10^{11} Hz in the frequency scale). The lower of these two, $\bar{E}$, serves as the upper laser level and the ground level $^4A_2$ serves as the lower laser level. The lifetime of the upper laser level, $\bar{E}$, is about 3 ms (for spontaneous decay to the ground level). This decay is accompanied by the emission of a photon of wavelength 0.6943 μm in the red region (denoted by the R_1 line). This time of 3 ms is very long by atomic standards. Therefore it is more than súfficient for building up a population at level $\bar{E}$ and, hence, achieving population inversion. Subsequent stimulated emission will produce radiation of wavelength 0.6943 μm. Thus, according to our discussion of Sec. 7.9, ruby may be considered a three-level laser.

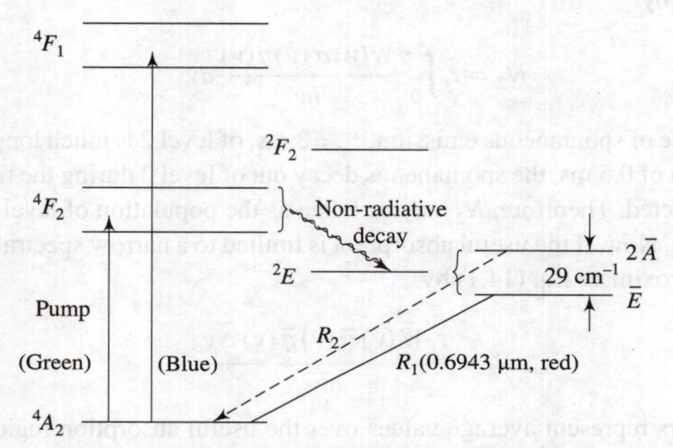

Fig. 14.4 Simplified energy-level diagram of Cr^{3+} ion in ruby, and pertinent transitions

Example 14.1 (a) For a pulsed ruby laser, calculate the energy in the pulse required to obtain threshold inversion. Assume that the density of Cr^{3+} atoms per cm^3 is 1.9×10^{19}. The flash lamp for pumping produces a pulse of light of duration $\tau_f = 0.5$ ms. The lifetime of spontaneous emission $\tau_{sp} = 3$ ms; the average absorption coefficient over the blue and green bands is $\bar{\alpha}(\nu) = 2\ \text{cm}^{-1}$; the average pump frequency absorbed by ruby, $\bar{\nu} \approx 5.45 \times 10^{14}$ Hz. (b) Calculate the threshold electrical energy input to the flash lamp per cm^2 of the ruby crystal surface. Assume that the efficiency of conversion from electrical energy to optical energy is 50%; the fraction of total optical output that is usefully absorbed by the active medium (ruby) is 15%; the fraction of the lamp light focused by the optical system onto the laser rod is 20%.

Solution

(a) For the sake of simplicity, let us assume the shape of the pulse to be rectangular. If we assume that a pulse of duration τ_f produces an optical flux of $W(\nu)$ watts per unit area per unit frequency at a frequency ν at the surface of the ruby crystal, then the amount of energy absorbed by the crystal per unit volume is given by (Yariv 1997)

$$\tau_f \int_0^\infty W(\nu)\alpha(\nu)d\nu$$

where $\alpha(\nu)$ is the absorption coefficient of the crystal. If we assume that the absorption quantum efficiency (that is, the probability that the absorption of a pump photon at a frequency ν results in transferring one atom into the upper laser level, say, $\bar{E}$) is $\eta(\nu)$, then the number of atoms pumped into level 2 (i.e., level $\bar{E}$) per unit volume will be given by

$$N_2 = \tau_f \int_0^\infty \frac{W(\nu)\alpha(\nu)\eta(\nu)}{h\nu} d\nu \tag{14.1}$$

As the lifetime of spontaneous emission, $\tau_{sp} = 3$ ms, of level 2 is much longer than the flash duration of 0.5 ms, the spontaneous decay out of level 2 during the time of flash may be neglected. Therefore, N_2 may be taken as the population of level 2 (i.e., $\bar{E}$) after the flash. Now, if the useful absorption is limited to a narrow spectral region $\Delta\nu$, one may approximate Eq. (14.1) by

$$N_2 = \frac{\tau_f \overline{W}(\nu)\overline{\alpha}(\nu)\overline{\eta}(\nu)\overline{\Delta}\nu}{h\overline{\nu}} \tag{14.2}$$

where the bars represent average values over the useful absorption region of width $\overline{\Delta}\nu$.

Since ruby is a three-level laser, the populations N_1 and N_2 at levels 1 and 2, respectively, should satisfy the condition

$$N_1 + N_2 = N_0 \tag{14.3}$$

where N_0 is the density of active atoms (Cr^{3+}). Of course, here we are assuming that the population N_3 at level 3 is negligible because of the very fast transition rate out of level 3. If the pumping level is high enough so that the population at level 2 becomes

$$N_2 = N_1 = N_0/2 \tag{14.4}$$

the optical gain will be zero. This means that roughly half of the chromium atoms must be raised to level $\bar{E}$ to achieve transparency (zero gain) on the R_1 line. Further pumping will yield gain and oscillation if appropriate feedback is supplied.

Using the data given here, we get

$$N_2 \approx \frac{N_0}{2} = \frac{1.9}{2} \times 10^{19} = 9.5 \times 10^{18} \text{ cm}^{-3}$$

Taking $\bar{\nu} = 5.45 \times 10^{14}$ Hz and $\bar{\eta}(\nu) \approx 1$, we obtain [from Eq. (14.2)] the following pump energy that must fall on each cm^2 of the ruby crystal surface in order to achieve threshold inversion:

$$\overline{W}(\nu)\overline{\Delta}\nu\tau_f = \frac{N_2 h\bar{\nu}}{\bar{\alpha}(\nu)\bar{\eta}(\nu)} = \frac{9.5 \times 10^{18} \times 6.6 \times 10^{-34} \times 5.45 \times 10^{14}}{2 \times 1}$$

$$= 1.7 \text{ J cm}^{-2}$$

(b) From the data given, the threshold energy input to the flash lamp per cm^2 of the ruby surface will be

$$\frac{1.7 \text{ J cm}^{-2}}{0.15 \times 0.20 \times 0.5} \approx 113 \text{ J cm}^{-2}$$

Note These are obviously rough estimates. Nevertheless, they give the order of magnitude of the powers involved in laser pumping.

14.2.2 The Nd^{3+}:YAG Laser

Pure yttrium aluminium garnet (YAG), $Y_3Al_5O_{12}$, is an optically isotropic crystal with a cubic structure. It is doped with trivalent neodymium (Nd^{3+}) ions (about 0.73% by weight) to make it a laser material. Nd^{3+} ions serve as active atoms. They introduce well-defined energy levels in YAG as shown in Fig. 14.5.

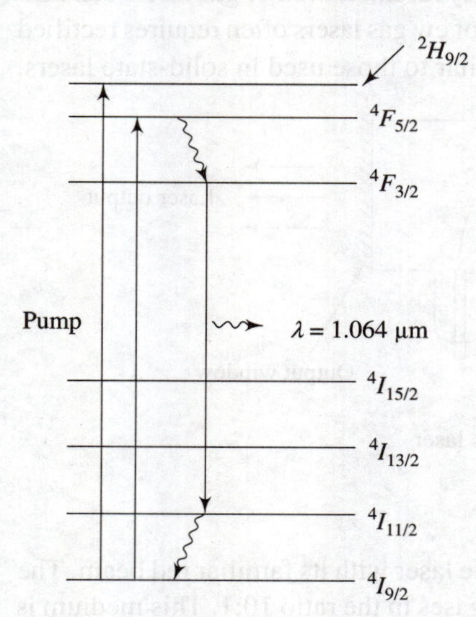

Fig. 14.5 Simplified energy-level diagram of Nd^{3+} in YAG. The typical pumping route and subsequent decays are shown by appropriate arrows.

Absorption of pump frequencies raises the atoms to higher levels: the $^4F_{3/2}$ level which forms the upper lasing level. The dominant laser transition is from $^4F_{3/2}$ (at 11,507 cm^{-1}) to $^4I_{11/2}$ (at 2110 cm^{-1}), giving a spectral line at $\lambda_0 = 1.064$ µm. The lower laser level $^4I_{11/2}$ decays non-radiatively. Thus, from our discussion in Sec. 7.9, the Nd^{3+}:YAG laser may be considered a four-level laser.

In this case, the optical gain is much greater than that of ruby. This causes the laser threshold to be very low and also makes cw operation much easier. As the absorption bands are narrow, krypton gas, whose spectrum matches the pumping bands, is normally used in the pumping lamp.

Nd^{3+}-doped glasses also make useful laser systems. However, because of the amorphous nature of glass, the absorption bands are much broader than those of YAG. This leads to wider fluorescent lines. Second, glass is a poor thermal conductor, and hence it is difficult to remove this waste heat, which in turn limits the repetition rate of glass lasers.

14.3 GAS LASERS

Although a number of gases have been shown to exhibit laser action, only a few have been exploited commercially. Of them, helium–neon, argon ion, and CO_2 lasers have been extensively studied and used. A typical gas laser is shown in Fig. 14.6. The pumping is normally achieved through the electrical discharge between a pair of electrodes. The discharge occurs along the axis of the laser cavity. In this case, some volume of the gas is not utilized. In order to achieve uniform excitation of a larger volume of the gas, transverse discharge is employed. Excitation of gas lasers can also be carried out by electron beams. The operation of cw gas lasers often requires rectified ac, while pulsed lasers use power supplies similar to those used in solid-state lasers.

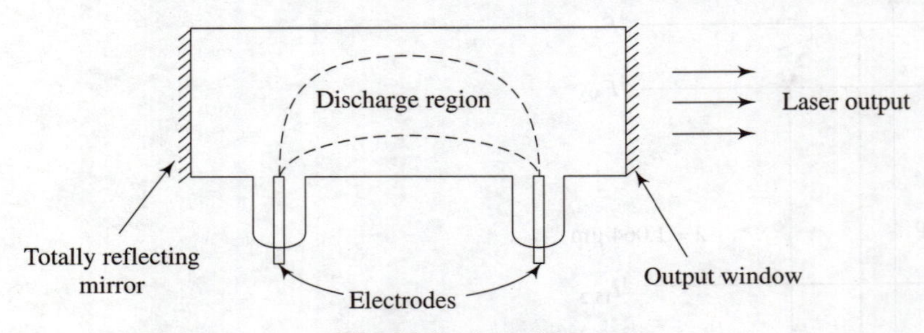

Fig. 14.6 A gas laser

14.3.1 The He–Ne Laser

The most popular and first cw laser is the He–Ne laser with its familiar red beam. The laser medium is a mixture of helium and neon gases in the ratio 10:1. This medium is excited by electrical discharge (either dc or rf current). The pumping action is rather complex and indirect. First the helium atoms are excited by the energetic electrons in the discharge into a variety of excited states. Most of these excited atoms accumulate in the long-lived metastable states 2^3S and 2^1S, whose lifetimes are 0.1 ms and 5 μs, respectively. As shown in Fig. 14.7, these two levels happen to be very close to the 2*S* and 3*S* levels of Ne. When the excited He atoms collide with the Ne atoms, energy is exchanged and the Ne atoms are pumped to the respective levels. The atoms at the Ne 3*S* level eventually decay to the 2*p* level as a result of stimulated emission, and a spectral line of $\lambda = 0.6328$ μm is emitted. The atoms at the 2*S* level drop to 2*p* by emitting light at 1.15 μm. However, stimulated emission tends to occur between the

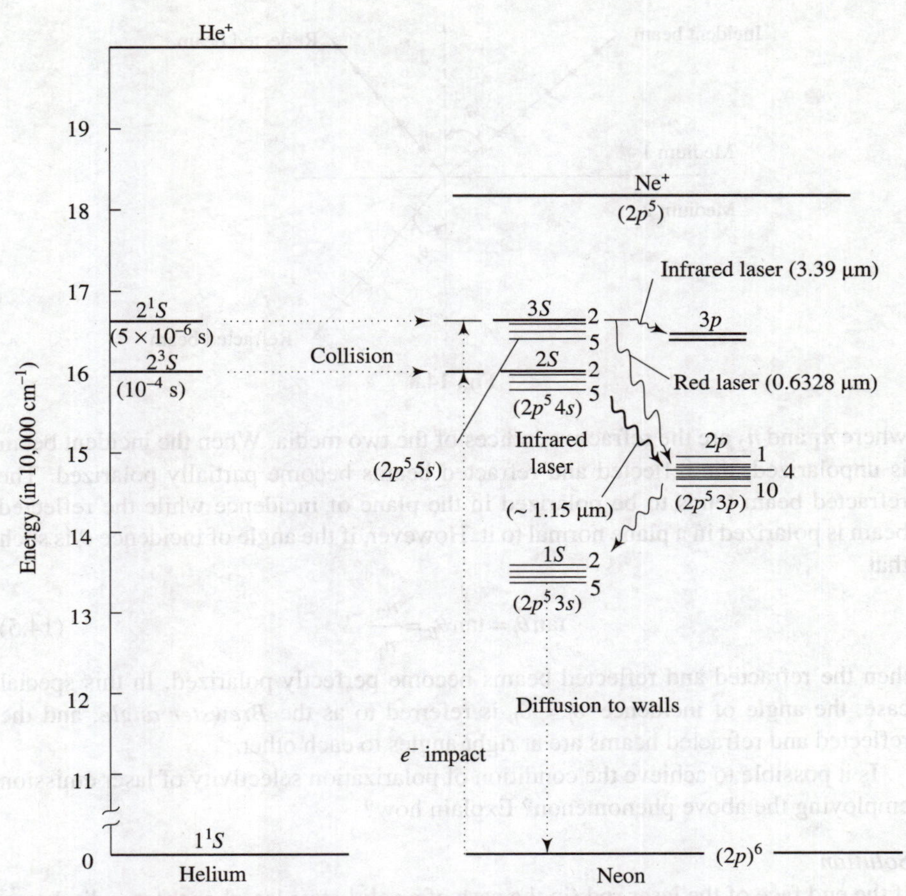

Fig. 14.7 He–Ne energy levels. The dominant excitation paths for the red and infrared laser transitions are shown (Bennett 1962).

3S and 3p levels emitting light at $\lambda = 3.39$ μm. Thus, laser action would normally take place at 3.39 μm (infrared) instead of the desired 0.6328 μm (red). This problem is overcome by attenuating the 3.39-μm line in the cavity. Normally, the output power of He–Ne lasers is in the range 0.5–5 mW.

Example 14.2

Brewster angle window It is well known that a beam of light traversing from one medium to another is divided into two beams at the interface as shown in Fig. 14.8. A part of the beam is reflected back into the same medium at an angle of incidence θ_i, and the remaining part is refracted at an angle θ_r. Referring to Snell's law, we have

$$n_1 \sin\theta_i = n_2 \sin\theta_r$$

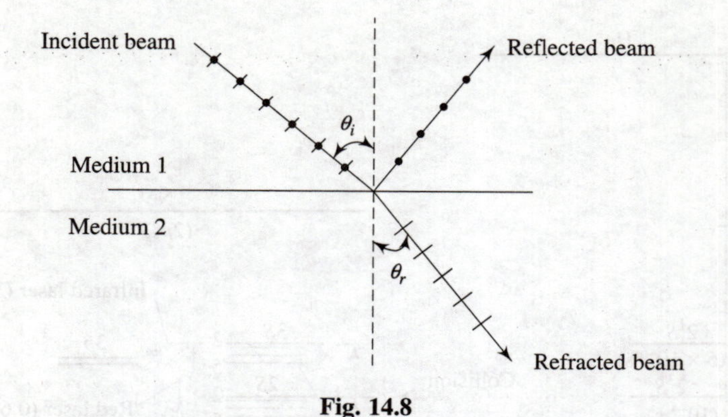

Fig. 14.8

where n_1 and n_2 are the refractive indices of the two media. When the incident beam is unpolarized, the reflected and refracted beams become partially polarized. The refracted beam tends to be polarized in the plane of incidence while the reflected beam is polarized in a plane normal to it. However, if the angle of incidence θ_i is such that

$$\tan\theta_i = \tan\theta_B = \frac{n_2}{n_1} \tag{14.5}$$

then the refracted and reflected beams become perfectly polarized. In this special case, the angle of incidence $\theta_i = \theta_B$ is referred to as the *Brewster angle*, and the reflected and refracted beams are at right angles to each other.

Is it possible to achieve the condition of polarization selectivity of laser emission employing the above phenomenon? Explain how?

Solution

If the end face of the laser rod (in the case of a solid-state laser) or the gas discharge tube (in the case of a gas laser) is tilted such that the normal to the end face and the optic axis OO' are at the Brewster angle corresponding to the refractive index of the material of the end face window, the emitted laser radiation will be plane-polarized. The relevant configuration is shown in Fig. 14.9.

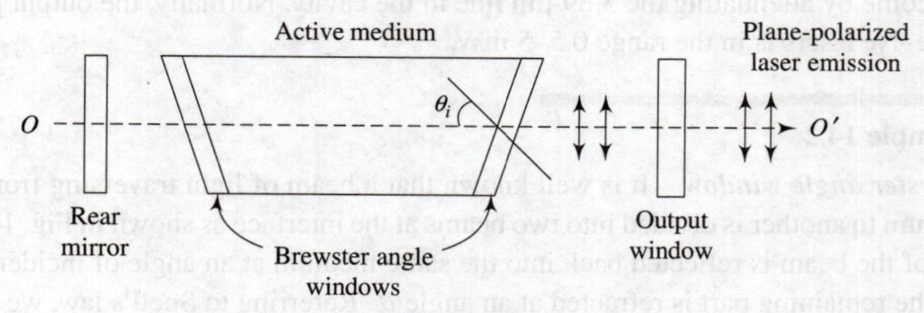

Fig. 14.9

14.3.2 The CO_2 Laser

The CO_2 laser is one of the most efficient, powerful, useful, and most studied lasers. A He–Ne laser, described earlier, can at the most produce an output of 100 mW for which the required length of the discharge tube would be quite large. On the other hand, even a small CO_2 laser can produce tens of watts. One can obtain hundreds of kilowatts in the cw mode and even terawatts in the pulsed mode. Therefore, this laser also has applications in diverse fields, e.g., trimming operations in industrial manufacturing, welding and cutting of many-inches-thick steel, weaponry, laser fusion, etc.

The lasers described earlier depend on electronic transitions between states in which the electronic orbitals are different. However, the CO_2 laser is a molecular laser in which the concerned energy levels involve the internal vibration of molecules, that is, relative motion of the constituent atoms. The atomic electrons remain in their lowest energy states.

The laser medium is a mixture of CO_2, N_2, and He gases in the pressure ratio 1:2:3. A pertinent energy-level diagram is shown in Fig. 14.10.

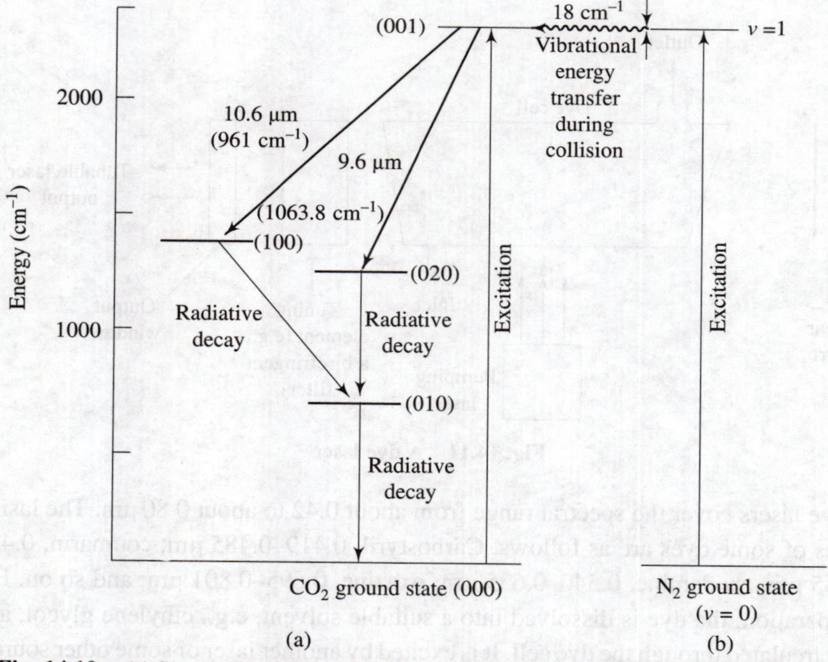

Fig. 14.10 (a) Some of the low-lying vibrational levels of the CO_2 molecule, including the upper and lower levels for the 10.6-μm and 9.6-μm laser transitions. (b) Ground state ($v = 0$) and first excited state ($v = 1$) of the nitrogen molecule, which plays an important role in the selective excitation of the (001) CO_2 level (Yariv 1997).

The pumping sequence for laser action is as follows: (i) Energy is transferred to the electrons in the discharge by the applied electric field. (ii) The electrons transfer this energy to neutral gas atoms. (iii) A large fraction of CO_2 molecules excited by the electron impact tend to accumulate in the long-lived (001) level. (iv) A very large fraction of N_2 molecules excited by the discharge tend to accumulate in the $v = 1$ level. (v) Most of the excited N_2 molecules collide with the ground-state CO_2 molecules and transfer their energy to the latter in the process. CO_2 molecules, therefore, get excited to the (001) state. Laser oscillations can occur at 10.6 μm and 9.6 μm as shown in Fig. 14.10. However, the gain is stronger at 10.6 μm.

14.4 DYE LASERS

There are many organic dyes which are suitable laser materials. The importance of dye lasers lies in the fact that they can be tuned to give a continuously variable output over a wide range of wavelengths. They are excited either by another laser (e.g., cw argon or pulsed nitrogen) or by flash tubes. A tunable dye laser is shown in Fig. 14.11.

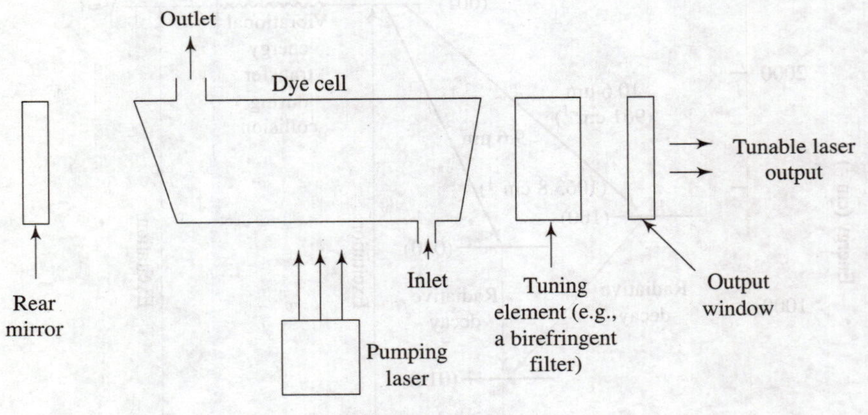

Fig. 14.11 A dye laser

Dye lasers cover the spectral range from about 0.42 to about 0.80 μm. The lasing ranges of some dyes are as follows. Carbostyril, 0.419–0.485 μm; coumarin, 0.435 –0.565 μm; rhodamine, 0.540–0.635 μm; oxazine, 0.695–0.801 μm; and so on. For cw operation, the dye is dissolved into a suitable solvent, e.g., ethylene glycol, and then circulated through the dye cell. It is excited by another laser or some other source. The tuning is achieved by rotating the birefringent filter placed inside the optical cavity.

14.5 *Q*-SWITCHING

According to their mode of oscillation, lasers have been divided into three groups.

(i) Continuous lasers, which emit a continuous light beam with constant power. Such devices require continuous steady-state pumping of the active medium.

(ii) Pulsed lasers with free oscillation, in which the emission takes the form of periodic light pulses. These devices require pulsed operation of the pumping system. The pumping achieves population inversion of the lasing levels periodically and for short periods only.

(iii) Pulsed lasers with controlled losses, in which the concentration of energy reaches a maximum so that they give rise to giant pulses of short duration. Their peak power is of the order of 10^8 W or more.

The giant-pulse mode of oscillation is realized by controlling the losses inside the optical cavity. This control is achieved through a device known as the *Q-switch*. Some examples of *Q*-switches are shown in Fig. 14.12.

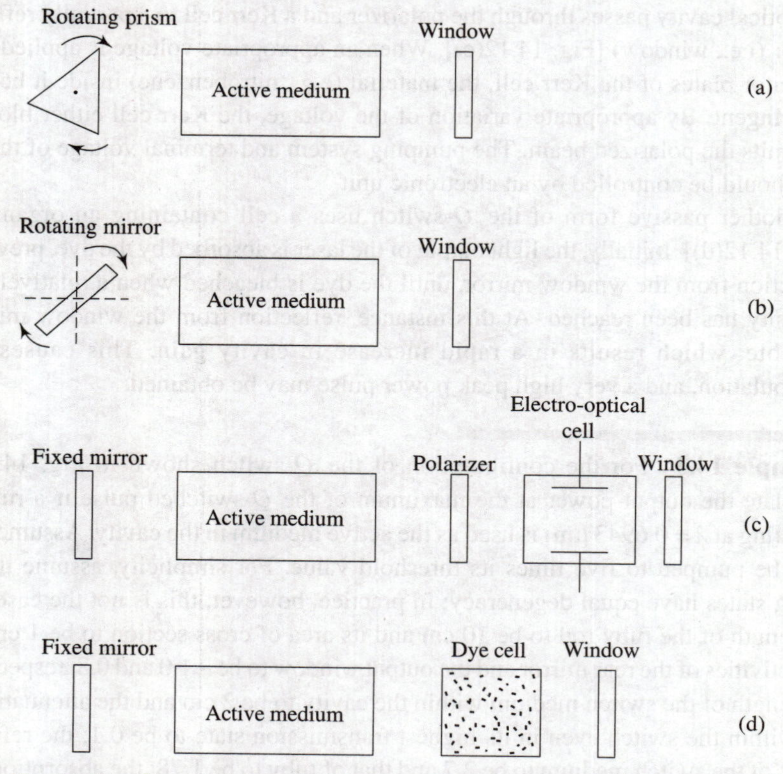

Fig. 14.12 Commonly used *Q*-switches: (a) rotating prism device, (b) rotating mirror device, (c) electro-optical switch based on the Kerr effect, and (d) dye cell

In the giant-pulse or *Q*-switched mode, the active medium is excited without feedback by blocking the reflection from one of the end mirrors of the optical cavity. The end mirror is then suddenly allowed to reflect, employing either a mechanical or an electro-optical switch. The suddenly applied feedback from the mirror causes a rapid population inversion of lasing levels, and this results in a very high peak power output pulse. The duration of the light pulse may be of the order of 0.1 μs.

A mechanically driven device such as a rotating deviation prism or a mirror, or a passive device such as an electro-optical cell or a dye cell may be used as an optical switch. In a rotating prism optical switch [Fig. 14.12(a)] the totally reflecting end mirror of the cavity is replaced by a deviating prism. This prism is rotated by a synchronous motor at a high speed (about 30,000 revolutions/min). When this optical switch is employed, the pulses of the flash lamp are electronically controlled and synchronized with the rotation of the optical switch.

In the rotating mirror optical switch, one of the end mirrors is rotated as shown in Fig. 14.12(b). In the electro-optical switch based on the Kerr effect, the light leaving the optical cavity passes through the polarizer and a Kerr cell to a partially reflecting mirror (i.e., window) [Fig. 14.12(c)]. When an appropriate voltage is applied to the capacitor plates of the Kerr cell, the material (e.g., nitrobenzene) inside it becomes birefringent. By appropriate variation of the voltage, the Kerr cell either blocks or transmits the polarized beam. The pumping system and terminal voltage of the Kerr cell should be controlled by an electronic unit.

Another passive form of the *Q*-switch uses a cell containing an organic dye [Fig. 14.12(d)]. Initially, the light output of the laser is absorbed by the dye, preventing reflection from the window mirror, until the dye is bleached when a relatively high intensity has been reached. At this instance, reflection from the window mirror is possible, which results in a rapid increase in cavity gain. This causes rapid depopulation, and a very high peak power pulse may be obtained.

Example 14.3 For the configuration of the *Q*-switch shown in Fig. 14.12(c), calculate the output power at the maximum of the *Q*-switched pulse if a ruby rod (emitting at $\lambda = 0.6943$ μm) is used as the active medium in the cavity. Assume that it is to be pumped to five times its threshold value. For simplicity, assume that the lasing states have equal degeneracy; in practice, however, this is not the case. Take the length of the ruby rod to be 10 cm and its area of cross section to be 1 cm^2; the reflectivities of the rear mirror and the output window to be ~ 1.0 and 0.8, respectively; the length of the switch medium within the cavity to be 2 cm and the attenuation per cm within the switch even in its highest transmission state to be 0.1; the refractive index of the switch medium to be 2.7 and that of ruby to be 1.78; the absorption cross section σ of Cr^{3+} ions in ruby at $\lambda = 0.6943$ μm to be 1.5×10^{20} cm^{-2}; and the total length of the cavity to be 14 cm. The transmission coefficients at the interfaces between the active medium and air and that between the switch and air may be taken to be (arbitrarily) same and equal to 0.98.

Solution

[1]The geometry of the above Q-switched laser may be drawn as shown in Fig. 14.13. For this system the condition for threshold may be written as follows:

The round-trip gain = 1

i.e., $R_1R_2(T_1T_2T_3T_4)^2\exp(-2\alpha_s l_s)\exp(2g_{th}l_m) = 1$

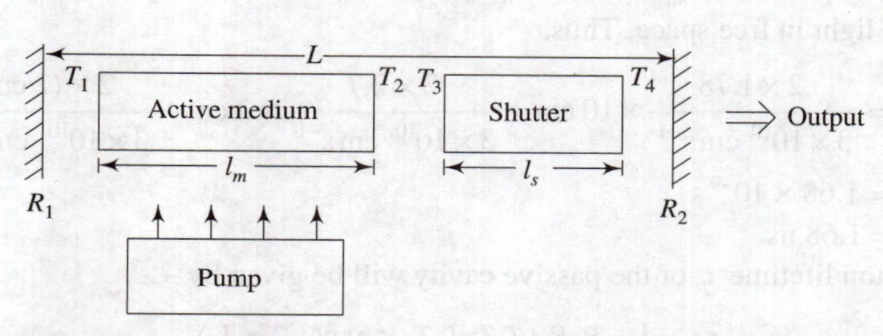

Fig. 14.13 The geometry of a Q-switched laser. T_1, T_2, T_3, and T_4 are the transmission coefficients at the interfaces shown; l_m is the length of the active medium; l_s is the length of the shutter medium; L is the cavity length; R_1 and R_2 are the reflectivities of the rear mirror and the output window, respectively.

Thus the threshold gain coefficient g_{th} will be given by

$$g_{th} = \frac{1}{2l_m}\ln\left[\frac{1}{R_1R_2(T_1T_2T_3T_4)^2}\right] + \frac{\alpha_s l_s}{l_m} \tag{14.6}$$

Substituting the values of the given parameters, we get

$$g_{th} = 2.948 \times 10^{-2}\ \text{cm}^{-1}$$

The threshold inversion density will then be given by

$$(N_2 - N_1)_{th} = \frac{g_{th}}{\sigma} = 1.965 \times 10^{18}\ \text{cm}^{-3}$$

and the total number of inverted atoms at the threshold will be

$$n_{th} = (N_2 - N_1)_{th}V$$

where V is the volume of the gain medium, or

$$n_{th} = (N_2 - N_1)_{th}Al_m$$

where A is the area of cross section of the ruby rod. Thus,

$$n_{th} = 1.965 \times 10^{18} \times 1 \times 10 = 1.965 \times 10^{19}\ \text{atoms}$$

[1]In analysing this system, we have followed Joseph T. Verdeyen (1993).

As required in the problem, it is to be pumped to five times its threshold; i.e., the initial inversion will have to be

$$n_i = 5n_{\text{th}} = 9.825 \times 10^{19} \text{ atoms}$$

Now the round-trip time τ_{RT} within the cavity may be obtained as follows:

$$\tau_{\text{RT}} = \frac{2\,n_g}{c} l_m + \frac{2\,n_s}{c} l_s + \frac{2\,l_{\text{air}}}{c} \tag{14.7}$$

where n_g and n_s are the refractive indices of the gain medium (ruby) and the switch medium, respectively, l_{air} is the free space left between the cavity mirrors, and c is the speed of light in free space. Thus,

$$\tau_{\text{RT}} = \frac{2 \times 1.78}{3 \times 10^{10} \text{ cm s}^{-1}} \times 10 \text{ cm} + \frac{2 \times 2.7}{3 \times 10^{10} \text{ cm s}^{-1}} \times 2 \text{ cm} + \frac{2 \times (2 \text{ cm})}{3 \times 10^{10} \text{ cm s}^{-1}}$$

$$= 1.68 \times 10^{-9} \text{ s}$$

$$= 1.68 \text{ ns}$$

The photon lifetime τ_p of the passive cavity will be given by

$$\frac{1}{\tau_p} = \frac{1 - R_1 R_2 (T_1 T_2 T_3 T_4)^2 \exp(-2\alpha_s l_s)}{\tau_{\text{RT}}} \tag{14.8}$$

This gives $\tau_p = 2.42$ ns.

Let us now calculate the maximum number of photons inside the cavity, which is given by the following relation:

$$N_p(\text{max}) = \frac{n_i - n_{\text{th}}}{2} - \frac{n_{\text{th}}}{2} \ln\left(\frac{n_i}{n_{\text{th}}}\right) = 2.35 \times 10^{19} \text{ photons}$$

Now $h\nu N_p(\text{max})$ represents the maximum optical energy stored in the cavity. Photons are lost due to various loss mechanisms inside the cavity, but only the part of the loss that is coupled through the output window represents the power available outside the cavity.

The total loss coefficient, prorated over the length of the cavity, say, $\langle \alpha_T \rangle$, can be written as

$$\langle \alpha_T \rangle = \frac{\alpha_s l_s}{L} + \left(\frac{1}{2L}\right) \ln\left\{\frac{1}{R_1 (T_1 T_2 T_3 T_4)^2}\right\} + \frac{1}{2L} \ln\left(\frac{1}{R_2}\right) \tag{14.9}$$

The sum of the first two terms on the RHS of Eq. (14.9) gives the internal loss coefficient $\langle \alpha_{\text{int}} \rangle$ and the third term gives the external loss coefficient $\langle \alpha_{\text{ext}} \rangle$.

Therefore, the maximum power emerging from the output window may be given by

$$P_{\text{max}} = \frac{\langle \alpha_{\text{ext}} \rangle}{\langle \alpha_T \rangle} \frac{h\nu N_p(\text{max})}{\tau_p} \tag{14.10}$$

In the present case, $\langle \alpha_T \rangle = 0.0280\ \text{cm}^{-1}$ and $\langle \alpha_{\text{ext}} \rangle = 0.00796\ \text{cm}^{-1}$. Substituting the values of other parameters, we get $P_{\max} = 7.87 \times 10^8$ W = 787 MW, which is quite large indeed.

14.6 MODE-LOCKING

An ideal homogeneously broadened laser can oscillate at a single frequency. However, all practical lasers are inhomogeneously broadened and hence may oscillate at a number of frequencies, which are separated by

$$\omega_q - \omega_{q-1} = \frac{\pi c}{l} \equiv \omega \tag{14.11}$$

where l is the length of the gain medium as well as the distance between the mirrors of the cavity, and c is the speed of light in free space. Here we have assumed that the refractive index of the gain medium $n = 1$ (as in the case of the He–Ne laser).

The total optical electric field resulting from such multimode oscillation at a particular point may be given by

$$e(t) = \sum_n E_n \exp[i\{(\omega_0 + n\omega)t + \phi_n\}] \tag{14.12}$$

where E_n is the amplitude of the nth mode, which is oscillating at an angular frequency of $(\omega_0 + n\omega)$; ω_0 here has been arbitrarily taken to be the reference frequency. ϕ_n is the phase of the nth mode. It is easy to prove that $e(t)$ is periodic in $T = 2\pi/\omega = 2l/c$, which is the round-trip transit time inside the resonator. Thus

$$e(t+T) = \sum_n E_n \exp\left[i\left\{(\omega_0 + n\omega)\left(t + \frac{2\pi}{\omega}\right) + \phi_n\right\}\right]$$

$$= \sum_n E_n \exp[i\{(\omega_0 + n\omega)t + \phi_n\}] \exp\left[i\left\{2\pi\left(\frac{\omega_0}{\omega} + n\right)\right\}\right] \tag{14.13}$$

Since ω_0 is a reference frequency, we can take it to be $\omega_0 = m\pi c/l$; m is an integer and $\omega = \pi c/l$ [from Eq. (14.11)], the ratio ω_0/ω is an integer m. Therefore,

$$\exp[i\{2\pi(m+n)\}] = 1$$

This reduces Eq. (14.13) to

$$e(t+T) = \sum_n E_n \exp[i\{(\omega_0 + n\omega)t + \phi_n\}] = e(t) \tag{14.14}$$

In order that $e(t)$ maintains a periodic nature, the phases ϕ_n should be fixed. However, in many lasers, the phases ϕ_n vary randomly with time. This causes the laser output power to fluctuate randomly, thus reducing the possibility of its application in cases where temporal coherence is an important consideration.

There are two ways in which the laser can be made coherent. These are (i) to allow the laser to oscillate only at a single frequency and (ii) to force the phases ϕ_n of the modes to maintain their relative values. The second method is called *mode-locking*, and this causes the oscillation intensity to consist of a periodic train with a period $T = 2l/c = 2\pi/\omega$.

In order to simplify things, let us lock each mode to a common origin of time and take each phase ϕ_n to be zero. Further, assume that there are N oscillating modes with equal amplitudes E_n, and that $E_n = 1$. Substituting these parameters in Eq. (14.12), we get

$$e(t) = \sum_{-(N-1)/2}^{(N-1)/2} e^{i(\omega_0 + n\omega)t} \tag{14.15}$$

$$= e^{i\omega_0 t} \frac{\sin(N\omega t/2)}{\sin(\omega t/2)} \tag{14.16}$$

The average laser output power

$$P(t) \propto e(t)e^*(t)$$

where $e^*(t)$ is the complex conjugate of $e(t)$,

or
$$P(t) \propto \frac{\sin(N\omega t/2)}{\sin(\omega t/2)} \tag{14.17}$$

Some evident features of $P(t)$ are a follows: (i) the peak power P_{peak} is equal to n times the average power P_{av}, where N is the number of modes locked together; (ii) the peak amplitude of the field is equal to N times the amplitude of a single mode; (iii) the individual pulse width τ_p, defined as the time from the peak to the first zero, can be estimated by the relation

$$(P_{\text{peak}})\tau_p \approx (P_{\text{av}})T = (P_{\text{av}})\left(\frac{2\pi}{\omega}\right) = (P_{\text{av}})\left(\frac{2l}{c}\right)$$

or
$$(NP_{\text{av}})\tau_p = (P_{\text{av}})\frac{2l}{c}$$

or
$$\tau_p = \frac{2l}{cN} \tag{14.18}$$

But the number of oscillating modes N is approximately given by the ratio of the transition line shape width $\Delta\omega$ to the frequency spacing ω between the modes; i.e.,

$$N = \frac{\Delta\omega}{\omega} = \frac{\Delta\omega}{\pi c/l} \tag{14.19}$$

[where we have used Eq. (14.11)].

From Eqs (14.18) and (14.19), we get

$$\tau_p = \frac{2l}{c}\frac{(\pi c/l)}{\Delta\omega} = \frac{2\pi}{\Delta\omega} = \frac{1}{\Delta\upsilon} \tag{14.20}$$

Thus the width of the mode-locked pulse is inversely proportional to the gain line width.

Example 14.4 Calculate the pulse width τ_p and the spatial length L_p of mode-locked pulses for the following cases: (a) a He–Ne laser for which $\Delta\upsilon = 1.5 \times 10^9$ Hz and (b) a ruby laser for which $\Delta\upsilon = 6 \times 10^{10}$ Hz. Comment on the results.

Solution

(a) $\tau_p = 1/(\Delta\upsilon) = 1/(1.5 \times 10^9)$ s = 0.66 ns

$L_p = c\tau_p = 3 \times 10^8 \times 0.66 \times 10^{-9}$ m = 19.8 cm

(b) $\tau_p = 1.66 \times 10^{-11}$ s = 16.6 ps

$L_p = 3 \times 10^8 \times 1.66 \times 10^{-11} = 5 \times 10^{-3}$ m = 5 mm

The results simply indicate that enormous number of photons can be packed into a very small space L_p occupied by these pulses.

14.7 LASER-BASED SYSTEMS FOR DIFFERENT APPLICATIONS

In the past two decades, there has been a tremendous increase in the applications of lasers in various fields, e.g., industry, medicine and surgery, communications, science and technology, defence, environmental monitoring, etc. In this section, we briefly describe some of the major laser-based systems for different applications.

14.7.1 Remote Sensing Using Light Detection and Ranging

The distance of a remote object can be determined by measuring the time t taken by a short laser pulse to reach the object and get reflected back to the observer. The distance d of the object will then be given by

$$d = \frac{ct}{2} \tag{14.21}$$

where c is the speed of light in free space. As this technique is similar to the radar technique using radio-frequency waves, it is called an optical radar or laser radar or *lidar* (light detection and ranging). Lidar has been used to determine the distance between points on the earth and the moon to an accuracy of a few inches. Such a system is shown in Fig. 14.14. The major components of this system are a pulsed laser, a photodetector with a timing circuit, and a collecting and focusing telescope [Fig. 14.14(b)] to collect the reflected light. In its simplest form, the power received at the telescope, P_z, may be given by Beer's law:

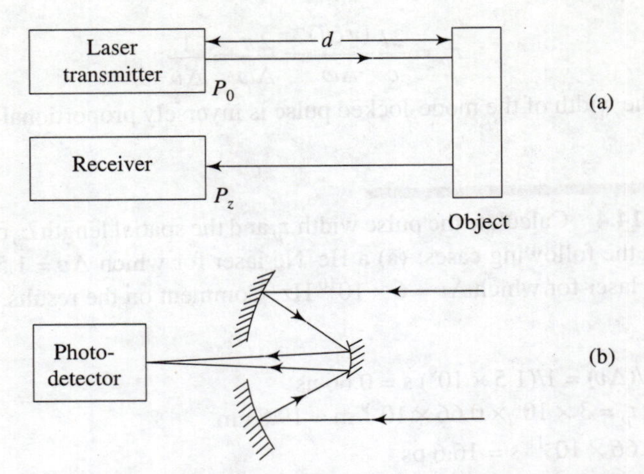

Fig. 14.14 (a) Lidar system and (b) collecting and focusing telescope

$$P_z = P_0 \exp(-\alpha 2d) \tag{14.22}$$

where P_0 is the power transmitted and α is the coefficient of attenuation per unit length. Attenuation of the transmitted signal may be caused by absorption, and scattering of light by the medium between the laser and the object.

Lidar has many applications, all of which derive from the high directionality and high power of laser radiation. These include monitoring of volcanic aerosols that may affect global climate, monitoring of atmospheric pollution, detection and characterization of fog layers, laser-based weaponry, etc. Lasers are also used in electro-optic counter measures (EOCMs) in the battlefield. The purpose of this system is to make ineffective similar systems used by the enemy on ground, air, or at sea, ensuring at the same time reliable and uninterrupted functioning of the system.

Example 14.5

Airborne laser bathymetry Lidar is also used in airborne bathymeters (to measure the depth of water or the location of any submerged object). For example, a typical airborne laser bathymeter may be mounted on a fixed-wing aircraft or a helicopter. Suppose it employs a 3-MW peak power *Q*-switched Nd^{3+}:YAG laser with 4-ns pulses at the rate of 200 Hz. Assume it is emitting a fundamental wavelength $\lambda_1 = 1.064$ µm (IR) and also its second harmonic at $\lambda_2 = 0.532$ µm (green). The IR pulse at λ_1 is reflected back from the sea surface, while the green pulse at λ_2 penetrates the sea water and is received by the airborne sensor after getting reflected by the bottom of the sea or any submerged object. The difference Δt in the arrival times of the two pulses will give the depth of the sea bottom or that of the submerged object. If we assume that the refractive index of sea water is approximately 1.33 and that the submerged object is located at a depth of 50 m from the surface of the sea, then the time difference between the two pulses will be

$$\Delta t = \frac{n \times 2 \times 50\,\text{m}}{c\,\text{ms}^{-1}} = \frac{1.33 \times 2 \times 50\,\text{m}}{3 \times 10^{8}\ \text{ms}^{-1}} = 0.44 \times 10^{-6}\ \text{s}$$

$$= 0.44\ \mu\text{s}$$

This is indeed a long duration as compared to the duration of the pulse, which is only 4 ns.

14.7.2 Lasers in Materials Processing in Industries

The use of lasers in industrial processing has been increasing gradually. Material processing with high-power lasers includes cutting, welding, drilling, marking, surface modification, prototyping/manufacturing, etc. These are, in fact, examples of the peaceful use of directed energy application. When an intense laser beam strikes a target, a part of it is reflected and the remainder is absorbed. The absorbed energy heats the surface. This heating can be very rapid, and its extent can be controlled for different applications. Some important industrial lasers and their potential applications are given in Table 14.1.

Table 14.1 Some important industrial lasers and their potential applications

Type of laser	Applications
Solid-state lasers	
Nd^{3+}:YAG Nd^{3+}:glass	• Light to heavy duty industrial drilling, cutting, welding, marking, etc.
Ho^{3+}:YLF Er^{3+}:YLF	• Industrial pollution monitoring, wireless initiation of thermal batteries, explosives, propellants, etc.
Gas lasers	
CO_2 N_2	• Light to heavy duty industrial drilling, cutting, welding, surface modification, etc.
Argon ion	• Industrial wire stripping
Semiconductor lasers	
Laser diodes	• Fiber-optic communication, compact disc drives, laser printers, bar code scanners, optoelectronic devices
Other lasers	
Dye lasers	• R & D, medical diagnostics
Excimer lasers (a combination of two gases: rare gas + halogen)	• Optical lithography and stereolithography, precision micro-machining, polishing, etc.
Chemical lasers	• Light to heavy duty jobs for material processing
Free-electron lasers X-ray lasers	• Mainly for directed energy applications

14.7.3 Lasers in Medical Diagnosis and Surgery

Nowadays lasers have become an indispensable tool in medical diagnosis and surgery. For the purpose of diagnosis the laser-induced fluorescence method is normally employed. For example, it is possible to distinguish between normal and diseased tissues by correlating their spectral features with other pathological data. Attempts have been made to distinguish between normal, benign, and malignant human breast tissues using this technique. Figure 14.15 gives the results of typical studies.

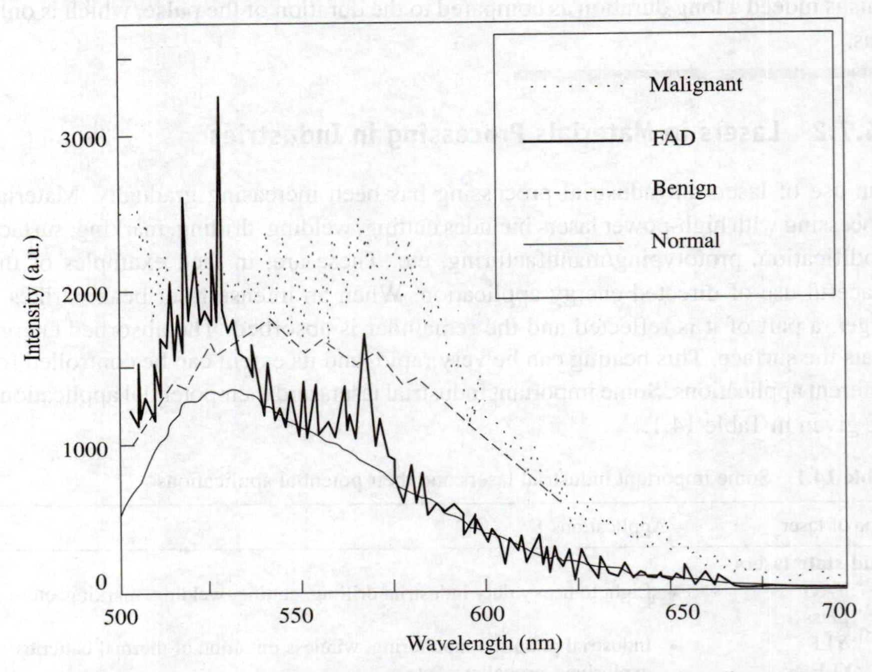

Fig. 14.15 Fluorescence spectral profiles of normal, benign, and malignant tumour tissues and aqueous flavin adenine dinucleotide (FAD) (Nair et al. 2001)

This technique has also been used for the detection of lung cancer. A chemical called *haematoporphyrin derivative* (or Hpd) is introduced into the patient's body. This chemical concentrates in the cancerous cells. The suspected areas are illuminated by light of 0.40 μm (blue laser), which is the absorption region of Hpd. The cancerous cells containing this chemical fluoresce in the range 0.60–0.70 μm and in the process reveal their presence.

It is also possible to analyse cells and their contents for the diagnosis of hereditary diseases using laser microfluorimetry. Cellular structures as small as 0.3 μm and cellular processes as fast as 0.2 ns can easily be recorded by this technique.

Lasers are now increasingly being used in medical surgery. The principal uses are to precisely cut, cauterize, and damage or destroy affected areas/tissues. The use of lasers provides the following major advantages over conventional surgery:

(i) Surgery can be performed under a microscope so that the affected areas can be located and treated accurately.
(ii) Sterilization is not required because no mechanical instruments are used.
(iii) There is less danger of haemorrhage and also less post-operative pain because photocoagulation is done by the laser.
(iv) Fiber-optic endoscopes can be used to locate, diagnose, and treat inaccessible areas of the body.
(v) Lasers can be used to weld blood vessels and, hence, reduce the number of sutures.
(vi) Surgeries can be controlled by computers.

Table 14.2 gives some typical applications of different types of lasers in surgery.

Table 14.2 Applications of lasers in surgery

Type of laser	Applications
Solid-state lasers	
Nd^{3+}:YAG laser	• Eye surgery, photocoagulation, spinal surgery, brain surgery, plastic surgery
Ruby laser	• Hair removal (cosmetics)
Gas lasers	
He–Ne laser	• Laser Doppler velocity meter
Argon ion laser	• Eye surgery (removal of cataract), photocoagulation, angioplasty, brain surgery
CO_2 laser	• Removal of cancerous growth, lesions, dermatology, spinal surgery, skin resurfacing (cosmetics)
Other lasers	
Metal–vapour lasers	• Fluorimetry (Hpd)
Excimer lasers	• Eye surgery, angioplasty
Dye lasers	• Removal of benign pigmented marks
Semiconductor lasers	
Diode lasers	• Cosmetics: hair removal and teeth whitening

14.7.4 Lasers in Defence

Lasers can be used in the battlefield for target range finding, target designation and tracking, and guidance. Nearly all battlefield tanks and armoured combat vehicles utilize the services of a laser range finder. These devices normally use the *Q*-switched Nd^{3+}:YAG laser. Recently, laser systems that can serve as EOCM devices have been developed. The purpose of these devices is to temporarily or even permanently disable electro-optic devices/sensors used by the enemy in the battlefield.

Another area of interest for use in defence is directed energy weapons (DEWs) or laser weapons. In fact, these are nothing but lasers which have high enough power and appropriate control mechanisms to disable the guidance system of warheads, or trigger an explosion of the fuel or warhead, or cause temporary or permanent damage to the target.

14.7.5 Lasers in Scientific Investigation

Lasers have been widely used in the investigation of atoms and molecules. In fact, there exists a separate branch of spectroscopy called laser spectroscopy. Because of high radiant power and narrow spectral width, a laser is an ideal source for selective excitation of atoms and molecules and for studying absorption and emission properties under such conditions. A laser microprobe is a powerful tool for studying materials available in lesser quantities (of the order of micrograms or less).

There are two ways in which a laser microprobe can be used in spectral analysis. In the first method, the energy of the laser beam is used for simultaneously evaporating the sample as well as exciting the erupted vapour and finally causing the emission. In the second method, the evaporation of the sample is produced by the laser pulse, but the excitation is caused by an auxiliary electrical discharge (cross-excitation).

In the first method (Fig. 14.16) a fairly powerful laser or Q-switched mode of operation is required. Further, a high-speed spectrograph or spectrometer is also needed to record the emission spectrum.

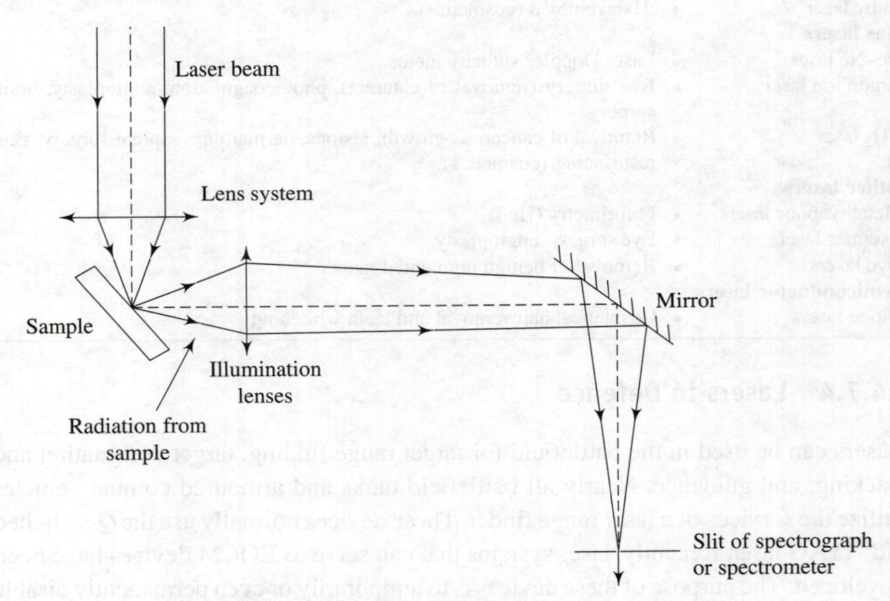

Fig. 14.16 Optics involved in laser-induced emission analysis

In the application of the second method, a spark gap is formed by two pointed carbon electrodes. The centre of the spark gap coincides with the optical axis of the laser beam, as shown in Fig. 14.17. The sample is kept at a distance of 1–2 mm below the centre of the spark gap. A dc voltage (1–5 kV) is applied to the electrodes. As soon as the laser pulse irradiates the sample, it produces a strongly ionized vapour. This vapour renders the spark gap conducting and the capacitor *C* discharges, which gives rise to the spark. The excitation of the vapour is caused by the spark. The value of the damping resistance *R* is such that no spark occurs following the discharge of the capacitor *C*. The characteristics of the spectrum can be modified by the inductor *L* within the discharge circuit. The amount of sample necessary for this technique is of the order of 0.1 μg.

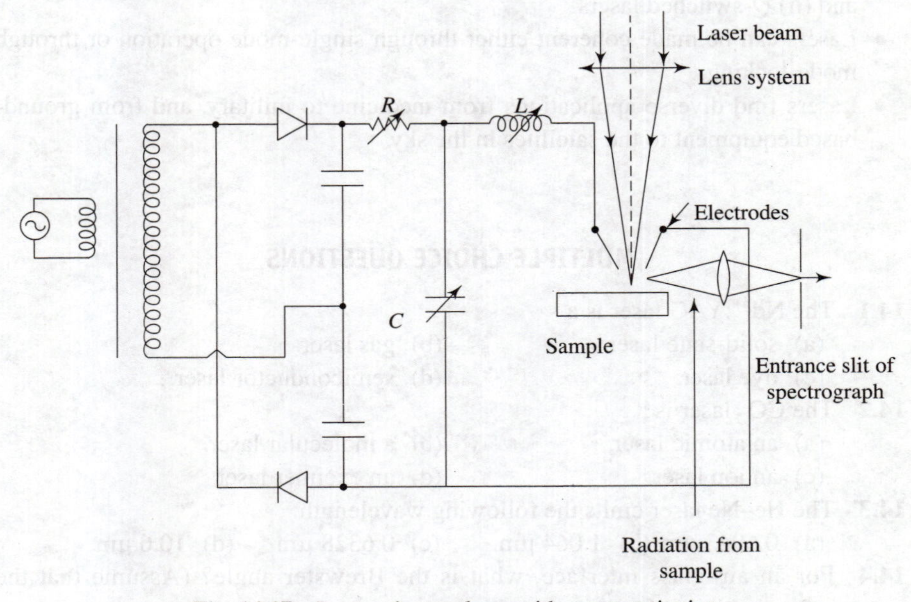

Fig. 14.17 Laser microanalyser with cross-excitation

Both the above-mentioned methods are generally employed for local analysis or microanalysis. However, they can also be used for macroanalysis, i.e., to know the average composition of a large homogeneous sample. One way is to focus an energetic laser beam onto a large area (with a diameter of several tenths of a millimeter) and the second way is to scan the specimen with a sequence of laser pulses. In the latter, two servomotors are employed for moving the sample in two mutually perpendicular directions. These motors are electronically controlled.

There are a number of other systems which have been developed for measurements of different kinds in various fields of scientific investigation.

SUMMARY

- Lasers may be classified according to the state of the active medium. There are four broad categories, namely, (i) solid-state lasers, (ii) gas lasers, (iii) dye lasers, and (iv) semiconductor lasers.
- The prime lasers in these categories are the ruby laser (emission wavelength $\lambda = 0.6943$ μm), Nd^{3+}:YAG laser ($\lambda = 1.064$ μm), He–Ne laser ($\lambda = 0.6328$ μm), CO_2 laser ($\lambda = 10.6$ μm), and tunable dye lasers.
- According to the mode of oscillation, lasers may be categorized in three groups, namely, (i) continuous-wave (cw) lasers, (ii) pulsed lasers with free oscillations, and (ii) *Q*-switched lasers.
- Lasers can be made coherent either through single-mode operation or through mode-locking.
- Lasers find diverse applications from medicine to military, and from ground-based equipment to the satellites in the sky.

MULTIPLE CHOICE QUESTIONS

14.1 The Nd^{3+}:YAG laser is a
(a) solid-state laser. (b) gas laser.
(c) dye laser. (d) semiconductor laser.

14.2 The CO_2 laser is
(a) an atomic laser. (b) a molecular laser.
(c) an ion laser. (d) an excimer laser.

14.3 The He–Ne laser emits the following wavelength:
(a) 0.6943 μm (b) 1.064 μm (c) 0.6328 μm (d) 10.6 μm

14.4 For an air–glass interface, what is the Brewster angle? (Assume that the refractive index for glass is 1.5.)
(a) 30° (b) 42.1° (c) 56.3° (d) 90°

14.5 Which of the following modes of operation should be used to achieve laser pulses of very high power?
(a) Continuous-wave operation (b) Pulsed operation with free oscillation
(c) *Q*-switched operation (d) All of these

14.6 What is the temporal width of a mode-locked rhodamine 6G dye laser emitting at 0.6 μm if $\Delta\nu = 10^{13}$ Hz?
(a) 6.6×10^{-10} s (b) 8.4×10^{-11} s
(c) 3.3×10^{-13} s (d) 10^{-13} s

14.7 What is the spatial length of a mode-locked Nd^{3+}:glass laser if $\Delta\nu = 3 \times 10^{12}$ Hz?
(a) 0.1 mm (b) 3.3 cm (c) 1 m (d) 7 m

14.8 CO_2 lasers can be used in
(a) light to heavy duty industrial jobs.
(b) medical surgery.
(c) military defence.
(d) all of the above.

14.9 Laser range finders use the following lasers:
(a) Excimer lasers (b) Semiconductor lasers
(c) He–Ne lasers (d) *Q*-switched Nd^{3+}:YAG lasers

14.10 Directed energy weapons (DEWs) can accomplish the following tasks in the battlefield:
(a) Disable EOCM of the enemy
(b) Trigger an explosion of the fuel or warhead
(c) Cause a temporary or permanent damage to the target
(d) All of the above

Answers:

14.1 (a)	14.2 (b)	14.3 (c)	14.4 (c)	14.5 (c)
14.6 (d)	14.7 (a)	14.8 (d)	14.9 (d)	14.10 (d)

REVIEW QUESTIONS

14.1 Suggest, at least, three ways of classifying lasers. Give example(s) in each category.

14.2 Consider a pulsed ruby laser with the following parameters: $N_0 = 1.5 \times 10^{19}$ atoms of Cr^{3+} per cm^3, $\tau_f = 0.4$ ms, $\tau_{sp} = 3$ ms, $\bar{\alpha}(\nu) = 2\ cm^{-1}$, $\bar{\eta}(\nu) = 0.95$, and $\bar{\nu} = 5.1 \times 10^{14}$ Hz. Calculate the pump energy that must fall on each cm^2 of the ruby crystal surface to achieve threshold inversion.
Ans: 1.328 J cm^{-2}

14.3 Estimate the critical population inversion $(N_2 - N_1)_t$ for a He–Ne laser emitting at 0.6328 μm. Assume that $\tau_{sp} \approx 0.1$ μs, length of the cavity = 10 cm, $R_1 = 1$, $R_2 = 0.99$, and the total loss coefficient $\alpha = 0$. The Doppler-broadened width of the laser transition, $\Delta\nu \approx 10^9$ Hz.
Ans: $6.28 \times 10^8\ cm^{-3}$

14.4 (a) Suggest methods of *Q*-switching lasers.
(b) Derive an expression for the maximum number of photons inside the cavity of a *Q*-switched ruby laser.

14.5 (a) Suggest a method for mode-locking lasers.
(b) Derive an expression for an individual temporal pulse width of a mode-locked laser.

14.6 (a) Discuss the generalized configuration of a laser range finder.
(b) Suggest civilian as well as military applications of laser range finders.

14.7 What is lidar? Suggest the design of an instrument based on lidar for monitoring the pollution in the atmosphere.

14.8 A typical airborne laser bathymeter employs a 3-MW peak power *Q*-switched Nd^{3+}:YAG laser with a temporal pulse width of 4 ns, both at the fundamental wavelength $\lambda_1 = 1.064$ μm and its second harmonic $\lambda_2 = 0.532$ μm. The bathymeter records a time difference of 0.53 μs between the two beams when they are directed towards an object under the sea. Calculate the depth of the object below the surface of the sea.
Ans: 60 m

14.9 (a) Suggest medical applications of lasers.
(b) How can you measure blood velocity using a fiber-optic catheter and a laser?

14.10 (a) What is a DEW?
(b) What are the advantages of DEWs over conventional weapons? What are their limitations?

Part IV: Projects

15 Lab-oriented Projects

After reading this chapter you will be able to do projects on the following:

- PC-based characterization of multimode and single-mode optical fibers.
- Characterization of optoelectronic sources (LED and ILD)
- Fiber-optic sensing mechanisms
- Several other topics

15.1 INTRODUCTION

This chapter provides an introduction to the hands-on experience needed to master the laboratory techniques using fiber optics and optoelectronics and related topics. The laboratory-oriented projects described here cover a wide range of applications involving the use of optical fibers in sensing and communications. First we discuss projects dealing with characterization of optical fibers and optoelectronic sources [light-emitting diode/injection laser diode (LED/ILD)]. Then we take up projects involving applications of fiber optics technology in sensing and communications.

In order to perform an experiment or design and fabricate a device, one needs a set of basic tools and apparatus that can be used again and again. Further, these days, the trend is towards automatic measurements. Gone are the days when the experimenter used to perform experiments manually, i.e., note down the readings, plot graphs, do necessary calculations, and then draw conclusions. Manual methods used to take from hours to even days in some cases. Therefore, keeping these facts in mind, we have described in this chapter how an experimenter can make his own kit for performing PC-based measurements. In fact, such measurements have their own advantages; e.g., the experiments can be conducted automatically and quickly, the data can be stored and retrieved at any later time, the data can be presented in different forms (depending upon the program used), the data can be shared with others, etc. The projects described here can also be done manually if some components or building blocks are not available. We will discuss both the ways of performing experiments.

15.2 MAKE YOUR OWN KIT

In order to perform any experiment with optical fibers, especially glass fibers, it is necessary to prepare the fiber ends. The first step in this process is to remove the outer sheath of the fiber. A small portion (about 2 inches) of the fiber is soaked for about 3–5 min in an appropriate solvent which dissolves the outer sheath. A single-edge razor blade may also be used for stripping the outer jacket of the fiber. The second step is to cleave the stripped end of the fiber. The method of cleaving has already been described, in brief, in Chapter 6. A *fiber cleaver* is required for this purpose. The third step is to examine the quality of the cleave. This is done under a high-power microscope. Good quality cleaving will produce a flat end face that is free of any defects and perpendicular to the optical axis of the fiber. While inspecting, if the fiber is illuminated at one end by the light from an incandescent lamp (e.g., a tungsten-filament lamp) and the other end is viewed, one can see the light shining through the central portion of the fiber. This is the core of the optical fiber. The region surrounding the core is the cladding. Both the ends of the fiber need to be prepared.

In all measurements involving optical fibers, there is a need to mount the fiber in a proper position, and there should be flexibility of minor adjustments in the x-, y-, and z-direction and also to position it at an appropriate angle (θ). Normally the bare cleaved end of the fiber is placed in a chuck (V-groove) and the latter is mounted on a micro-positioner which allows x, y, z, and θ variation. An optical bench, which may simply be a mechanical breadboard on which the micro-positioners or other devices may be mounted firmly, is also a prerequisite.

A suitable monochromatic source of light, e.g., a He–Ne laser (1–3 mW), a photodetector with an amplifier, and a meter or simply a power meter with attendant power supplies are other essential items. Table 15.1 gives the list of essential items needed to make one's own kit.

Table 15.1 Essential items of a fiber-optic kit

	Item (purpose)	Minimum quantity required
1	Fiber cleaver (for cleaving optical fibers)	1
2	Fiber chucks or V-grooves (for mounting optical fibers)	10
3	Micro-positioners (for mounting chucks or V-grooves)	6
4	Mounting posts (for mounting micro-positioners or other components)	10
5	Mechanical breadboard, normally $2' \times 2'$ area with threaded holes, equally spaced (for mounting the post, source, detector etc.), or a suitable optical bench	1

(contd)

Table 15.1 *(contd)*

	Item (purpose)	Minimum quantity required
6	Screw kit (set of screws for mounting different components)	1
7	Screwdriver set	1
8	He–Ne laser (1–3 mW) with mount	1
9	Power meter	1
10	High-power microscope	1
11	Multimode glass fiber (1 km)	1 spool
12	Single-mode glass fiber	20 m
13	Jacket solvent or razor blade	1
14	10 × and 20 × lenses for focusing	1 each

If the measurements are planned to be done automatically using a PC, then some extra modules are needed. These are mentioned in Table 15.2.

Table 15.2 Extra modules needed for PC-based measurements

	Item (purpose)	Minimum quantity required
1	PC (preferably Pentium)	1
2	Data acquisition card [Normally this will have a built-in analog-to-digital converter (ADC) and input and output (I/O). If this is not so, one can use separate ADC and I/O cards.]	1
3	Stepper card (for rotating the stepper motor)	1
4	Stepper motor 2.5 kg cm (for making rotation as well as translation stages)	2
5	Set of gears (for increasing the resolution in measurements)	2–4
6	Multiport (0–12 V)/(5 A) power supply	1

15.3 SOME SPECIFIC PROJECTS

In this section we will discuss some specific projects that can be accomplished manually and/or automatically. These are only illustrative examples. One can do many more experiments using the same kit. It depends solely on the experimenter to properly plan the experiments and conduct them successfully. In Sec. 15.4, we have listed some more projects which can be performed using the same kit.

Project 1: PC-based Measurement of the Numerical Aperture of a Multimode Step-index (SI) Optical Fiber

Project 2: PC-based Measurement of the MFD of a Single-mode fiber

Project 3: Characterization of Optoelectronic Sources (LED and ILD)

Project 4: Fiber-optic Proximity Sensor

Project 5: PC-based Fiber-optic Reflective Sensor

Project 6: PC-based Fiber-optic Angular Position Sensor

Project 7: Fiber-optic Differential Angular Displacement Sensor

PROJECT 1
PC-BASED MEASUREMENT OF THE NUMERICAL APERTURE OF A MULTIMODE STEP-INDEX (SI) OPTICAL FIBER

Principle of Measurement

The numerical aperture (NA) determines the light-gathering capacity of an optical fiber. It also determines the number of modes guided by a fiber. Therefore, NA represents an important characterization parameter. It is given by the following relation (see Sec. 2.3):

$$\mathrm{NA} = n_a \sin\alpha_m = (n_1^2 - n_2^2)^{1/2} = n_1\sqrt{2\Delta} \tag{15.1}$$

where the symbols have their usual meaning.

If the fiber is kept in air, $n_a = 1$; hence the measurement of NA requires measuring the angle of acceptance, α_m. In the present method, light from the laser is launched into the fiber as shown in Fig. 15.1. The diameter of the laser beam is roughly 1 mm and that of the multimode fiber is 50–100 µm, so nearly all the light launched into the fiber has the same incident angle α. In this launch condition, if the end face of the fiber is rotated about point O in Fig. 15.1, one can measure the amount of power accepted by the fiber as a function of the incident angle α. The angle at which the accepted power falls to 5% of its peak accepted power (corresponding to $\alpha = 0$) gives the measure of α_m. This is a standard practice (prescribed by Electronic Industries Association, Washington, DC).

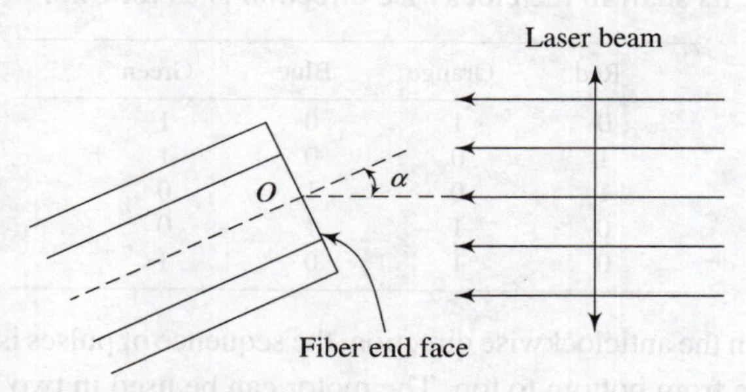

Fig. 15.1 Plane wave launch of a laser beam into an optical fiber

Apparatus Required

Most of the items mentioned in Tables 15.1 and 15.2 are required to assemble the apparatus. The block diagram of the system design is shown in Fig. 15.2, and the stepper card circuit is shown in Fig. 15.3.

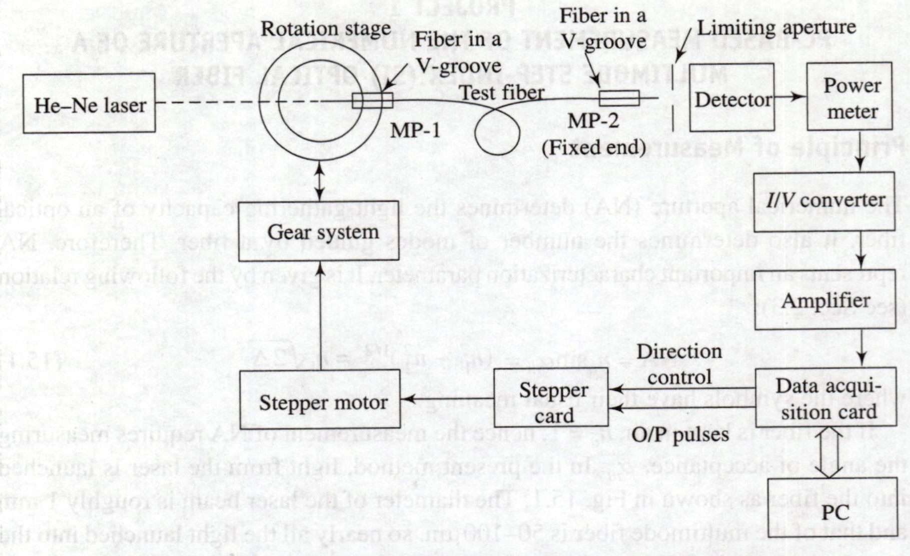

Fig. 15.2 Block diagram of a PC-based system for measuring NA

Hardware Description[1]

The stepper card forms the backbone of the hardware for this experiment. The stepper motor used has four phases. The sequence of pulses to the four phases of the stepper motor to rotate its shaft in the clockwise direction is as follows:

Red	Orange	Blue	Green
0	1	0	1
1	0	0	1
1	0	1	0
0	1	1	0
0	1	0	1

To rotate it in the anticlockwise direction, the sequence of pulses is given by reading the above table from bottom to top. The motor can be used in two modes:

(a) single stepping mode in which the step size is 1.8°
(b) multiple stepping mode in which the step size is 0.9°

In this experiment the stepper motor was used in the single stepping mode. Three gears were used to reduce the incremental angle to 0.5°. The motor was run with the help of a stepper card.

---

[1]The hardware needed for this kit was designed and fabricated by Mr Rajesh Purohit.

The IC 7474 is a dual *D* flip-flop integrated circuit (IC). It is a negative-edge-triggered one. When a negative edge of a going pulse is given to the clock input of the stepper card, the four outputs follow the sequence shown earlier. Initially, when the power is switched on, the four outputs will be 0101. With each clock pulse the sequence follows.

The stepper motor requires 1.5 A current but the IC 7474 can supply current only of the order of milliamperes, so the outputs of IC 7474 are connected to darlington pairs of transistors to increase the sourcing current capacity. Diodes are used for the freewheeling action. When an output goes from logic 1 to 0, because the current in the coil, which is inductive, cannot change suddenly, diodes are used to provide an alternative path for the current. Here we must recall that the induced electromotive force in an inductance will be of opposite polarity to the source voltage. If diodes are not there, the change in current (*di*) with respect to time (*dt*) of the inductance will be very high during the transition of an output. The voltage developed increases proportionally. This very high transient voltage may damage other parts of the circuit.

Software Description

For measurement of NA (done in our laboratory), the software is written in C. A typical program is given in Appendix A15.1.[2]

This program has separate functions for direction control, reading data from an ADC, initializing graphics, plotting the graph, and calculating the NA. The direction of the motor is controlled by writing 1 for the clockwise and 0 for the anticlockwise rotation. ADC channel 0 has been configured as the ADC input port in the function 'initialise'. The data from the ADC is stored in an array (sensed [ ]), which is then used for plotting the graph and also for calculating the numerical aperture.

Procedure

(i) Make the connections to the stepper card as per the circuit diagram shown in Fig. 15.3, i.e., the white and black wires of the stepper motor to +12 V and the other four wires (red, orange, blue, and green) to the stepper card. [In a typical set-up, the clock input to the stepper card is taken from the DACO OUT pin of the extension board (PCLD 8115), and the direction input to the stepper card is taken from the ADC (PCL 818HG) digital port.] Once the connections are made, run the motor to check its functioning.

(ii) Take a multimode step-index fiber about 1–2 m long and cleave both its ends properly. Mount both ends of the fiber in the chucks (or V-grooves) to be placed in the two micro-positioners (MP-1 and MP-2). MP-1 is mounted on the rotation

[2]This program code has been written by Ms Shweta Gaur and Ms K. Priya.

stage and MP-2 is mounted near the power meter. Switch on the laser and align it with the fiber end such that, for $\alpha = 0$, the power meter reads maximum. You may need to put two or three drops of glycerine, which will act as a cladding mode stripper, near the launching end of the fiber.

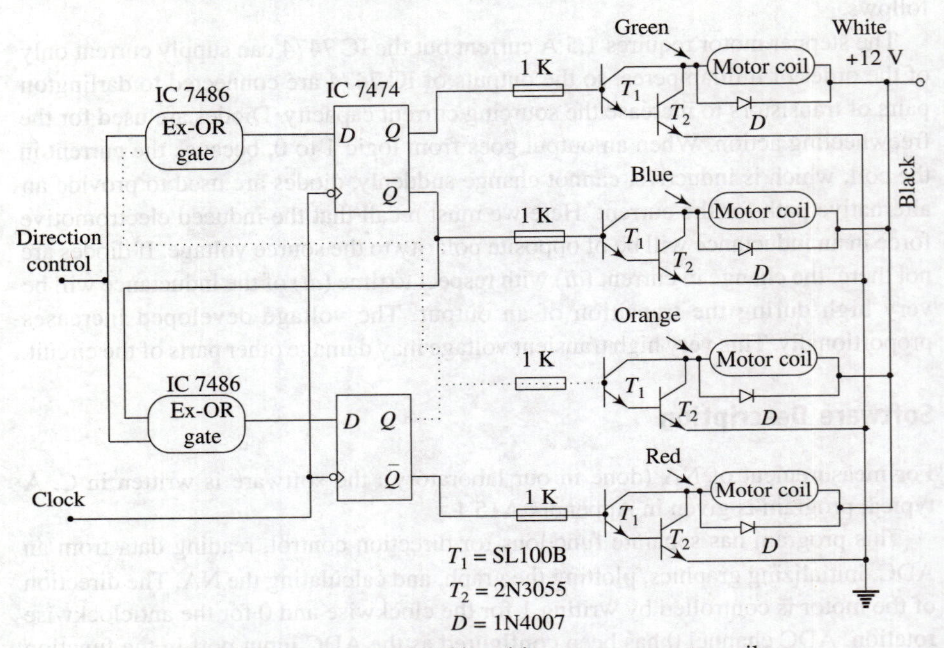

Fig. 15.3 Stepper card and its wiring to stepper motor coils

(iii) Now switch on the power supply and run the program (C:\tc\aper.c). The motor makes the fiber end move from, say, –15° to +15° (these values can be changed). The corresponding power at the other end of the fiber is detected by the detector (power meter), the output photocurrent of the detector is converted into voltage by the *I-V* converter, amplified, and fed to a PC via the ADC in the data acquisition card. The output in the form of either the graph of variation of optical power with incident angle α or the calculated value of the NA can be seen on the monitor.

Exercise 15.1 Modify the program code to include the calculation of the *V*-parameter of the fiber and the number of modes guided by the fiber.

Hint: We know that the *V*-parameter is given by [recall Eq. (4.16)]

$$V = \frac{2\pi a}{\lambda}(n_1^2 - n_2^2)^{1/2} = \frac{2\pi a}{\lambda}(\text{NA}) \tag{15.2}$$

The diameter '2*a*' of the fiber core is normally specified by the manufacturer and the wavelength of the source is known (for the He–Ne laser, $\lambda = 0.6328$ μm).

Further, the total number of modes, M, supported by the SI fiber [from Eq. (4.67)] is given by

$$M_S = \frac{V^2}{2} = \frac{1}{2}\left[\frac{2\pi a}{\lambda}(\text{NA})\right]^2 \tag{15.3}$$

Therefore, knowing the value of NA, one can calculate both V and M_s.

PROJECT 2
PC-BASED MEASUREMENT OF THE MFD OF A SINGLE-MODE FIBER

Principle of Measurement

In single-mode fibers, the radial distribution of the optical power in the propagating fundamental mode plays an important role. Therefore, the mode field diameter (MFD) of the propagating mode constitutes a characteristic parameter of a single-mode fiber. Recall (see Sec. 5.3.1) that the field distribution of the fundamental mode can be approximated by the Gaussian function

$$\psi(r) = \psi_0 \exp(-r^2/w^2)$$

where $\psi(r)$ is the electric or magnetic field at a radius r, ψ_0 is the axial field (at $r = 0$), and w is known as the mode field radius, which is the radial distance from the axis at which ψ_0 drops to ψ_0/e. Thus the MFD is $2w$. The above equation gives the field distribution at the output end of the fiber. This is also called the near-field pattern. However, it can be shown that the far-field pattern at a distance of z from the fiber end is also Gaussian and that the corresponding intensity is given (Ghatak & Shenoy 1994) by

$$I(r) = I_0 \exp\left[-\frac{2r^2}{w_z^2}\right] \tag{15.4}$$

where I_0 is a constant and w and w_z are the Gaussian MFDs of the near-field and far-field distributions. If $z \gg w_z/\lambda$,

$$w_z \approx \frac{\lambda z}{\pi w} \tag{15.5}$$

Substituting for w_z from Eq. (15.5) in Eq. (15.4), we get

$$I(r) = I_0 \exp\left[-\frac{2\pi^2 r^2 w^2}{\lambda^2 z^2}\right]$$

or

$$I(\theta) = I_0 \exp\left[-\frac{2\pi^2 w^2}{\lambda^2}\tan^2\theta\right] \tag{15.6}$$

where $\tan\theta = r/z$, θ being the far-field radiation angle. The angle θ at which the intensity $I(r)$ drops to $1/e^2$ of its maximum value I_0, at $\theta = 0$, is given by

$$\tan\theta_e = \frac{\lambda}{\pi w} \tag{15.7}$$

This gives the MFD, $2w$, as

$$2w = \frac{2\lambda}{\pi \tan\theta_e} \tag{15.8}$$

Thus, by measuring the angle θ_e, it is possible to calculate the MFD. The variation of $I(\theta)$ with θ is shown schematically in Fig. 15.4.

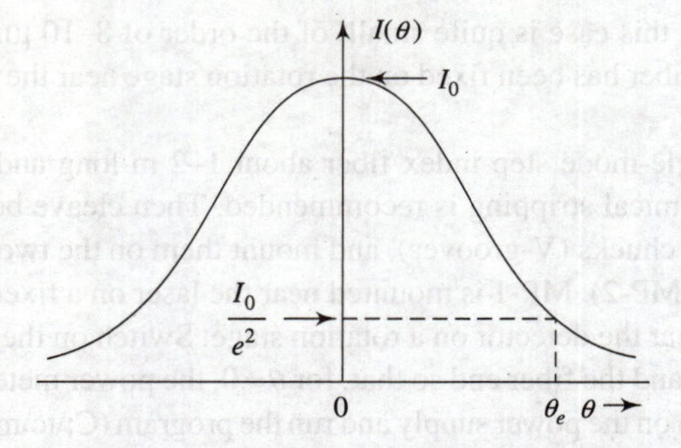

Fig. 15.4 Far-field pattern of a single-mode fiber

Apparatus Required

Most of the items mentioned in Tables 15.1 and 15.2 will be required to assemble the apparatus. The block diagram of the system design is shown in Fig. 15.5.

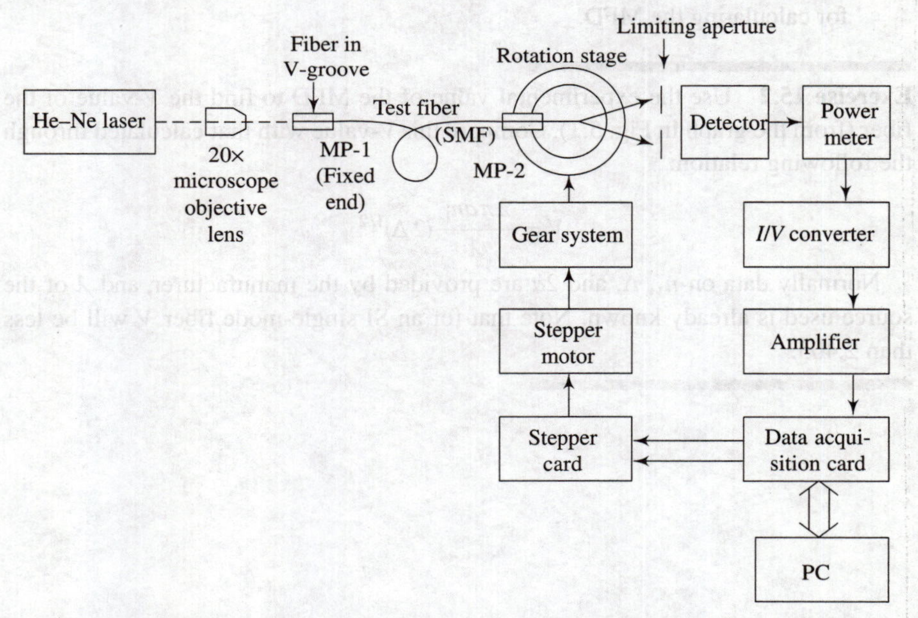

Fig. 15.5 Block diagram of a PC-based system for measuring the MFD of an SMF

Procedure

(i) Assemble the apparatus as shown in Fig. 15.5. If you compare this figure with Fig. 15.2, you will find two changes. First a 20× microscope objective lens has been used to focus the laser beam onto the tip of the SMF (because the fiber diameter in this case is quite small, of the order of 8–10 μm). Second, the far end of the fiber has been fixed on the rotation stage near the detector (or power meter).

(ii) Take a single-mode step-index fiber about 1–2 m long and remove the outer jacket. Chemical stripping is recommended. Then cleave both the ends, place them in the chucks (V-grooves), and mount them on the two micro-positioners (MP-1 and MP-2). MP-1 is mounted near the laser on a fixed post and MP-2 is mounted near the detector on a rotation stage. Switch on the laser and align the laser beam and the fiber end so that, for $\theta = 0$, the power meter reads maximum.

(iii) Now switch on the power supply and run the program (C:\tc\mfd.c). The program code is almost similar to that used in Project 1 except for minor changes that have to be made for calculating the MFD. As the fiber end on the rotation stage rotates, say, from $-10°$ to $+10°$, the intensity of the light incident on the detector gradually increases to a maximum (say, I_0) for $\theta = 0$ and then falls off again as θ increases. The difference between angular positions (on the negative and positive sides), for which I_0 falls to I_0/e^2, gives the value of $2\theta_e$. Use Eq. (15.8) for calculating the MFD.

Exercise 15.2 Use the experimental value of the MFD to find the *V*-value of the fiber (from the graph in Fig. 5.1). Compare this *V*-value with that calculated through the following relation:

$$V = \frac{2\pi a n_1}{\lambda}(2\,\Delta)^{1/2}$$

Normally data on n_1, Δ, and $2a$ are provided by the manufacturer, and λ of the source used is already known. Note that for an SI single-mode fiber *V* will be less than 2.405.

PROJECT 3
CHARACTERIZATION OF OPTOELECTRONIC SOURCES (LED AND ILD)

Principle of Measurement

In Chapter 7, we have already discussed the theory of optoelectronic sources in detail. From that discussion, we know that light-emitting diodes (LEDs) and injection laser diodes (ILDs) are generally used in fiber-optic communication. They are also used in sensors, displays, and other optoelectronic circuitry. In this project our objective is to study some of the optical and electrical characteristics of these sources and learn the difference between them.

In fact an optoelectronic source is characterized by the distribution of optical power radiated from its surface. Thus they are divided into two categories: (i) Lambertian source, which emits light in all possible directions, and (ii) collimated source, which emits light in a very narrow range of angles about the normal to its emitting surface. A surface-emitting LED closely resembles a Lambertian source and the ILD approximates a collimated source. In general, the angular distribution of the power radiated per unit solid angle in a direction at an angle θ to the normal to the emitting surface is given by

$$P(\theta) = P_0(\cos\theta)^m \tag{15.9}$$

where P_0 is the power radiated per unit solid angle normal to the surface. For a diffuse source, $m = 1$ [recall Eq. (A7.1)]; and for a collimated source, m is large (typically $m = 20$ for an ILD).

Another property of interest, which distinguishes an LED from an ILD is the variation of output power of the source with the drive current. For communication grade sources, rise time, spectral half-width, and the modulation response are prime factors.

Apparatus Required

LED and ILD sources, variable dc power supply (0–5 V), a digital multimeter, function generator (20 MHz), cathode ray oscilloscope (CRO) (20 MHz) [if a digital storage oscilloscope (DSO) is available, it is better], and all the items of Table 15.2 are required. The block diagram of the system design is shown in Fig. 15.6.

Procedure

The experiment is to be conducted in three parts. Part I involves the measurement of *P-I* characteristics of the LED, part II involves the measurement of $P(\theta)$ as a function of θ, and part III involves the measurement of the frequency response of the LED.

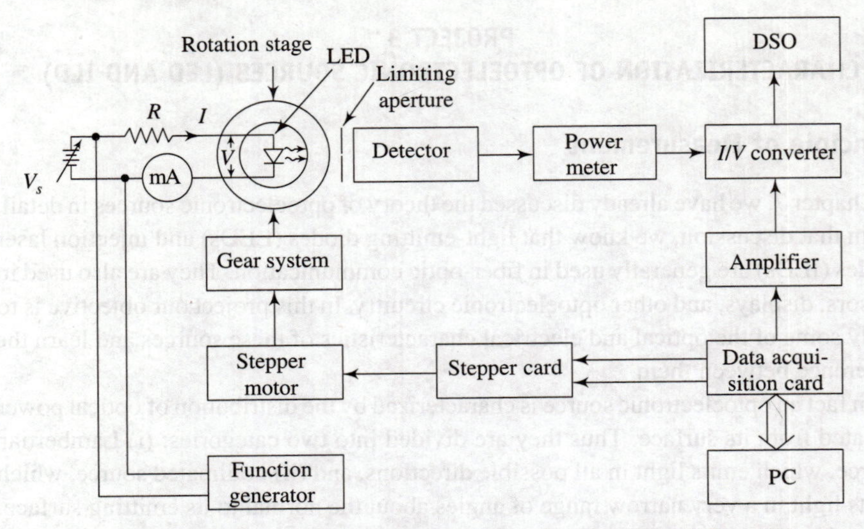

Fig. 15.6 Experimental set-up for characterization of an LED or ILD

(i) Assemble the experimental set-up as shown in Fig. 15.6. Connect the variable voltage dc power supply V_s, resistor R, and milliammeter (mA) and mount the LED (under test) on the rotation stage. Now vary the voltage V_s in suitable steps and with the help of a multimeter measure the voltage V across the diode and read the current I in the milliammeter. Normally the current must not exceed ~ 120 mA and the voltage should not exceed ~ 1.5 V.

Simultaneously read the optical power in the power meter. Plot power P as a function of current I. A typical plot is shown in Fig. 15.7. From this plot determine the operational position (bias point I_B) for the LED. Normally this is the current setting that yields half the maximum output power from the source.

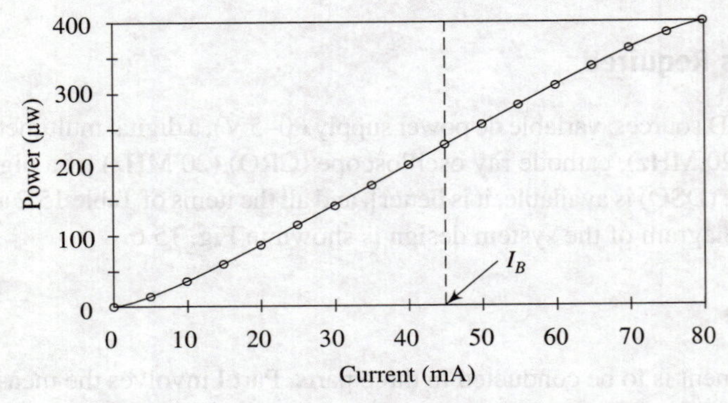

Fig. 15.7 LED power vs current for a typical LED

It is important to note that the detector should be placed near the LED. For optimum results, mount the LED at the centre of the rotation stage and the detector just outside the rim of the rotation stage and align them properly.

(ii) Now set the LED at the bias point I_B for the rest of the experiment. Now energize other blocks for PC-based measurement of the radiation pattern of the LED. Prior to this measurement, suitably modify the program code for measuring $P(\theta)$ and name the file (e.g., C:\tc\led.c). As in the previous two projects, run the program. The stepper motor will rotate the LED from, say, –10° to 10°, and the corresponding output will be recorded and plotted on the screen of the monitor.

(iii) The third part is again manual. Switch off the power supply to the stepper motor/card, etc. Again align the LED with the detector, with the LED at the dc bias current I_B. Now connect a sinusoidal input to the LED from the function generator using a T-connector. The same signal should be given to the reference channel of the CRO (or DSO) and the detector output should be given to the measuring channel. Vary the frequencies from about 20 Hz to about 20 MHz in suitable steps. At each frequency note the peak-to-peak voltage of the measuring channel. Plot the output as a function of frequency on a semilog graph paper. The highest frequency at which the response drops to half of its maximum value gives the modulation bandwidth of the LED.

(iv) Repeat steps (i), (ii), and (iii) for an ILD and compare the results.

Exercise 15.3 In part (iii) of the above experiment, instead of sinusoidal input, connect a square wave signal and measure the rise times (for the LED and ILD). The rise time is the time difference between the 10% and 90% responses.

PROJECT 4
FIBER-OPTIC PROXIMITY SENSOR

Principle of Sensing

As we have already discussed in Chapter 13, there are mainly three types of fiber-optic sensors, namely, (i) intensity-modulated sensors, (ii) phase-modulated sensors, and (iii) spectrally modulated sensors. This project belongs to the first category. In the present technique of sensing, an optical signal is launched through one arm of the bifurcated fiber bundle onto a target element (essentially a reflector), which is coupled to the object whose position is to be monitored. The signal after a reflective modulation is collected by the second arm of the fiber bundle and sent to the detector. The output of the latter is a function of the position of the target element with respect to the end of the fiber bundle. Twin fibers or a single fiber with a directional *y*-coupler can also be used in place of a multimode fiber bundle. This technique seems to offer low-cost production of sensors for a wide range of applications, e.g., measurement of displacement, force, pressure, temperature, level, etc.

Apparatus Required

An LED and a *p-i-n* photodiode, one bifurcated fiber bundle, a mechanical breadboard with accessories for mounting components, a front surface reflector, five posts, one translational stage. The experimental set-up is shown in Fig. 15.8(a) and the circuit diagram is given in Fig. 15.8(b). We see that the IR LED (used in the circuit) is forward-biased using a +5 V supply and 1 K current-limiting resistor. The *p-i-n* photodiode is reverse-biased, again by using a + 5 V supply and 1 K current-limiting resistor. The first stage of amplification is a current-to-voltage converter with an amplification factor of 1510, i.e., the voltage output of the first amplifier is

$$V_1 = -IR$$

where I is the detector current (typically of the order of 0.0006 μA) and R = 1510 K. This V_1 is further amplified by an inverting amplifier with an amplification factor of 1500 to finally give V_0. The following is the list of electronic components required: three 1 K resistors, two 1500 K resistors, one 10 K resistor, and two μA 741 operational amplifiers.

Procedure

Once the experimental set-up is ready, move the reflector so that it almost touches the tip of the bifurcated fiber bundle. We take this as d = 0. Then move the reflector away from the tip of the fiber bundle in suitable steps (say, 0.1 mm) and take the

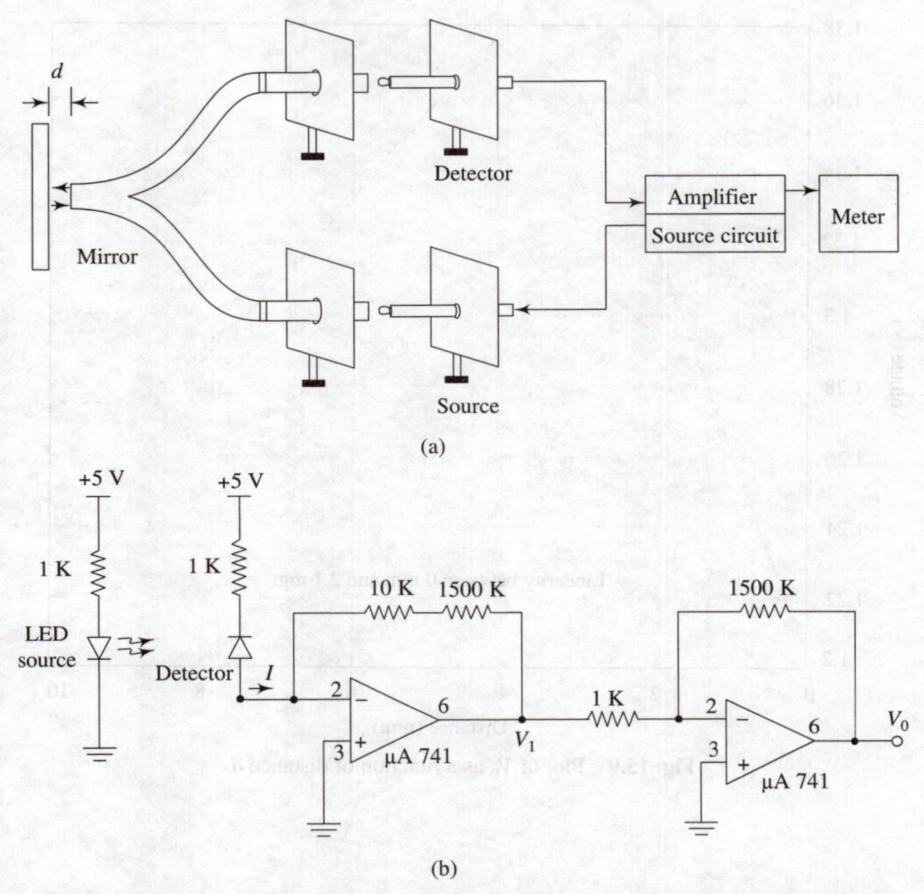

Fig. 15.8 (a) Experimental set-up for position sensing, (b) circuit diagrams for source and detector

corresponding reading, i.e., V_0 (in volts). Plot voltage V_0 as a function of distance d. A typical plot is shown in Fig. 15.9.

We see from this graph that initially the voltage increases with distance up to 3.8 mm, at which it is maximum, and then decreases up to about 9.2 mm. The interesting point is that there is a linear region between 0 and 2.1 mm. Therefore, this region can be used for sensing the position of an object.

Exercise 15.4 (a) Plot voltage V_0 as a function of $(\text{distance})^{-1}$ and find the linear region. (b) Plot voltage V_0 as a function of $(\text{distance})^{-2}$ and again find the linear region. (c) Compare these linear ranges with that plotted in Fig. 15.9.

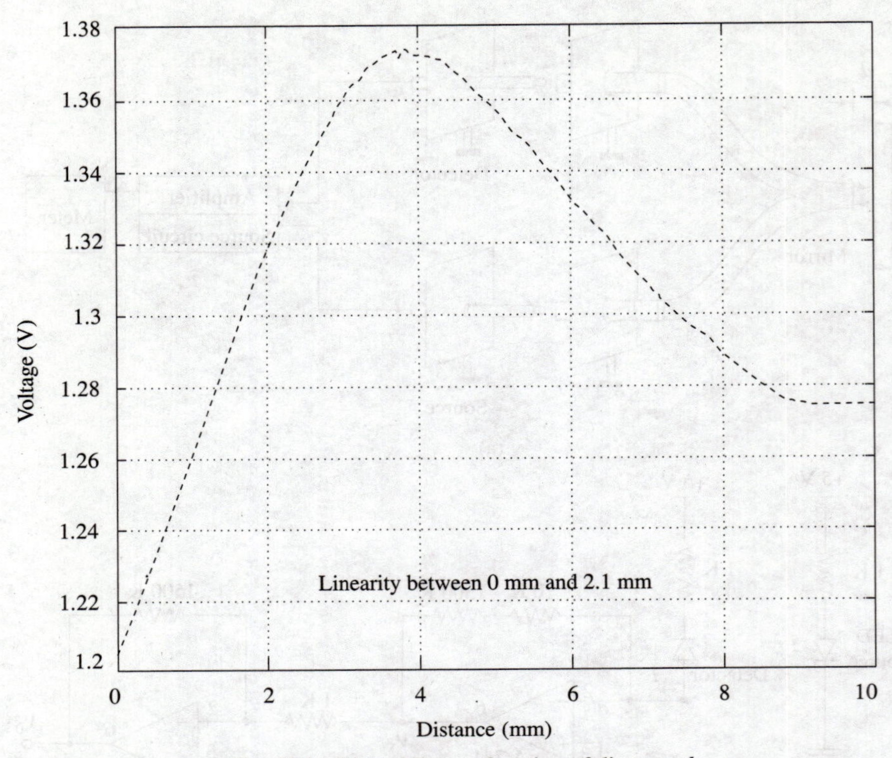

Fig. 15.9 Plot of V_0 as a function of distance d

PROJECT 5
PC-BASED FIBER-OPTIC REFLECTIVE SENSOR

Principle of Sensing

The design of the sensor is shown in Fig. 15.10(a). Herein, light from an optical source (a continuous-wave He–Ne laser) is launched through one arm of the bifurcated fiber bundle, which makes it fall onto the sensing element. The reflected flux of light is collected by the second arm of the bundle and sent to the Si photodetector. The output of the latter is a photocurrent, which is converted into voltage and given to an analog-to-digital converter card on the PC.

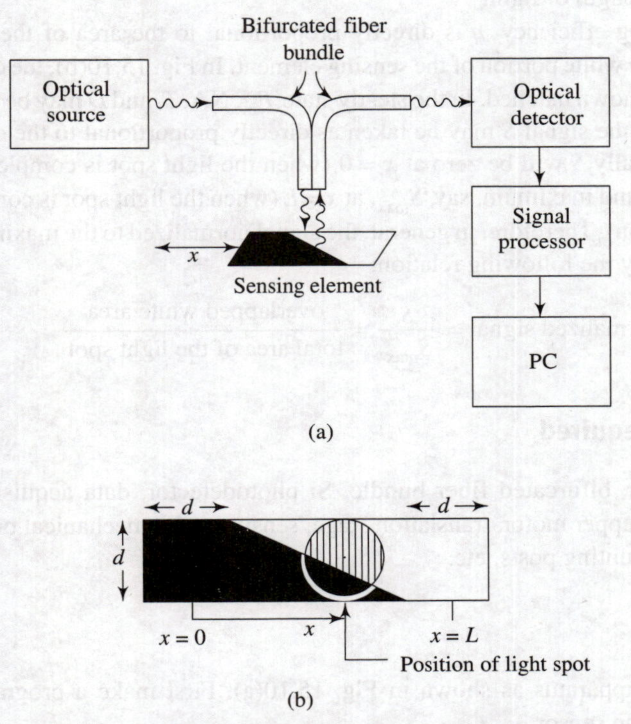

Fig. 15.10 (a) Block diagram of sensor, (b) enlarged view of the sensing element

The heart of the sensor is the sensing element shown in Fig. 15.10(b). It is a strip of length $L + d$ and width d, which is diagonally painted with rough black in one half and reflecting white in the other half. The diameter of the spot size of the light incident on the strip is made equal to d and the distance between the fiber end and the element is then kept fixed. If the element is moved in a direction (x) transverse to the direction of the incident light, the flux of light coupled back into the bundle will depend on the position of the element because the black portion reflects minimum flux of light whereas the white portion reflects maximum flux of light.

The signal S developed by the detector may, approximately, be given by the following relation:

$$S = P_{in}(\mathrm{NA})\eta TD \tag{15.10}$$

where P_{in} is the optical power supplied by the source to the transmitting arm of the fiber bundle, NA is the numerical aperture of the bundle, η is the coupling efficiency at which the optical power is coupled to the receiving arm of the bundle, T is the transmission efficiency of the bundle (which takes into account the absorption or scattering losses within the fibers), and D is the responsivity of the detector for the incident wavelength of light.

The coupling efficiency η is directly proportional to the area of the light spot overlapping the white portion of the sensing element. In Fig. 15.10(b), the overlapped area has been shown hatched. In the steady state, P_{in}, NA, T, and D may be taken to be constant. Thus the signal S may be taken as directly proportional to the overlapped white area. Ideally S will be zero at $x = 0$ (when the light spot is completely in the black portion) and maximum, say, S_{max}, at $x = L$ (when the light spot is completely in the white portion). Therefore, in general, the signal normalized to the maximum value will be given by the following relation:

$$\text{Normalized signal} = \frac{S}{S_{max}} = \frac{\text{overlapped white area}}{\text{total area of the light spot}} \tag{15.11}$$

Apparatus Required

A He–Ne laser, bifurcated fiber bundle, Si photodetector, data acquisition card, stepper card, stepper motor, translation stage, sensing strip, mechanical breadboard, appropriate mounting posts, etc.

Procedure

Assemble the apparatus as shown in Fig. 15.10(a). First make a program for the calibration of the sensor.

For calibration, the PC controls the mechanical movement of the sensing element through a separate mechanical assembly (not shown). The element is made to move by an increment of Δx (whose value depends on the resolution required) and the voltage corresponding to the light flux reflected by the element is read by the computer. Typically the values of Δx, d, and L may be taken to be 1, 5, and 40 mm, respectively. The voltage readings corresponding to the displacement from $x = 0$ to $x = L$ are stored and a calibration curve is plotted on the screen of the monitor. A typical curve depicting the variation of the normalized signal (experimental value of S/S_{max}) as a function of normalized displacement (x/L) is shown in Fig. 15.11. It may be noted that the experimental curve almost coincides with the theoretical curve (which is a plot of the normalized overlapped white area as a function of x/L).

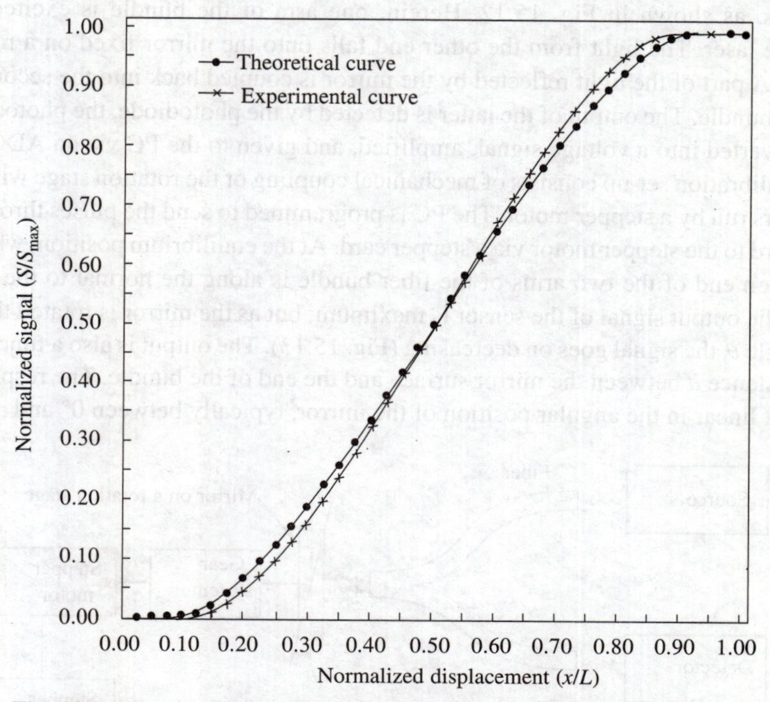

Fig. 15.11 Experimental and theoretical variations of the normalized signal as a function of x/L for a typical set-up

Once calibrated, the sensing element is coupled to the object whose displacement is to be monitored. As the object moves, a change in the flux of reflected light occurs, which is communicated to the PC. The latter reads the corresponding voltage value and compares it with the calibration data and also with the previous voltage value, thus giving both the magnitude and direction of displacement.

Exercise 15.5 Compare this reflective sensor with that of Project 4. Mention the merits and demerits of this sensor. List the possible applications of this sensor.

PROJECT 6
PC-BASED FIBER-OPTIC ANGULAR POSITION SENSOR

Here also, reflective modulation is used for sensing the angular position of the object. It utilizes a bifurcated fiber bundle as a medium for transmission and reception of signals, as shown in Fig. 15.12. Herein, one arm of the bundle is excited by a He–Ne laser. The light from the other end falls onto the mirror fixed on a rotating stage. A part of the light reflected by the mirror is coupled back into the second arm of the bundle. The output of the latter is detected by the photodiode, the photocurrent is converted into a voltage signal, amplified, and given to the PC via an ADC card. The calibration set-up consists of mechanical coupling of the rotation stage with a set of gears run by a stepper motor. The PC is programmed to send the pulses through an I/O card to the stepper motor via a stepper card. At the equilibrium position, when the common end of the two arms of the fiber bundle is along the normal to the mirror face, the output signal of the sensor is maximum; but as the mirror is rotated through an angle θ, the signal goes on decreasing (Fig. 15.13). The output is also a function of the distance d between the mirror surface and the end of the bundle. The response is almost linear in the angular position of the mirror, typically between 0° and ±10°.

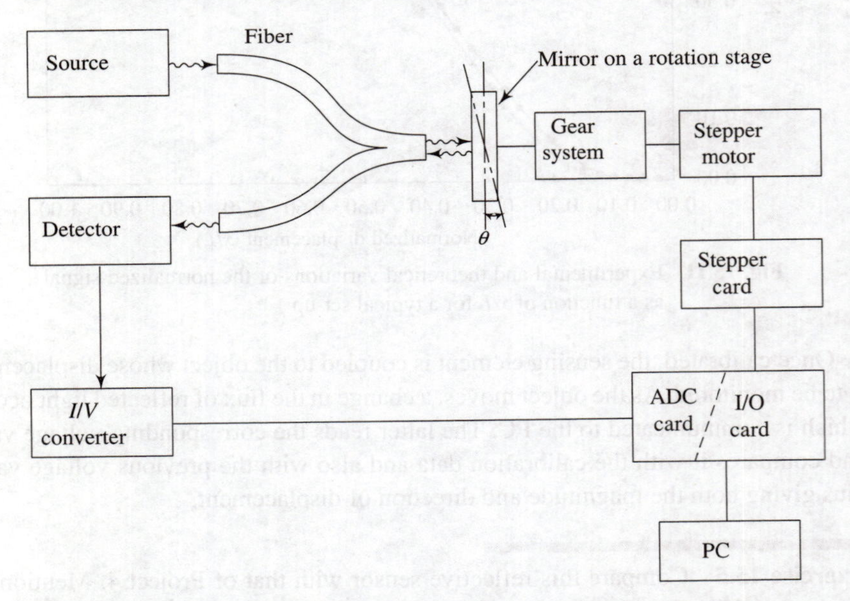

Fig. 15.12 Block diagram of the calibration set-up for an angular position sensor using a bifurcated fiber bundle

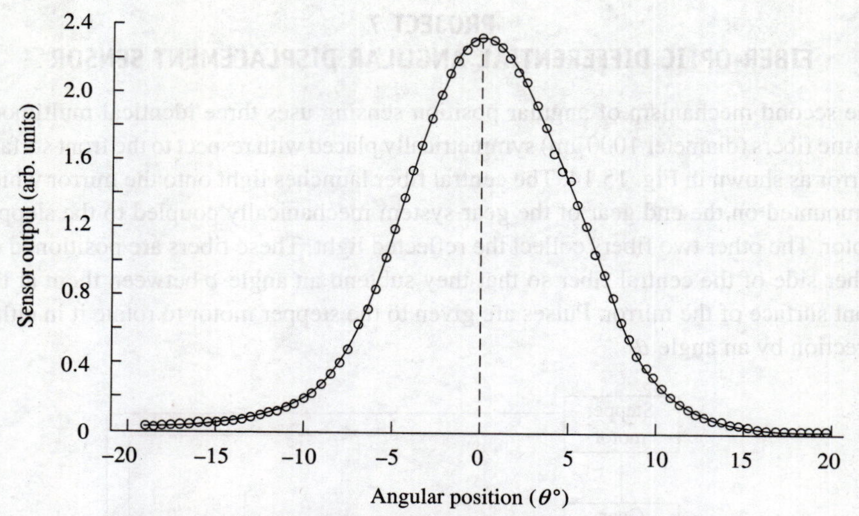

Fig. 15.13 Response (output of a sensor) as a function of angle θ

PROJECT 7
FIBER-OPTIC DIFFERENTIAL ANGULAR DISPLACEMENT SENSOR

The second mechanism of angular position sensing uses three identical multimode plastic fibers (diameter 1000 μm) symmetrically placed with respect to the front surface mirror as shown in Fig. 15.14. The central fiber launches light onto the mirror which is mounted on the end gear of the gear system mechanically coupled to the stepper motor. The other two fibers collect the reflected light. These fibers are positioned on either side of the central fiber so that they subtend an angle ϕ between them at the front surface of the mirror. Pulses are given to the stepper motor to rotate it in either direction by an angle θ.

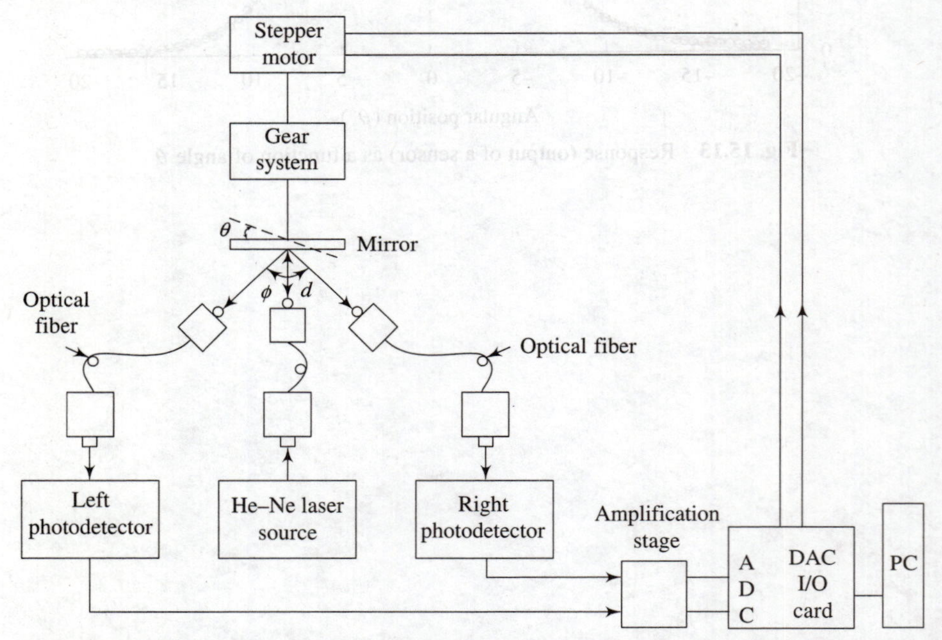

Fig. 15.14 Block diagram of the calibration set-up for an angular position sensor using three multimode fibers

In the equilibrium position ($\theta = 0$) both the outer fibers collect equal amounts of light. Thus the difference ΔV of the output of the detectors placed behind these fibers is zero at this position. Any change θ in the angular position of the mirror causes a change in ΔV, which is amplified and fed to a PC via an ADC card. The circuit diagram of the amplification stage is shown in Fig. 15.15. The response of the sensor is dependent on the settings of the three fibers relative to each other. Figure 15.16 depicts a typical case, where $\phi = 30°$. As can be seen in the figure, the response is almost linear in the range from $\theta = 0°$ to $\pm 10°$ rotation of the mirror. For other settings (e.g.,

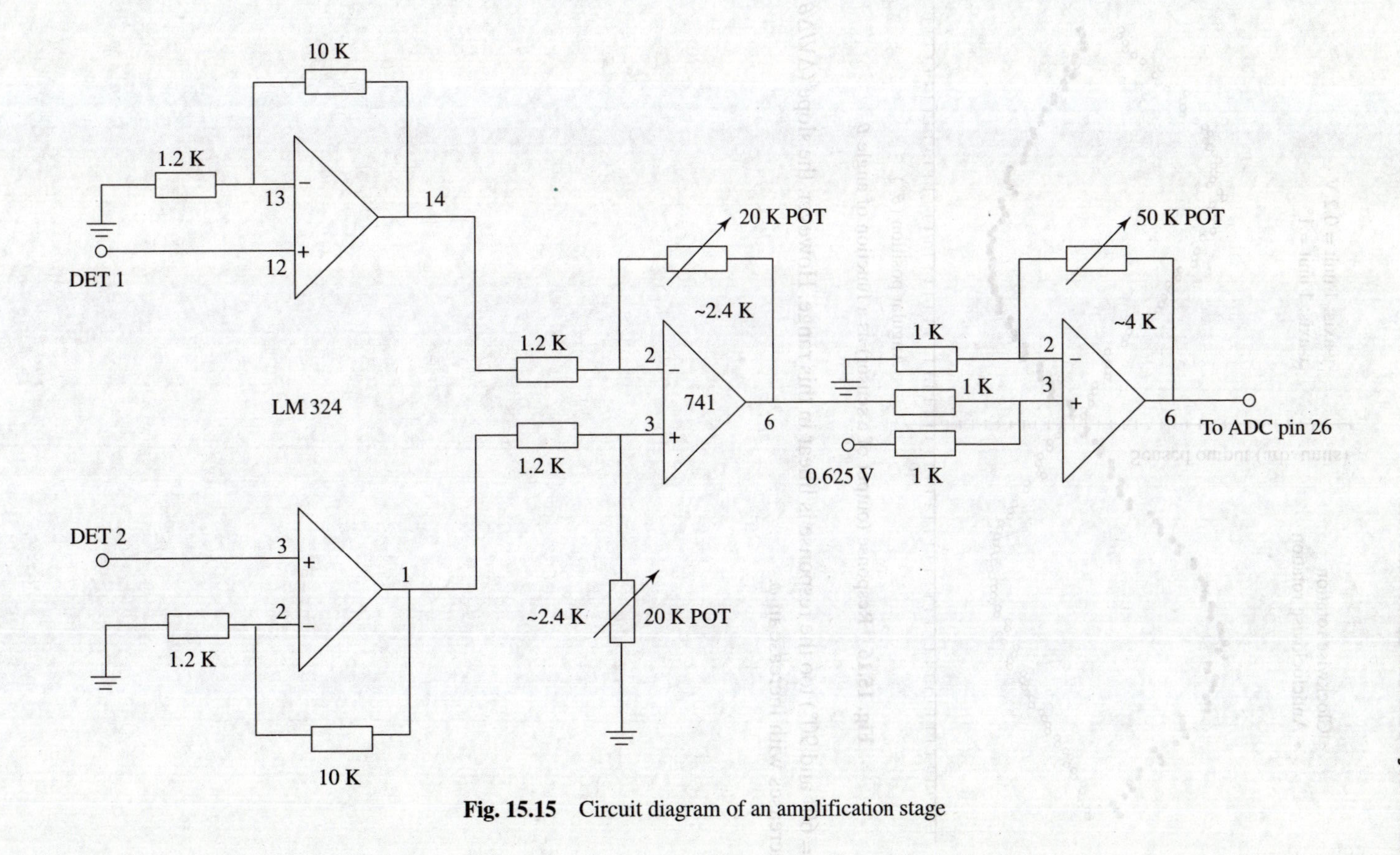

Fig. 15.15 Circuit diagram of an amplification stage

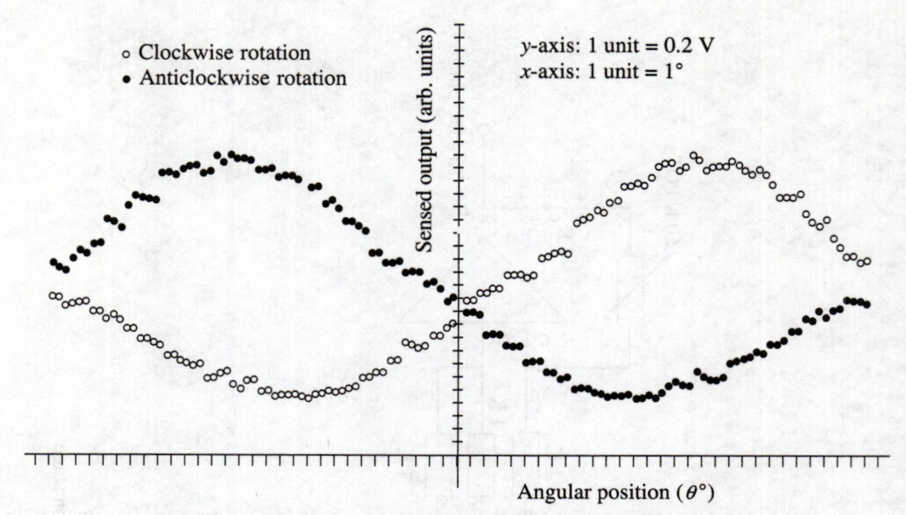

Fig. 15.16 Response (output of a sensor) as a function of angle θ

$\phi = 60°$ and 90°) too the response is linear in this range. However, the slope ($\Delta V/\Delta\theta$) increases with increase in ϕ.

15.4 MORE PROJECTS

It is possible to do some more projects using the above-mentioned kit. One may need some extra modules, which are normally available in college laboratories. However, it is not possible to describe them in detail here. We will mention only the titles of projects and the relevant references of the section/chapter of this book.

Project 8: Measurement of Fiber-to-fiber Misalignment Losses
[Refer to Fiber-to-fiber misalignment losses in Sec. 6.5.1.]

Project 9: Measurement of Spectral Attenuation of a Multimode Optical Fiber
[Refer to Sec. 6.8.1.]

Project 10: Measurement of Intermodal Dispersion of a Multimode Optical Fiber
[Refer to Sec. 6.8.2.]

Project 11: Measurement of the Refractive Index Profile of an Optical Fiber
[Refer to Sec. 6.8.4.]

Project 12: Measurement of Emission Spectra of (a) LED and (b) ILD
[This is in fact an extension of Project 3.]

Project 13: Measurement of Responsivity of the *p-i-n* Detector
[Refer to Chapter 8. An experiment similar to Project 3 can easily be done.]

Project 14: Measurement of Optical Bandwidth and Rise-time Budget Analysis of a Fiber-optic Communication System
[In the apparatus for Project 3, introduce an appropriate length (say, about 1 km) of the fiber and make a simplex link and repeat that experiment. Also refer to Rise-time budget analysis in Sec. 12.2.1.]

Project 15: Differential Fiber-optic Sensing and Switching
[Refer to Fig. 13.6 and related discussion.]

Project 16: Microbend Sensor for Measurement of Force, Pressure, and Stress
[Refer to Fig. 13.7 and related discussion.]

Project 17: Fiber-optic Mach–Zehnder Interferometric Sensor
[Refer to Fig. 13.8 and related discussion.]

Project 18: Fiber-optic Bar Code Scanner

[This is the same as Project 5, except that the sensing element will be replaced by a bar code and the program code has to be suitably modified.]

Project 19: Fluoro-optic Temperature Sensor

[Refer to Sec. 13.6.1.]

APPENDIX A15.1: A TYPICAL C PROGRAM FOR THE MEASUREMENT OF NA

```
/* Program to calculate Numerical Aperture of
   optical fiber */
#  include <stdio.h>
#  include <conio.h>
#  include <dos.h>
#  include <graphics.h>
#  include <math.h>
#  define BASE 0x300          /* ADC_LO byte */
#  define ADC_HI BASE+1
#  define MUX BASE+2          /* Start and stop channel no. */
#  define DIR BASE+3          /* Port for controlling direction
                                 of stepper motor */
#  define DAC_LO BASE+4
#  define DAC_HI BASE+5
#  define CNTRL BASE+9
#  define TD 900
#  define HI 0xff  /* value corresponding to +5 volts */
#  define PI 3.141
void rotatec(void);
void rotatea(void);
void initialise(int chno);
float ADC(void);
void AVG(float);
void max(int);
void init(int angle);
void plot(int angle);
float sensed[120];
void main()
{
int i,angle;
int gd,gm;
int a=0;
char dir;
clrscr();
initialise(0x00);
printf("ENTER ANGLE\n");
scanf("%d",&angle);
fflush(stdin);
printf("ENTER THE DIRECTION--'c' for clockwise, 'a' for
anticlockwise\n");
scanf("%c",&dir);
switch(dir)
{
case 'a':
```

```
rotatec();
delay(TD);
for(i=0; i < 2*angle; i++)
{
outportb(DAC_LO,0x00);
outportb(DAC_HI,0x00);
delay(TD);                              /* for sending pulses to
                                           stepper card */
outportb(DAC_LO,0x00);
outportb(DAC_HI,0xff);
delay(TD);
sensed[a] = ADC();
a++;
}
a=0;
delay(TD);
rotatea();
delay(TD);
printf("anti clockwise rotation\n");
for(i=0; i < 4*angle; i++)
{
outportb(DAC_LO,0x00);
outportb(DAC_HI,0x00);
delay(TD);
outportb(DAC_LO,0x00);
outportb(DAC_HI,0xff);
delay(TD);
sensed[a] = ADC();
printf("angle:%.1f\t volt:%.2f\n",(i-2.0*angle)/2,sensed[a]);
a++;
}
delay(TD);
break;

case 'c':
rotatea();
delay(TD);
for(i=0;i<2*angle;i++)
{
outportb(DAC_LO,0x00);
outportb(DAC_HI,0x00);
delay(TD);
outportb(DAC_LO,0x00);
outportb(DAC_HI,0xff);
delay(TD);
sensed[a] = ADC();
a++;
}
```

```
a=0;
delay(TD);
rotatec();
delay(TD);
printf("clockwise rotation\n");
for(i=0; i < 4*angle; i++)
{
outportb(DAC_LO,0x00);
outportb(DAC_HI,0x00);
delay(TD);
outportb(DAC_LO,0x00);
outportb(DAC_HI,0xff);
delay(TD);
sensed[a] = ADC();
printf("angle:%.1f\t volt:%.2f\n",(i-2.0*angle)/2,sensed[a]);
a++;
}
delay(TD);
break;
}
max(angle);
init(angle);
plot(angle);
}
void rotatea(void)
{
outportb(DIR,0xff);    /* sending high to rotate anticlockwise */
}
void rotatec(void)
{
outportb(DIR,0x00);    /* sending low to rotate clockwise */
}
void initialise(int chno)
{
outportb(CNTRL,chno);  /*select software trigger */
outportb(MUX,chno);
outportb(ADC_HI,0x05);  /* select voltage range 0-1V */
outportb(MUX,0x00);
return;
}
float ADC(void)
{
unsigned char  lo, hi;
unsigned int   data;
float volt;
outportb(BASE,0x00);            /* Triggering the ADC conversion */
```

```
delay(TD);
lo=inportb(BASE);
hi=inportb(ADC_HI);
data=(unsigned int)((hi<<4) | (lo >> 4));
volt=((data)*(4.5/1024));     /* calibration of o/p voltage */
return(volt);
}
void max(int ang)
{
signed float pos_min=0,pos_max=0;
int i;
float maximum;
maximum = sensed[0];
for(i=1;i<120;i++) maximum = (maximum>sensed[i])?(maximum):
(sensed[i]);
for(i=0;i<120;i++)
{
sensed[i] = sensed[i]/maximum;    /* normalizing data */
}
for(i=1;i<120;i++) if (sensed[i]>(0.05)) break;
pos_min=(((i-1)*0.5)-ang);
for(i=0;i<120;i++) if (sensed[i] == 1.0) break;
for(i=i;i<120;i++) if (sensed[i] < (0.05)) break;
pos_max=((i-1)*0.5)-ang);
printf("\n Maximum Intensity          : %2f", maximum);
printf("\n Minimum Position           : %.1f", pos_min);
printf("\n Maximum Position           : %.1f", pos_max);
printf("\n Light Gathering Angle      : %.3f degrees", ((pos_max-pos_
min)*0.5));
printf("\n Numerical Aperture is      : %.4f  ",  sin(PI/180*((pos_
max-pos_min)*0.5)));
getch();
return;
}
void init(int angle)
{
int i,j;
int gd,gm;
gd = DETECT;
clrscr();
initgraph(&gd,&gm,"c:\\tc\\bgi");
setcolor(YELLOW);
line(10,400,630,400);
setcolor(YELLOW);
line(310,400,310,10);
j=(600/(2*angle));
```

```
 if(j<10)
 j=10;
 for(i=10;i<600;i+=j)
 {
 putpixel(i,400,RED);
 putpixel(i-1,400,RED);
 putpixel(i+1,400,RED);
 }
 outtextxy(410,440,"X Axis:ANGLE");
 printf("PLOT OF ANGLE Vs POWER\n\n\n");
 printf("X Axis Scale :1 unit=1 degree\n");
 printf("Y Axis Scale :1 unit=0.1 volt");
 for(i=10;i<400;i+=20)
 {
 putpixel(310,i,RED);
 putpixel(310,i-1,RED);
 putpixel(310,i+1,RED);
 }
 outtextxy(320,100,"Y Axis:POWER");
 outtextxy(310,420,"0");
 outtextxy(410,420,"+10");
 outtextxy(510,420,"+20");
 outtextxy(610,420,"+30");
 outtextxy(210,420,"-10");
 outtextxy(110,420,"-20");
 outtextxy(10,420,"-30");
 outtextxy(285,100,"1.0");
 outtextxy(285,250,"0.5");
 }

 void plot(int angle)  /* for plotting the graph with stored data
  */
 {
 int x,i=10,j,a=28;
 int y;
 j=(600/(4*angle));
 if (j<10)
 j=10;
 y=(((1-(sensed[a]))*300)+100);
 a++;
 putpixel(i,y,GREEN);
 moveto(i,y);
 for(a=a;a<100;a++)
 {
 setcolor(GREEN);
    i+=j;
 y=(((1-(sensed[a]))*300)+100);
 lineto(i,y);
 }
```

```
setcolor(YELLOW);
line(10,385,600,385);
outtextxy(25,370,"0.05 Maximum Intensity");
}

/* end of file */
```

References

Agrawal, G.P. 2002, *Fiber-Optic Communication Systems*, 3rd edn, John Wiley, New York.

Barnoski, M.K. and S.M. Jensen 1976, *Appl. Opt.*, vol. 15, p. 2112.

Bennett, W.R. 1962, *Appl. Opt.*, vol. 1, suppl. 1: Optical masers, p. 24.

Buck, J.A. 1995, *Fundamentals of Optical Fibers*, John Wiley, New York.

Culshaw, B. and J. Dakin (eds) 1989, *Optical Fibers Sensors: Systems and Applications*, Artech House, Norwood, MA, vol. 2.

Desurvire, E. 1994, *Erbium-Doped Fiber Amplifiers: Principles and Applications*, John Wiley, New York.

Einstein, A. 1917, *Phys. Z.*, vol. 18, p. 121.

Ghatak, A.K. and M.R. Shenoy (eds) 1994, *Fiber Optics Through Experiments*, Viva Books, New Delhi.

Ghatak, A.K. and K. Thyagarajan 1999, *Introduction to Fiber Optics*, First South Asian edn, Cambridge University Press, Cambridge.

Giles, C.R. and E. Desurvire 1991, *J. Lightwave Technol.*, vol. 9, p. 271.

Gloge, D. 1971, *Appl. Opt.*, vol. 10, p. 2252.

Gloge, D. 1972, *Appl. Opt.*, vol. 11, p. 2506.

Gloge, D. and E.A.J. Marcatili 1973, *Bell Syst. Tech. J.*, vol. 52, p. 1563.

Gowar, John 2001, *Optical Communication Systems*, 2nd edn, Prentice Hall of India, New Delhi.

Grattan, K.T.V. and B.T. Meggitt (eds) 1998, *Optical Fiber Sensor Technology*, Chapman and Hall, London, vol. 2.

Henry, P.S., R.A. Linke, and A.H. Knauck 1988, in S.E. Miller & I.P. Kaminov (eds), *Optical Fiber Telecommunications*, Academic Press, San Diego, CA, vol. II, ch. 21.

Hussey, C.D. and F. Martinez 1985, *Electron lett.*, vol. 21, p. 1103.

Keck, D.B. 1985, *IEEE Commun. Mag.*, vol. 23, p. 17.

Keiser, Gerd 2000, *Optical Fiber Communications*, 3rd edn, McGraw-Hill, Singapore.

Khare, R.P. 1993, *Analysis Instrumentation: An Introduction*, CBS Publishers, New Delhi.

Khare, R.P., Aditi Sharma, and Sartaj Singh 1995, *Indian J. Eng. Mater. Sci.*, vol. 2, p. 58.

Khare, R.P., Vandana Prabhu, and T.V. Jaishankar 1996, *Indian J. Pure Appl. Phys.*, vol. 34, p. 823.

Kressel, H. and J.K. Butler 1977, *Semiconductor Lasers and Heterojunction LEDs*, Academic Press, New York.

Lee, T.P. and T. Li 1979, in S.E. Miller & A.G. Chynoweth (eds), *Optical Fiber Telecommunications*, Academic Press, San Diego, CA, vol. I, ch. 18.

Marcuse, D. 1977, *Bell Syst. Tech. J.*, vol. 56, p. 703.

Marcuse, D. 1979, *Appl. Opt.*, vol. 18, p. 2930.

Miller, C.M. 1978, *Bell Syst. Tech. J.*, vol. 57, p. 75.

Nair, Maya S. et al. 2001, *Laser Horiz.*, vol. 5, no. 1, p. 19.

O'Mahony, M.J. 1988, *J. Lightwave Technol.*, vol. 6, p. 531.

Okamoto, K. and T. Okoshi 1976, *IEEE Trans. Microwave Theory Tech.*, vol. MTT-24, p. 416.

Ovren, M. Adoleson and B. Hok 1983, *Proceedings of the International Conference on Optical Techniques in Process Control*, BHRA Publishers, Bedford, UK, p. 67.

Peterman, K. 1983, *Electron Lett.*, vol. 19, p. 712.

Pyne, D.N. and W.A. Gambling 1975, *Electron Lett.*, vol. 11, p. 176.

Ramaswami, Rajiv and Kumar N. Sivarajan 1998, *Optical Networks: A Practical Perspective*, Morgan Kaufmann, San Francisco, CA.

Rao, Y.J. 1998, 'Fiber Bragg grating sensors: Principles and applications', in K.T.V. Grattan & B.T. Meggitt (eds), *Optical Fiber Sensor Technology*, Chapman and Hall, London, vol. 2.

Shimada, S. and H. Ishio (eds) 1994, *Optical Amplifiers and Their Applications*, John Wiley, Chichester.

Smit, M.K. and C. Van Dam 1996, *IEEE J. Sel. Top. Quantum Electron.*, vol. 2, p. 236.

Sze, S.M. 1981, *Physics of Semiconductor Devices*, John Wiley, New York.

Tewari, R. and K. Thyagarajan 1986, *J. Lightwave Technol.*, vol. 4, p. 386.

Tsuchiya, H., H. Nakagome, N. Shimizie, and S. Ohara 1977, *Appl. Opt.*, vol. 16, p. 1323.

Udd, Eric (ed.) 1991, *Fiber Optic Sensors: An Introduction for Engineers and Scientists*, John Wiley, New York.

Udd, Eric (ed.) 1995, *Fiber Optic Smart Structures*, John Wiley, New York.

Verdeyen, Joseph T. 1993, *Laser Electronics*, 2nd edn, Prentice Hall of India, New Delhi.

Wickersheim, K.A. 1978, US Patent no. 4,075, 493.

Yariv, A. 1997, *Optical Electronics in Modern Communications*, 5th edn, Oxford University Press, New York, ch. 7.

Yariv, A. 1997a, *Optical Electronics in Modern Communications*, 5th edn, Oxford University Press, New York, ch. 15.

Index